Coherent

T. Meier   P. Thomas   S.W. Koch

Coherent Semiconductor Optics

From Basic Concepts
to Nanostructure Applications

With 1 3 9 Figures

 Springer

Privatdozent Dr. Torsten Meier
Professor Dr. Peter Thomas
Professor Dr. Stephan W. Koch
Fachbereich Physik, Philipps Universität
Renthof 5, D-35032 Marburg, Germany
E-mail: torsten.meier@physik.uni-marburg.de
         peter.thomas@physik.uni-marburg.de
         stephan.w.koch@physik.uni-marburg.de

ISBN 978-3-642-06896-6   e-ISBN 978-3-540-32555-0

Springer is a part of Springer Science+Business Media.

springer.com

© Springer-Verlag Berlin Heidelberg 2010

Cover design: *design & production* GmbH, Heidelberg

To our families,

as well as former, present, and future students

Preface

This book could not have been written without the extensive work of many diploma and Ph.D. students in our Theoretical Semiconductor Physics Group at the Philipps-Universität in Marburg. They have contributed to the fundamental understanding and to many applications in the area of coherent semiconductor optics. The one-dimensional tight-binding model, which is exclusively treated in the present book, has been the basis of many of their diploma and Ph.D. theses. The reader will find references to their results and also their names as authors of the publications listed in the sections "Suggested Reading".

In particular, the authors wish to thank Irina Kuznetsova, who prepared a large number of the figures and recalculated the underlying data on the basis of the equations presented in this book in cases where parameters or presentation had to be changed and/or optimized.

Some of the problems, in particular, those connected to the more introductory chapters, were solved by Swantje Horst and Joachim Kalden. They made valuable suggestions for improved formulation of the problems and pointed out a number of hints we should give our readers in order to help them with the solutions.

Furthermore, we wish to thank all our numerous collaborators, together with whom we have performed research in the area of coherent semiconductor optics in the past and present, for many valuable discussions. In particular, without the close cooperation between experiment and theory this research field would not have advanced to the present level.

In the time preceding the finalization of this book, sketches of the text have been used several times as notes for lectures given in the context of the Deutsche Forschungsgemeinschaft (DFG) Research Training Group "Optoelectronics of Mesoscopic Semiconductors" in Marburg and also during guest lectures in the context of the DFG European Research Training Group "Electron–Electron Interactions in Solids" (Marburg/Budapest) in Marburg and Budapest. We would like to thank the students attending these lectures for numerous corrections and suggestions which helped to improve the presentation.

We wish to thank the DFG for continued support of our research. The topics connected with the content of this book are currently funded by projects in the priority program SPP 1113 "Photonic Crystals" (Koch and Meier, KO 816/7), in the research group FOR 483 "Metastable Compound Semiconductor Systems and Heterostructures" (Koch, Meier, and Thränhardt, KO 816/9), and in the research group FOR 485 "Quantum Optics in Semiconductor Nanostructures" (Koch, KO 816/10). During the work on this book, Torsten Meier was supported by a fellowship of the Heisenberg program of the DFG (ME-1916/1) which is gratefully acknowledged.

We thank Renate Schmid, Marburg, for her excellent work on the text and her valuable suggestions regarding the proper (American) English. Of course, the authors of this book are to blame for misprints and errors which we have overlooked.

Marburg
June 2006

Torsten Meier
Peter Thomas
Stephan W. Koch

Contents

Part I

Preliminaries

1

Introduction

Mankind has always been fascinated by optical properties of solids. Besides manufacturing and availability aspects it has been the color of metals like gold, silver, and bronze that defined their outstanding ornamental and decorative value in history. The same is true for gemstones and minerals due to their colorful appearance and their intense reflections caused by a high index of refraction.

Whereas most gemstones are insulators, the precious metals are more or less good conductors. Only fairly recently another class of materials, the semiconductors, have caught the interest of researchers, engineers, and technicians. Semiconductors are situated somewhere in between metals and insulators, even though the exact boundaries are often not that well defined. Many aspects of our modern civilization are based on semiconductors. Virtually all technical equipment for everyday use contains semiconducting material in one form or another.

The first semiconductor applications used their unique transport properties. Contacts of semiconductors with metals were used in rectifiers and diodes. Then the invention of the transistor led to fabulous developments, most prominently in computing and information technology. The application of optical and optoelectronic semiconductor properties was somewhat delayed, mainly because of the fact that the initially dominant semiconductors, silicon and germanium, have indirect optical gaps and are thus not well suited for optical devices. Today, however, compound semiconductors like GaAs and its relatives are basis materials for optoelectronic applications, e.g., in light-emitting diodes, semiconductor lasers, etc. Besides the GaAs-like materials that are composed of elements from groups III and V of the periodic system, also II–VI compounds, as well as ternary, or even quaternary systems are becoming increasingly relevant.

Over the years, fundamental research on the optical properties of semiconductors has revealed a large number of fascinating effects. However, a real boost came with the development of techniques that allow for controlled atomic growth and thus nanoscale structuring of semiconductor materials.

Using epitaxial growth techniques and modern lasers that emit ultrashort light pulses in the range of femtoseconds makes it now possible to study optoelectronic linear and nonlinear processes on ultrafast time scales in structures characterized by length scales down to a few nm. Using structuring on nanometer (nm) scales the spatial dimension became a relevant physical parameter. Effectively zero-, one-, and two-dimensional systems can be fabricated and studied by growing suitable semiconductor heterostructures. These nanostructures may resemble atoms (quantum dots), wires (quantum wires), films (quantum wells), superlattices (periodic arrays of quantum wells), etc.

Using ultrafast laser pulses, electronic processes can be monitored on time scales that are comparable to typical interaction times among the electronic excitations. It is also possible to investigate the dynamics of the coherently excited many-particle system on time scales shorter than typical decoherence times. This coherent regime has attracted the interest of researchers most recently, as the concepts of "ultrafast switching", "coherent control", or "quantum computing" rely on processes which are destroyed by phase-breaking interactions.

By now, a large amount of experimental and theoretical work on optical properties of semiconductors in the coherent ultrafast time regime exists. The interpretation of the experimental findings requires a thorough treatment of the semiconductor as a many-particle system, where the often dominant processes result from the Coulomb interaction between the optically excited charge carriers. In addition, realistic semiconductor nanostructures always show a certain degree of disorder, which has a profound influence on their optoelectronic properties. One often applies external ac or dc electrical or magnetic fields to semiconductor structures. Such fields lead to new dynamical processes which might be useful for future applications.

1.1 Coherence

The dynamics of a classical plane wave with given wave vector and frequency is determined by the solution of the relevant wave equation. Generally, the wave is described by a spatially and temporally varying field $\Psi(\mathbf{r}, t)$. The propagation of this wave function in free space is called *coherent* if, by knowing the amplitude and phase at a given point in space at a given time, we can immediately tell the amplitudes and phases at all other times and space points.

In this book, we are dealing exclusively with temporal, not spatial coherence. A formal definition of temporal coherence of an arbitrary function $f(t)$ is given by the autocorrelation function

$$F(t) = \lim_{T \to \infty} \frac{1}{T} \int_0^T f^*(t')f(t'+t)dt. \tag{1.1}$$

The complex conjugate has been used in this definition in case the function is complex valued. If we find that $F(t)$ decays in time, we say that the coherence of the function $f(t)$ is characterized by a finite coherence time, i.e., the function $f(t)$ is subject to *dephasing*.

In most experiments, it is not the function $f(t)$ which describes, e.g., an electric field, that is measured. In a statistical ensemble, the expectation value of the field is often zero, i.e.,

$$\langle f(t) \rangle = 0. \tag{1.2}$$

In such a case, one may still record the related intensity I_f, which is given by

$$I_f(t) = \langle |f(t)|^2 \rangle. \tag{1.3}$$

The intensity is an equal-time autocorrelation function and can be nonzero even if $\langle f(t) \rangle = 0$.

Consider that $f(t)$ is the superposition of two functions $f_1(t)$ and $f_2(t)$.

$$f(t) = f_1(t) + f_2(t). \tag{1.4}$$

One then measures

$$\langle |f_1(t) + f_2(t)|^2 \rangle = \langle |f_1(t)|^2 \rangle + \langle |f_2(t)|^2 \rangle + 2\Re\langle f_1^*(t)f_2(t) \rangle. \tag{1.5}$$

Here, the third term on the right-hand side describes the interference of the two fields f_1 and f_2. This term is finite only if f_1 and f_2 are at least to a certain degree coherent to each other. Thus, in order to observe interference the temporal evolution of the phases of f_1 and f_2 must be determined by the coherent dynamics governed by, e.g., a wave equation. If, however, the dynamics of either f_1 or f_2 is dominated by incoherent processes, e.g., scattering events, interference is absent.

In nature, particles and waves interact with the surrounding media. The resulting *scattering processes* depend both on the nature of the particle or wave and that of the medium. In order to determine the trajectory of a classical particle, one has to insert all the (possibly time-dependent) forces into Newton's equation of motion and solve for $r(t)$, i.e., compute the particle's position as a function of time. In many cases one does not have the full information about all the acting forces, but only knows some statistical properties. In such a case, one has to resort to a statistical description yielding distributions of the dynamical variables of the particle.

In the case of waves we can distinguish two different cases:

(i) Imagine a source that emits a stationary wave into free space. At some point, this wave enters a medium that is characterized by static scattering centers. The wave is then scattered in such a way that a distorted wave pattern is formed. For a stationary situation this wave pattern is also stationary and is therefore still coherent. It can be computed from the wave equation, where the stationary scattering processes have to be included. This means, in particular, that the phases in all points are rigidly correlated. Examples are Bloch waves of electrons in atomic lattices and electron waves in disordered solids as well as light waves in photonic crystals and in disordered dielectrics. Also, waves in random stationary media are coherent in this sense, although their wave pattern can be very complicated.

(ii) If the medium possesses scattering centers that have some temporal dynamics, e.g., such as those of a gas of particles, we could still try to include this dynamics into the wave equation and attempt a solution for the scattered wave. However, often the trajectories of the scattering particles are not known in detail. Only their statistical properties can be specified, e.g., by giving the temperature and the density if the scattering medium is a thermal (heat) *bath*. Since we do not have full knowledge of the dynamics of the bath particles, we cannot hope to obtain all possible information about the scattered wave. Only some average quantities are accessible in such a situation. As the intensity of the scattered wave is related to some kind of local energy density, it seems reasonable to assume that we can still obtain decent information about the spatial and temporal behavior of the amplitude. However, the information about the phases is partially or completely lost. In this case, one says that the phase has suffered phase-breaking interactions or dephasing. In other words, the wave-like excitation is no longer coherent. Since the information about the intensity is less sensitive, we arrive at a description where some local object, e.g., a maximum of the intensity, moves in space. This is strongly reminiscent of the dynamics of a classical particle. In fact, dephasing destroys the wave-like features that are introduced into the description by using quantum-mechanical concepts, and we are left with a classical description of particle motion.

In the coherent regime, on the other hand, we retain the full coherent nature of the excitation, which this book focuses on. We will see, however, that even in this coherent regime *optical phase coherence* can be limited if this notion is defined in terms of certain properties of experimentally determined transient signals.

1.2 Basic Optical Principles

The measurement of optical properties of solids often follows the scheme where one obtains information about the material system by applying an external field and recording the response of the system to this field. In many cases, one is interested in the equilibrium properties of the material system. These properties can be investigated by experiments in the linear response regime, where the response linearly depends on the excitation field. Examples are linear optical absorption spectra, refractive index, etc.

Modern semiconductor optics, however, often investigates nonlinear optical properties. In particular, if dynamical processes are of interest, one often applies sequences of ultrashort laser pulses to the system and records the sub-sequent dynamical response. Examples are pump–probe and four-wave-mixing experiments. The nonlinear response can either be measured in the frequency domain (bleaching, induced absorption, etc.) or in the time domain (decay, echoes, quantum beats, etc.).

An interesting application of ultrafast optical measurements is related to transport properties. As the laser field generates populations of electrons and holes, the subsequent spatio-temporal dynamics of these particles can be investigated. For these studies, local excitation and detection schemes can be applied, e.g., using near-field optical microscopy. Such studies yield informa-tion about the interesting and intricate interplay between structural disorder and many-particle interactions and their influence on the spatio-temporal dy-namics of the optically generated excitations. Alternatively, coherent-control schemes make it possible to induce particle currents on an ultrafast time scale by homogeneous optical excitation of the semiconductor system.

Besides optical excitation, also externally applied electric and magnetic ac and dc fields lead to interesting dynamical signatures. Examples are Bloch oscillations in dc-biased superlattices, dynamical localization in an ordered semiconductor induced by an ac electric field, and Aharonov–Bohm oscil-lations of excitons and biexcitons in semiconductor rings subjected to a magnetic flux. Since these dynamical effects are related to moving charges, the system emits electromagnetic radiation. The upper time limit at which the mentioned coherent dynamical processes can be observed is determined by phase-breaking interactions. As these occur typically on time scales in the picosecond range for the case of excitonic excitations in typical semi-conductors, parameters have to be chosen such that the relevant periods of the coherent oscillations are shorter than this time limit. Considering picosecond periods, the emitted signals thus have frequencies in the terahertz range.

1.3 Relevant Material Systems

Most of the experiments mentioned so far have been performed mainly on semiconductor nanostructures with reduced effective spatial dimensions. Here, one-dimensional quantum wires and two-dimensional quantum films and ar-rays composed of these structures are widely used examples. The III–V system GaAs and similar III–V systems are prototype materials. However, II–VI sys-tems and some wide-gap III–V compounds like GaN have larger excitonic binding energies. Therefore, these materials are considered to be useful candi-dates for some applications. However, the majority of experiments have been performed on GaAs-type semiconductor heterostructures. These also domi-nate most applications in information technology. In this book, we therefore predominantly take fundamental electronic properties of this class of mate-rials as a basis for the parameters of our schematic model. These include effective electron and hole masses, exciton binding energies, optical selection rules, etc. However, extensions to other material parameters are often straightforward.

In the main part of our calculations, we concentrate on a one-dimensional model, the evident realization of which would be quantum wires. However, in many cases the predictions of the model do not strongly depend on the dimensionality of the system. Therefore, they can, *mutatis mutandis*, be taken as a guideline also for experiments on two-dimensional or three-dimensional heterostructures like quantum wells and superlattices.

1.4 Related Systems

As mentioned above, the results obtained on the basis of the schematic one-dimensional model can be easily transferred to one-dimensional heterostructures like quantum wires. One could also attempt to apply the predictions to other, more natural one-dimensional systems like polymeres, liquid crystals composed of disc-shaped molecules (discotic crystals), and even to biological systems. However, these systems belong to the class of highly correlated electron systems, and the theoretical approach used here is often not directly applicable. The linear optical spectra of such systems are studied in ongoing research and the investigation of dynamical processes of optical excitations is a fascinating field that appears to be far from being settled yet.

1.5 Aim of the Book

Although considerable progress has been achieved in the theoretical description of coherent optical properties of semiconductor structures, a full theoretical treatment taking into account all details of the real heterostructure, the full many-particle Coulomb interaction, disorder, coupling to lattice vibrations and external fields is beyond the capability of even the largest modern computing facilities. Therefore, one is forced to simplify the theoretical model by neglecting certain complications or by considering a simplified electronic or atomic structure. Even then, a numerical simulation of the dynamical processes often presents a formidable problem. In particular, the many-particle interaction leads to a hierarchy of equations or to an infinite number of relevant terms.

In this book, we present a discussion of coherent semiconductor optics starting from the simplest possible model of a semiconductor. The advantage of this approach, besides leading to mathematically transparent equations and numerically tractable simulation schemes even for situations including external electric and magnetic fields and disorder, lies in the fact that the physical principles of the various dynamical processes can be clarified and introduced without being distracted by numerous details necessary to describe a real semiconductor structure. On the other hand, one should not expect that such a schematic model quantitatively accounts for all of the detailed features observed in experiments on a particular structure.

Nevertheless, for the time being, a number of new predictions and interpretations concerning dynamical processes of optical excitations in the coherent regime is only possible on the basis of such a schematic model. Although it is conceivable that these theories will be substantiated for realistic situations once the next generation of computers is available, the underlying general physical principles can most easily be perceived on the basis of the schematic model.

The presentation of this book is based on a one-dimensional tight-binding model for a semiconductor with finite length. If periodic boundary conditions are applied, this model represents a ring-like structure. The theoretical treatment will in most cases make use of a real-space representation, which allows us to incorporate disorder in a most natural way and to consider structures with a finite length.

There is also a tutorial reason for working with such a model system. Coherent optical properties were first observed for atomic and molecular systems because of their much longer phase relaxation times compared to solids. The theoretical models used in this context were based on systems having only a few single-particle energy levels. These few-level systems allow for an easy and transparent introduction of many dynamic processes initiated by ultrashort laser pulses. The transfer of the developed concepts to a solid is then relatively straightforward. Complications and fascinating differences to few-level systems arise *mainly* from the many-particle Coulomb interaction which has to be implemented into the model. Coulomb effects are responsible for the mutual repulsion of carriers with equal charges and for new resonances due to the attractive interaction between the oppositely charged electrons and holes.

When introducing students to the field of semiconductor optics, we realized that, although there is a large amount of theoretical work on dynamical processes of optical excitations in the coherent regime, a presentation that summarizes the fundamental principles and explanations without going into material-specific details is lacking. In this book, the material is therefore presented at a level useful for students intending to work in the field of semiconductor optics. It is also suitable for researchers more interested in applications of optoelectronic devices as it provides a basis for a fundamental understanding of optical properties of semiconductor heterostructures in the coherent regime.

This book consists of three parts. In Part I we develop the theoretical concepts and the equations of motion, which form the basis of the following two parts. Part II deals with applications of the theory to level systems and the semiconductor model, while Part III is devoted to special dynamic properties of semiconductors that can be studied by coherent optical experiments.

1.6 Necessary Prerequisites

In the first place this book is written for students intending to learn the principles of modern semiconductor optics. They should have some knowledge of quantum mechanics, including the basics of second quantization. In the introductory chapters describing few-level systems we have, however, avoided using second quantization and apply the Dirac bra and ket notation. This is sufficient since many-particle interactions are not considered here. However, for the discussion of many-particle interaction effects in the chapters dealing with the semiconductor system, second quantization techniques are needed.

Some background knowledge of solid state physics is helpful, but is not absolutely necessary. As far as mathematics is concerned, the reader should be familiar with simple differential equations and with the Fourier transformation.

1.7 Suggested Reading

1. L. Allen and J.H. Eberly, *Optical Resonance and Two-Level Atoms* (Wiley, New York 1975)
2. N. Bloembergen, *Nonlinear Optics* (Benjamin, New York, 1965).
3. M. Born and E. Wolf, *Principles of Optics* (Pergamon, New York 1970)
4. C. Cohen-Tannoudji, J. Dupont-Roc, and G. Grynberg, *Photons and Atoms* (Wiley, New York 1989)
5. H. Haug and A.-P. Jauho, *Quantum Kinetics in Transport and Optics of Semiconductors* (Springer, Berlin 1996)
6. H. Haug and S.W. Koch, *Quantum Theory of the Optical and Electronic Properties of Semiconductors*, 4th edn. (World Scientific, Singapore 2004)
7. S. Glutsch, *Excitons in Low-Dimensional Semiconductors*, Springer Series in Solid-State Sciences, Vol. 141 (Springer, Berlin 2004)
8. J.D. Macomber, *The Dynamics of Spectroscopic Transitions* (Wiley, New York 1976)
9. S. Mukamel, *Principles of Nonlinear Optical Spectroscopy* (Oxford, New York 1995)
10. N. Peyghambarian, S.W. Koch, and A. Mysyrowicz, *Introduction to Semiconductor Optics* (Prentice Hall, Englewood Cliffs, New Jersey 1993)
11. W. Schäfer and M. Wegener, *Semiconductor Optics and Transport Phenomena* (Springer, Berlin 2002)
12. A. Stahl and I. Balslev, *Electrodynamics of the Semiconductor Band Edge*, Springer Tracts in Modern Physics, Vol. 110 (Springer, Berlin 1987)

2

Experimental Techniques

Various experimental techniques that are used to obtain insight into dynamical processes apply pulsed excitation by light. The light field is usually characterized by a photon energy $\hbar\omega$ close to the fundamental optical gap of the semiconductor, E_g. We start by discussing the measurement of the linear optical response. From linear response theory this regime is known to yield information about the equilibrium properties of the system. It thus forms the basis for the interpretation of the nonequilibrium experiments, which we discuss below. These include pump–probe and wave-mixing experiments. Finally, we briefly discuss some experimental approaches which allow us to induce and detect electronic transport all-optically.

2.1 Linear Optical Spectra

We assume the following situation: The surface of an otherwise homogeneous, isotropic semiconductor defines the x-y plane of a coordinate system and the semiconductor fills the half-space with positive z. A stationary electric light field with a wave vector parallel to the z direction, $\boldsymbol{k} = (0, 0, k_0)$,

$$E(z,t) = E_0 e^{ik_0 z - i\omega t} + c.c., \qquad (2.1)$$

and with photon energy $\hbar\omega$ propagates for $z < 0$ in free space in the z direction. Here,

$$k_0 = \frac{\omega}{c} \qquad (2.2)$$

is the wave number and c the velocity of light in free space. In (2.1), c.c. indicates the complex conjugate. For the present discussion, it is sufficient to consider only the first term.

At $z = 0$, the amplitude of the field is $\boldsymbol{E}(z = 0) = \boldsymbol{E}_0$ and for positive z, i.e., inside the semiconductor, its Fourier components are phenomenologically given by

$$E(z,\omega) = E_0 e^{i(k(\omega)+i\kappa(\omega))z}. \tag{2.3}$$

Here, the wave number

$$k(\omega) = \frac{\omega}{v(\omega)} \tag{2.4}$$

describes the dispersion of the medium, which is due to the frequency dependence of the phase velocity $v(\omega)$ there. The extinction coefficient $\kappa(\omega)$ describes the exponential decay of the light field amplitude according to the Lambert–Beer law.

The index of refraction $n(\omega)$ is defined for the present system as

$$n(\omega) = \frac{k(\omega)}{k_0} = \frac{ck(\omega)}{\omega} \tag{2.5}$$

and the absorption coefficient describing the decay of the light intensity according to

$$|E(z)|^2 = |E_0|^2 e^{-\alpha(\omega)z} \tag{2.6}$$

is

$$\alpha(\omega) = 2\kappa(\omega). \tag{2.7}$$

In order to make contact with a microscopic description, one first has to write down the macroscopic Maxwell's equations which can be combined into a wave equation if we neglect magnetic interactions. For a homogeneous medium without external sources within the medium, it reads

$$\Delta E(r,t) - \frac{1}{c^2}\frac{\partial^2}{\partial t^2} D(r,t) = 0. \tag{2.8}$$

The Fourier transform into the frequency domain yields

$$\Delta E(r,\omega) + \frac{\omega^2}{c^2} D(r,\omega) = 0. \tag{2.9}$$

Here, D is the electrical displacement which is related to the electric field by

$$D(r,\omega) = E(r,\omega) + 4\pi P(r,\omega) \tag{2.10}$$

(we are using Gaussian cgs units here).

The polarization P is the central quantity in optics and, therefore, also in this book. It contains all the information about the coherent optical excitations. In the coherent regime, the ultrafast dynamics of the excitations induced by the external light field did not yet suffer any phase-breaking interactions and relaxations.

Linear response means that the optical polarization is related to the electric field by the linear relation

$$P(r,\omega) = \chi(\omega) E(r,\omega). \tag{2.11}$$

Here, we assume for simplicity that the medium is isotropic and that we do not have to consider spatial dispersion. Otherwise, the linear optical susceptibility χ is a tensorial quantity and the relation (2.11) would be nonlocal. This would then lead to a susceptibility that depends on wave number. As long as we do not consider light propagation effects, it is safe to use the local relation (2.11).

The validity of this model depends on the ratio of the wavelength λ of the exciting light field relative to other relevant length scales of the system. If λ is much larger than the Bohr radius of an exciton or the diameter of the unit cell, the spatially local susceptibility is a good approximation.

Instead of the linear susceptibility $\chi(\omega)$, one may also define the dielectric function $\epsilon(\omega)$ by

$$
\begin{aligned}
\boldsymbol{D}(\boldsymbol{r},\omega) &= \epsilon(\omega)\boldsymbol{E}(\boldsymbol{r},\omega) \\
&= (\epsilon'(\omega) + i\epsilon''(\omega))\boldsymbol{E}(\boldsymbol{r},\omega),
\end{aligned} \tag{2.12}
$$

thus

$$
\epsilon(\omega) = 1 + 4\pi\chi(\omega). \tag{2.13}
$$

Both functions are complex and their real and imaginary parts $\epsilon'(\omega) = 1 + 4\pi\chi'(\omega)$ and $\epsilon''(\omega) = 4\pi\chi''(\omega)$, respectively, are intimately related due to causality via the Kramers–Kronig relations.

Inserting (2.3) and (2.12) into the wave equation, i.e., into (2.8), we obtain for the real part and for the imaginary part

$$
k^2(\omega) - \kappa^2(\omega) = \frac{\omega^2}{c^2}\epsilon'(\omega),
$$

$$
2k(\omega)\kappa(\omega) = \frac{\omega^2}{c^2}\epsilon''(\omega), \tag{2.14}
$$

respectively. The absorption coefficient $\alpha(\omega)$ is then given by

$$
\alpha(\omega) = \frac{\epsilon''(\omega)\omega}{n(\omega)c} = \frac{4\pi\chi''(\omega)\omega}{n(\omega)c}. \tag{2.15}
$$

In most semiconductors, the index of refraction $n(\omega)$ is dominated by strong optical resonances in the ultraviolet energy range, i.e., energetically much higher than the fundamental absorption edge E_g. For optical transitions close to E_g, the index of refraction can therefore be replaced by a so-called background index n_b, which can be assumed to be independent of ω in the frequency range of interest. The optical absorption spectrum $\alpha(\omega)$ is then essentially determined by the imaginary part of the dielectric function $\epsilon''(\omega)$ which is up to 4π equal to the imaginary part of the linear susceptibility, $\chi''(\omega)$. Therefore, this function is usually simply referred to as the optical absorption. (From linear response theory, one deduces that in fact the energy dissipated in the medium is proportional to $\omega\chi''(\omega)$.)

For the index of refraction, we obtain

$$n(\omega) = \sqrt{\frac{1}{2}(\epsilon'(\omega) + \sqrt{\epsilon'(\omega)^2 + \epsilon''(\omega)^2})}. \qquad (2.16)$$

For frequencies where there is no absorption, i.e., $\hbar\omega < E_g$, we find the well-known result $n = \sqrt{\epsilon'} = \sqrt{\epsilon}$.

2.2 Pump–Probe Experiments

In pump–probe experiments, one typically first excites the system by a pulse No. 1 (pump) having wave vector $\mathbf{k}_1$. After a delay time τ, the linear response with respect to a second pulse No. 2 (probe) with wave vector $\mathbf{k}_2$ is measured. This yields the absorption $\alpha(\omega, \tau)$. Usually, the *differential absorption* modified by the first pulse

$$\delta\alpha(\omega, \tau) = \alpha(\omega, \tau) - \alpha_0(\omega) \qquad (2.17)$$

is plotted as a function of frequency. Here, $\alpha_0(\omega)$ is the linear response to pulse No. 2 if the system is in equilibrium before pulse No. 2 arrives, i.e., without pulse No. 1. The detector is placed either in the direction $\mathbf{k}_2$ of the transmitted light of pulse No. 2 (see Fig. 2.1), or it detects the signal reflected by the sample in the specular direction. In the latter case, one studies the differential reflectivity.

The differential optical absorption or reflectivity monitors the changes in the optical response induced by the pump pulse. The delay can be taken to be zero; then both pulses arrive at the same time. It can even be taken to be negative, i.e., the probe pulse precedes the pump pulse. In this case, the differential absorption displays interesting coherent spectral oscillations, as is shown below.

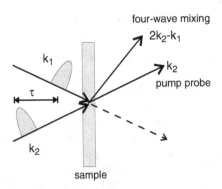

Fig. 2.1. Schematic kinematics of four-wave-mixing and pump–probe experiments which are performed with two optical laser pulses having a time delay τ

2.3 Wave-Mixing Experiments

Wave-mixing experiments are a variant of pump–probe experiments, however, the kinematics is different. Again, a series of pulses with wave vectors k_i impinges on the sample. But now, instead of monitoring the signal in one of the directions k_i, one detects the signal diffracted into one of the possible directions given by the kinematics of the experiment. If, e.g., two pulses are applied, the optical excitation (i.e., the polarization) due to the first and the field of the second delayed pulse forms an interference pattern in the sample. The optical properties of the material system are then modulated spatially, forming a grating, which diffracts off the second pulse into various directions. In this four-wave-mixing geometry, which is called the self-diffraction geometry, one usually monitors the signal in the direction $2k_2 - k_1$ (first-order diffraction), see Fig. 2.1. Since the first pulse is involved once, while the second one enters twice, and we observe a fourth pulse, this experiment is called four-wave mixing.

Other pulse sequences are also possible. For example, one often monitors the diffracted signal using a third pulse which is delayed with respect to both the first and the second one.

As diffraction is involved, it is obvious that wave-mixing experiments rely on coherence. If the polarization induced in the material by the first pulse underwent random phase variations, the formation of a grating would not be possible and there would be no diffracted signal. Therefore, wave-mixing experiments are frequently used to study the decay of the coherent optical polarization, due to various interactions.

2.4 Transport Phenomena

In various chapters of this book, we discuss electrical transport phenomena in the coherent regime. Coherence in semiconductors tends to be destroyed on very short time scales, depending on the particular situation, in the 0.1–10 ps range. In certain cases, e.g., for mesoscopic systems or disordered samples, these times can be much longer, extending up to nanoseconds. Generally, however, the generation of the species that move in space (electrons, holes, excitons), and the detection of their movement, both require techniques that work on an ultrafast time scale. Therefore, optical generation and detection should preferably be applied.

A fairly obvious method is the following: One excites the material system locally at a certain point in space. This generates particles or excitations locally, which then spread out in space. In general, these particles or excitations can be described by wave packets. At a later time, one monitors the arrival of the particles or excitations at a different point in space. This is a kind of time-of-flight experiment. In principle, it can be used to distinguish ballistic from diffusive transport. These two transport modes are limiting cases, which apply if there is no scattering (ballistic) or if the mean free path is short compared to the distance the particle has moved (diffusion). The length scales of these transport modes are in general quite short, often even shorter than the wavelength of light. Therefore, the optical excitation has to be generated and monitored on extremely small length scales. This requires sophisticated methods, like scanning-near-field-optical microscopy (SNOM), which suffer from inherent limitations like limited temporal resolution. For the analysis of spatial dynamics on longer length scales, solid immersion lenses are useful tools to generate local excitations.

Fortunately, alternative methods have been developed to generate currents and to detect them on ultrafast time scales that do not rely on high spatial resolution. Optical excitation by suitably designed light pulses can produce net currents of the optically excited electrons and holes rising on the time scale of the optical pulses. This scheme is discussed in detail in Chap. 18.

3

Few-Level Systems

Historically, nonlinear frequency- or time-resolved optical experiments were first performed on atomic and molecular systems and on isolated centers in otherwise transparent solids. On the length scale of the incident light, the spatial extension of the excited species is negligible, such that one can consider effectively zero-dimensional objects which are characterized by a discrete set of electronic levels. In thermodynamic equilibrium, i.e., before the light arrives at the sample, the energetically lowest levels are occupied according to the Pauli exclusion principle. Pioneering theoretical approaches have been developed for these systems. They have the advantage of yielding relatively simple equations and physically intuitive interpretations for the various nonlinear phenomena.

In solids, the discrete electronic levels are merged into energy bands and the different electronic states are often strongly correlated due to Coulombic and other many-body interactions, such that the theoretical analysis is rather involved. In order to gain some first insights into the optical properties of such solids, the few-level systems can often be viewed as a useful starting point. The most simple model for a solid is a linear arrangement of point-like systems, denoted as "sites", with a sufficiently small spatial separation allowing the electrons to tunnel from site to site. Restricting the tunneling to nearest neighbors in the chain of sites leads to a one-dimensional tight-binding model of a solid, which is discussed in the next chapter.

In this chapter, however, we start at an even simpler level and analyze two-level absorbers. This concept is then generalized to three- and M-level absorbers. Finally, we investigate a quasi-continuum by assuming $M \to \infty$ in a finite energy interval.

For the theoretical description of dynamical processes of few-level systems, we choose the Schrödinger picture. In this chapter, we use the bra–ket notation, which allows us to write the Hamiltonian, the relevant observables, and the equations of motions for their expectation values in a very transparent way.

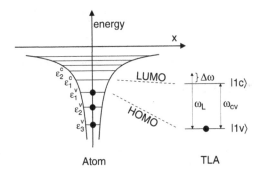

Fig. 3.1. The atomic multilevel system is reduced to a two-level absorber. The two-level absorber represents the optical excitations from the HOMO to the LUMO, i.e., the excitation of the ground state with the lowest frequency

3.1 Two-Level Absorbers

Let us consider an atom in its ground state, see Fig. 3.1. For simplicity, we assume for the moment that all levels are nondegenerate and neglect the electron spin.

Starting from the lowest electronic level, all levels are filled with a single electron up to the energetically highest occupied level. To comply with the notation used in later chapters, we count the occupied levels starting from the highest one downwards and denote them by superscripts v. The highest occupied level has therefore the energy ϵ_1^v and the corresponding state vector is written as $|1v\rangle$. In molecules, the highest occupied levels are called "HOMO"s (highest occupied molecular orbitals).

The lowest unoccupied level has the energy ϵ_1^c and state vector $|1c\rangle$. In molecules, the lowest unoccupied levels are called "LUMO"s. Additionally, there are also higher unoccupied levels. For simplicity, we use later the notion HOMO and LUMO referring to all occupied and unoccupied levels, respectively.

Let us concentrate on these two levels with energies ϵ_1^v and ϵ_1^c, respectively. Their energy separation is

$$\hbar\omega_{cv} = \epsilon_1^c - \epsilon_1^v. \tag{3.1}$$

We assume that the external light field has a central frequency ω_L close to ω_{cv} and a spectral width smaller than the spacing of the atomic levels. The frequency difference $\Delta\omega = \omega_L - \omega_{cv}$ is called detuning. When we discuss the situation of pulsed excitation, we have to keep in mind that pulses have a certain spectral width determined by the Fourier transform of their temporal envelope.

For sufficiently small detuning, the optical excitation predominantly involves the states $|1v\rangle$ and $|1c\rangle$. The other levels are not unaffected under pulsed excitation since they adiabatically follow the pulse amplitude to an extent that is determined by their detuning, i.e., $\hbar\omega_L - (\epsilon_i^c - \epsilon_j^v)$. However, without further interactions, these off-resonant excitations return to the un-excited state as soon as the pulse is turned off.

Therefore, if we are interested in resonant excitation and do not consider additional interactions, we may omit all levels other than $|1v\rangle$ and $|1c\rangle$ from the model and are left with the two-level absorber model, see Fig. 3.1.

Mathematically, this system can be described in a two-dimensional Hilbert space. A natural basis is given by the states $|1v\rangle$ and $|1c\rangle$, which are ortho-normal

$$\langle 1v|1v\rangle = 1,$$
$$\langle 1c|1c\rangle = 1,$$
$$\langle 1v|1c\rangle = 0, \tag{3.2}$$

and form a complete set

$$|1v\rangle\langle 1v| + |1c\rangle\langle 1c| = 1. \tag{3.3}$$

The material Hamiltonian $\hat{H}_0$ is then

$$\hat{H}_0 = |1v\rangle\epsilon_1^v\langle 1v| + |1c\rangle\epsilon_1^c\langle 1c|, \tag{3.4}$$

with the site energies

$$\epsilon_1^v = \langle 1v|\hat{H}_0|1v\rangle,$$
$$\epsilon_1^c = \langle 1c|\hat{H}_0|1c\rangle. \tag{3.5}$$

The strength of the interaction with the light field of this two-level absorber is determined by the dipole moment between the two states. The corresponding dipole operator is

$$\hat{d} = e\hat{r}, \tag{3.6}$$

where e is the electronic charge and $\hat{r}$ the position operator. In the two-dimensional Hilbert space, this operator can be expressed by, see (3.3),

$$\begin{aligned}
\hat{d} = 1 \cdot \hat{d} \cdot 1 \\
= |1v\rangle\langle 1v|\hat{d}|1v\rangle\langle 1v| \\
+ |1c\rangle\langle 1c|\hat{d}|1c\rangle\langle 1c| \\
+ |1v\rangle\langle 1v|\hat{d}|1c\rangle\langle 1c| \\
+ |1c\rangle\langle 1c|\hat{d}|1v\rangle\langle 1v|.
\end{aligned} \tag{3.7}$$

The first two lines contain matrix elements $\langle 1v|\hat{d}|1v\rangle$ and $\langle 1c|\hat{d}|1c\rangle$ which we call intraband dipole moments in anticipation of the situation in solids discussed in later chapters. In the simple atomic case considered here, these intraband quantities vanish due to symmetry.

The optical interband dipole matrix element of the two-level absorber is defined by

$$\boldsymbol{\mu}_{vc} = \langle 1v|\hat{\boldsymbol{d}}|1c\rangle. \tag{3.8}$$

The optical interband polarization operator is then given as

$$\hat{P} = \boldsymbol{\mu}_{vc}|1v\rangle\langle 1c| + H.C. \tag{3.9}$$

This operator describes transitions between the lower to the upper state of our system. Since we want to analyze the effects of optical excitation in the energy range close to ω_{cv}, we use the light–matter interaction Hamiltonian

$$\hat{H}_L = -\boldsymbol{E}(t) \cdot \hat{\boldsymbol{P}}(t). \tag{3.10}$$

The light field consists of components that travel in different directions and is given by

$$\boldsymbol{E}(t) = \sum_{l=1}^{n}(\boldsymbol{E}_l(t)\exp[i\boldsymbol{k}_l \cdot \boldsymbol{r} - i\omega_L t] + c.c.). \tag{3.11}$$

In this book, we work entirely in the semiclassical limit, i.e., we treat the light field as classical, but fully include the quantum nature of the material excitations. In various of the equations, we use $H.C.$ or $c.c.$ which stand for Hermitian conjugate and complex conjugate, respectively.

Generally, there is also the possibility that light couples to higher multipole moments of the atom. However, in this book we exclusively treat optical dipole transitions. If the dipole moment is zero, the optical transition is called (dipole-) forbidden.

The field $\boldsymbol{E}(t)$ may consist of a series of n pulses arriving with wave vectors $\boldsymbol{k}_l$, polarization vectors $\boldsymbol{e}_l$, and temporal envelopes $E_l(t)$

$$\boldsymbol{E}_l(t) = E_l(t)\boldsymbol{e}_l. \tag{3.12}$$

The unit vectors $\boldsymbol{e}_l$ can be used to describe linear or circular polarization of the exciting light field. If the light field propagates in free space in the z direction, $\boldsymbol{e}_l = \boldsymbol{e}_x, \boldsymbol{e}_y$ denote linear polarization in the x and y direction, respectively, while $\boldsymbol{e}_l = (\boldsymbol{e}_x \pm i\boldsymbol{e}_y)/\sqrt{2}$ denote left- and right-handed circular polarization.

The vector character of the electric field and of the optical dipole matrix element has to be taken seriously if realistic semiconductors are to be modeled. On the other hand, in order to gain insights into many problems of fundamental interest, the vector character of these quantities can often be ignored.

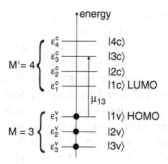

Fig. 3.2. A few-level system consisting of $M = 3$ lower and $M' = 4$ upper states. The optical dipole matrix element μ_{13} for the transition $|1v\rangle \rightarrow |3c\rangle$ is indicated

3.2 Three- and More-Level Absorbers

In some situations, the laser pulse spectrum is sufficiently wide, such that more than just two levels are involved in the resonant excitation. We also have to consider models, where the HOMOs and the LUMOs are degenerate.

The single site in the ground state is then characterized by M occupied levels with energies ϵ_1^v to ϵ_M^v and M' unoccupied levels ϵ_1^c to $\epsilon_{M'}^c$, see Fig. 3.2. The corresponding dipole matrix elements are $\mu_{\nu\nu'}$ with $\nu = 1, \ldots, M, \nu' = 1, \ldots, M'$. Accordingly, we now have an $(M + M')$-dimensional Hilbert space. The material Hamiltonian reads

$$\hat{H}_0 = \sum_{\nu=1}^{M} \epsilon_\nu^v |\nu v\rangle\langle\nu v| + \sum_{\nu'=1}^{M'} \epsilon_{\nu'}^c |\nu'c\rangle\langle\nu'c|. \tag{3.13}$$

The light–matter interaction in this case is

$$\hat{H}_L = -\boldsymbol{E}(t) \cdot \hat{\boldsymbol{P}}, \tag{3.14}$$

with the electric field from (3.11) and the polarization operator

$$\hat{\boldsymbol{P}} = \sum_{\nu=1}^{M} \sum_{\nu'=1}^{M'} \boldsymbol{\mu}_{\nu\nu'} |\nu v\rangle\langle\nu'c| + H.C. \tag{3.15}$$

3.3 Continua

It is also possible to use a system of M-level absorbers to model a continuum of states. Consider a site with a single occupied ground state $|1v\rangle = |g\rangle$ at the energy $\epsilon_g = 0$ and M' unoccupied states $|\epsilon_{\nu'}^c\rangle$, $\nu' = 1, \ldots, M'$, see Fig. 3.3. Imagine this system to be excited with a laser pulse that has a spectral width $\delta\omega$. In order to correctly model the excitation of a continuum, there have to be sufficiently many levels within the interval $\delta\omega$.

In reality, there is always some interaction not described explicitly by our model Hamiltonian. These additional interactions lead to some broadening of the optical resonances of the various closely spaced, but discrete, levels. This effect is called *homogeneous broadening*. In the calculations, this broadening is represented by phenomenological damping, i.e., a decay rate γ (homogeneous width) is inserted into the equations. The individual resonance lines then assume a Lorentzian spectrum with some width δ_ω. If one wants these Lorentzians to represent a smooth continuous spectrum, the level density has to be so large that the spacing of nearest-neighbor resonances is smaller than δ_ω.

Whereas the linear optical spectra of ideal discrete level systems consist of $\delta(\omega - \omega_i)$-like peaks, weighted by the square of the respective optical dipole moments $|\mu_i|^2$, the excitation into a continuum yields a smooth spectrum proportional to $|\mu(\epsilon)|^2$, see Fig. 3.4.

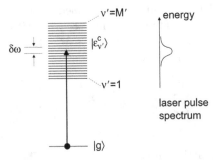

Fig. 3.3. A quasi-continuum is represented by M' discrete, closely spaced levels. The light pulse excites transitions with a weight proportional to the laser spectrum

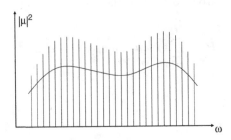

Fig. 3.4. Assigning a finite, but small width to the individual transitions leads to a smooth continuum

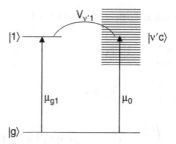

Fig. 3.5. From a common ground state $|g\rangle$, optical transitions are given by matrix elements μ_{g1} into a single discrete state $|1\rangle$ and into an unstructured continuum with constant density of states g and constant dipole matrix element μ_0. The discrete state is coupled to all continuum states with a coupling matrix element $V_{\nu'1}$. [After T. Meier et al., Phys. Rev. **B51**, 13977 (1995)]

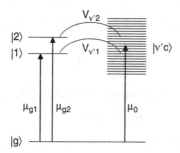

Fig. 3.6. The same situation as in the previous figure, but with two discrete states. [After T. Meier et al., Phys. Rev. **B51**, 13977 (1995)]

3.4 Fano Situations

Interesting spectral and temporal signatures arise in systems that have a common ground state, $|1v\rangle = |g\rangle$ at energy $\epsilon_g = 0$, and one or more $(M' = 1, 2, \ldots)$ discrete optical excitations into states $|\lambda'\rangle$ with energies $\epsilon_{\lambda'}^c$ that are degenerate with an excitation into a continuum, see Figs. 3.5 and 3.6. Following the previous paragraph, the continuum is represented by a quasi-continuum having a dense spectrum of discrete levels at energies $\epsilon_{\nu'}$ and states $|\nu'c\rangle$. While the true continuum has a continuous density of states distribution $\rho(\epsilon)$, the quasi-continuous version consists of a dense sequence of peaks $\sum_{\nu'} \delta_L(\epsilon - \epsilon_{\nu'})$, where $\delta_L(\epsilon - \epsilon_{\nu'})$ denotes Lorentzian lines.

Let us assume that there is coherent (for a precise definition of this notion see the next chapter) tunneling from the discrete excited states to the continuum states described by a coupling $V_{\nu'\lambda'} = \langle \nu'c|H_0|\lambda'\rangle$. The Hilbert space is spanned by the states $|g\rangle$, $|\nu'c\rangle$, and $|\lambda'\rangle$, i.e.,

$$1 = |g\rangle\langle g| + \sum_{\lambda'=1}^{M'} |\lambda'\rangle\langle\lambda'| + \sum_{\nu'} |\nu'c\rangle\langle\nu'c| \qquad (3.16)$$

and the material Hamiltonian expressed in this discrete basis $|g\rangle, |\nu'c\rangle, |\lambda'\rangle$ is

$$\hat{H}_0 = \sum_{\lambda'=1}^{M'} \epsilon_{\lambda'}^c |\lambda'\rangle\langle\lambda'|$$
$$+ \sum_{\nu'} \epsilon_{\nu'} |\nu'c\rangle\langle\nu'c|$$
$$+ \left(\sum_{\lambda'=1}^{M'} \sum_{\nu'} V_{\nu'\lambda'} |\nu'c\rangle\langle\lambda'| + H.C. \right). \tag{3.17}$$

The light–matter interaction is in this basis given by

$$\hat{H}_L = -\boldsymbol{E}(t) \cdot \left[\sum_{\lambda'=1}^{M'} (\boldsymbol{\mu}_{g\lambda'} |g\rangle\langle\lambda'| + H.C.) + \sum_{\nu'} (\boldsymbol{\mu}_{g\nu'} |g\rangle\langle\nu'c| + H.C.) \right]. \tag{3.18}$$

In 1961, U. Fano showed how the material Hamiltonian H_0 can be diagonalized, i.e.,

$$\hat{H}_0 = \sum_{\kappa'} \epsilon_{\kappa'} |\kappa'\rangle\langle\kappa'|. \tag{3.19}$$

In the new basis $|\kappa'\rangle$, the light–matter interaction Hamiltonian becomes

$$H_L = -\boldsymbol{E}(t) \cdot \hat{\boldsymbol{P}}, \tag{3.20}$$

with the polarization operator

$$\hat{\boldsymbol{P}} = \sum_{\kappa'} \boldsymbol{\mu}_{g\kappa'} |g\rangle\langle\kappa'| + H.C. \tag{3.21}$$

and new dipole matrix elements $\boldsymbol{\mu}_{g\kappa'}$ which determine the optical response.

In Chap. 9, we calculate the linear optical spectra numerically on the basis of (3.17) and (3.18) as an application of the equation of motion method, which constitutes the central method in this book. There, we also present the analytical result in terms of the basis $|\kappa'\rangle$ as derived by Fano for some simple situations, which yields precisely the same results as our numerical approach.

3.5 Suggested Reading

1. L. Allen and J.H. Eberly, *Optical Resonance and Two-Level Atoms* (Wiley, New York 1975)
2. U. Fano, "Effects of configuration interaction on intensities and phase shifts", Phys. Rev. **124**, 1866 (1961)
3. S. Glutsch, *Excitons in Low-Dimensional Semiconductors*, Springer Series in Solid-State Sciences, Vol. 141 (Springer, Berlin 2004)
4. T. Meier, A. Schulze, P. Thomas, H. Vaupel, and K. Maschke, "Signature of Fano-resonances in four-wave mixing experiments", Phys. Rev. **B51**, 13977 (1995)

4

Coherent Tunneling

In this chapter, we discuss the dynamical process of coherent tunneling. The notion "coherent" means that we omit all interactions with the environment. The dynamics is then governed by the Hamiltonian of the system alone which describes the motion of a particle in the basis of a low-dimensional (here two-dimensional) Hilbert space.

It is not obvious why in a presentation devoted to optics a process like tunneling is treated in detail. However, it is shown that the theoretical approach to coherent tunneling leads to equations of motion that closely resemble those needed to describe coherent optical excitation in discrete-level systems and semiconductors.

We start the discussion in terms of the eigenstates of the system. The results obtained are then rederived on the basis of the equation of motion of the density matrix $\rho(t)$. It is the latter approach that leads us to a deeper understanding of coherent time-dependent phenomena.

4.1 Analysis of Eigenstates

Consider two single-level objects at the positions (called sites in the following) x_l and x_r, where r refers to "right" and l to "left", see Fig. 4.1. The eigenvalues are ϵ_r and ϵ_l, respectively. The normalized eigenstates of the isolated sites are $|r\rangle$ and $|l\rangle$. We assume sufficient separation of the two sites, such that

$$\langle r|l \rangle = 0 \tag{4.1}$$

and

$$|r\rangle\langle r| + |l\rangle\langle l| = 1. \tag{4.2}$$

Strictly speaking, (4.1) is not exactly correct. However, for a sufficiently large separation the finite overlap of the states is small, $\langle r|l \rangle \ll 1$, and may, therefore, be neglected.

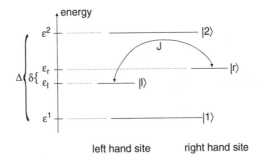

Fig. 4.1. Two states $|l\rangle$ (*left-hand site*) and $|r\rangle$ (*right-hand site*) are coupled by a coupling matrix element J. The delocalized eigenstates of this system are $|1\rangle$ and $|2\rangle$

We now introduce a coupling J between these sites, such that the Hamiltonian reads

$$
\begin{aligned}
\hat{H}_0 &= 1 \cdot \hat{H}_0 \cdot 1 \\
&= \epsilon_r |r\rangle\langle r| + \epsilon_l |l\rangle\langle l| + J(|l\rangle\langle r| + |r\rangle\langle l|),
\end{aligned} \tag{4.3}
$$

where

$$
\begin{aligned}
\epsilon_r &= \langle r|\hat{H}_0|r\rangle, \\
\epsilon_l &= \langle l|\hat{H}_0|l\rangle, \\
J &= \langle l|\hat{H}_0|r\rangle = \langle r|\hat{H}_0|l\rangle.
\end{aligned} \tag{4.4}
$$

We assume that J is negative. It is shown below that in this case the lower (upper) eigenstate of the coupled system has even (odd) symmetry, provided the two sites have equal symmetry.

This Hamiltonian is a good model for, e.g., two quantum wells separated by a barrier. In this case, $\epsilon_{r,l}$ are the levels in the isolated wells on the left-hand and right-hand sides and J describes tunneling through the barrier between both wells.

The coupled system must have two eigenstates $|1\rangle$ and $|2\rangle$ with energies ϵ_1 and ϵ_2, respectively, i.e.,

$$
\begin{aligned}
\hat{H}_0|1\rangle &= \epsilon_1|1\rangle, \\
\hat{H}_0|2\rangle &= \epsilon_2|2\rangle.
\end{aligned} \tag{4.5}
$$

Since the eigenfunctions $|1\rangle$ and $|2\rangle$ have to be orthonormal,

$$\langle 1|2\rangle = 0,$$
$$\langle 1|1\rangle = 1,$$
$$\langle 2|2\rangle = 1, \tag{4.6}$$

the diagonalization corresponds to a rotation in the two-dimensional Hilbert space, such that H_0 is diagonal in the new basis $|1\rangle$ and $|2\rangle$. Multiplying (4.5) by $\langle r|$ and $\langle l|$ from the left and using (4.4), we obtain

$$(\epsilon_r - \epsilon_1)\langle r|1\rangle + J\langle l|1\rangle = 0,$$
$$(\epsilon_l - \epsilon_1)\langle l|1\rangle + J\langle r|1\rangle = 0,$$
$$(\epsilon_r - \epsilon_2)\langle r|2\rangle + J\langle l|2\rangle = 0,$$
$$(\epsilon_l - \epsilon_2)\langle l|2\rangle + J\langle r|2\rangle = 0. \tag{4.7}$$

In order for this set of linear homogeneous equations to be nontrivially solvable, the secular determinant has to vanish, i.e.,

$$(\epsilon_r - \epsilon)(\epsilon_l - \epsilon) - J^2 = 0. \tag{4.8}$$

The two solutions for the eigenvalues are

$$\epsilon_{2,1} = \frac{\epsilon_r + \epsilon_l}{2} \pm \frac{1}{2}\sqrt{(\epsilon_r - \epsilon_l)^2 + 4J^2}. \tag{4.9}$$

We define, see Fig. 4.1,

$$\delta = \epsilon_r - \epsilon_l \tag{4.10}$$

and

$$\Delta = \epsilon_2 - \epsilon_1 = \sqrt{\delta^2 + 4J^2}, \tag{4.11}$$

which leads to

$$\epsilon_{2,1} = \frac{\epsilon_r + \epsilon_l}{2} \pm \frac{1}{2}\Delta. \tag{4.12}$$

Since the diagonalization is a rotation, it yields the eigenstates in the form

$$|1\rangle = \cos\phi|l\rangle + \sin\phi|r\rangle,$$
$$|2\rangle = -\sin\phi|l\rangle + \cos\phi|r\rangle, \tag{4.13}$$

where

$$\langle l|1\rangle = \langle r|2\rangle = \cos\phi,$$
$$\langle r|1\rangle = -\langle l|2\rangle = \sin\phi. \tag{4.14}$$

From the secular equation, (4.7), we obtain

$$\tan\phi = \frac{J}{\epsilon_1 - \epsilon_r} = \frac{\epsilon_r - \epsilon_2}{J} = \frac{J}{\epsilon_l - \epsilon_2} = \frac{\epsilon_1 - \epsilon_l}{J}. \tag{4.15}$$

Furthermore, since from (4.10)–(4.12) we have

$$2(\epsilon_r - \epsilon_2) = \delta - \Delta,$$
$$2(\epsilon_l - \epsilon_2) = -\delta - \Delta, \tag{4.16}$$

we obtain from the secular equation, (4.7),

$$(\delta - \Delta)\cos\phi = 2J\sin\phi,$$
$$(-\delta - \Delta)\sin\phi = 2J\cos\phi. \tag{4.17}$$

This can be combined multiplying the first equation with $\sin\phi$ and the second one with $\cos\phi$ and adding up, which gives

$$\cos\phi\sin\phi = -\frac{2J}{2\Delta}, \tag{4.18}$$

i.e.,

$$\sin^2\phi\cos^2\phi = \frac{J^2}{\Delta^2}. \tag{4.19}$$

Using similar manipulations we obtain

$$\sin^4\phi = \frac{(\delta - \Delta)^2}{4\Delta^2} \tag{4.20}$$

and

$$\cos^4\phi = \frac{(\delta + \Delta)^2}{4\Delta^2}. \tag{4.21}$$

For $\delta = 0$ we have $\sin\phi = \cos\phi = 1/\sqrt{2}$, thus $|1\rangle$ ($|2\rangle$) is symmetric (antisymmetric) as stated above, cf. (4.13). In the basis of the eigenstates of the coupled system, the individual states are

$$|l\rangle = \cos\phi|1\rangle - \sin\phi|2\rangle,$$
$$|r\rangle = \sin\phi|1\rangle + \cos\phi|2\rangle. \tag{4.22}$$

4.2 Population Dynamics

Coherent tunneling denotes a situation where at time $t = 0$ a particle is placed by some means in, say, the left well. Its state is then

$$|t = 0\rangle = |l\rangle, \tag{4.23}$$

which is not an eigenstate and thus shows temporal dynamics for larger times. This dynamics can be calculated by writing down the state of the particle for all times in terms of the eigenstates of the double well, which under this particular initial condition is given by

$$|t\rangle = \cos\phi\, e^{-i\epsilon_1 t/\hbar}|1\rangle - \sin\phi\, e^{-i\epsilon_2 t/\hbar}|2\rangle. \tag{4.24}$$

The probability of finding the particle at time t in one of the wells is given by the diagonal elements of the density matrix operator

$$\hat{\rho}(t) = |t\rangle\langle t| \tag{4.25}$$

taken in the basis of the states $|l\rangle$ and $|r\rangle$, i.e.,

$$\rho_{ll}(t) = \langle l|\rho(t)|l\rangle = |\langle l|t\rangle|^2,$$
$$\rho_{rr}(t) = \langle r|\rho(t)|r\rangle = |\langle r|t\rangle|^2. \tag{4.26}$$

Inserting (4.24), (4.19), (4.20), (4.21), and (4.11) leads to

$$\rho_{ll}(t) = 1 - \frac{4J^2}{\Delta^2}\sin^2\left(\frac{\Delta}{2\hbar}t\right),$$
$$\rho_{rr}(t) = 1 - \rho_{ll}(t). \tag{4.27}$$

This result describes periodic sinusoidal motion of the excitation between the left and the right state. In general, complete transfer is only obtained for identical sites, i.e., $\epsilon_r = \epsilon_l$, thus $\delta = 0$ and $4J^2/\Delta^2 = 1$, see Fig. 4.2. The period of the motion is given by $T = 2\pi\hbar/\Delta$, where Δ, see (4.11), is the

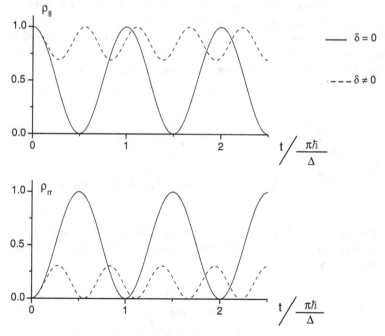

Fig. 4.2. Coherent oscillations for resonant ($\delta = 0$, *solid line*) and off-resonant ($\delta \neq 0$, *dashed line*) tunneling. Initially, at $t = 0$, the electron is situated at the *left-hand site*

energetic separation of the two eigenstates. The same is also true for $\delta \neq 0$. In this case, i.e., if the energies of the two isolated sites differ, there is an incomplete transfer to the right site for a particle having started at the left site, and the period of this motion is given in part also by the energy difference δ. In the limit $\delta \gg J$, the particle essentially stays at the left well, i.e., it is effectively localized there.

4.3 Equation of Motion Approach

More insight into the physics of coherent tunneling can be gained by studying the equation of motion for the density matrix $\hat{\rho}$. It determines the temporal evolution of an ensemble which we assume here to be composed out of identical two-level tunneling systems. One advantage of the density matrix theory is that we obtain a number of relations which are relevant also for optical excitation of two-level systems.

The Liouville-von Neumann equation of motion of the density matrix reads

$$i\hbar \frac{d}{dt}\hat{\rho} = [\hat{H}_0, \hat{\rho}]. \tag{4.28}$$

We are interested in the matrix elements ρ_{ll} and ρ_{rr}, whose equation of motion follows from the respective matrix elements of (4.28). When we write down these equations, we note that they couple to nondiagonal density matrix elements, ρ_{rl} and ρ_{lr}, appearing on the right-hand side of the equations for the diagonal elements ρ_{ll} and ρ_{rr}. Hence, in order to obtain a closed set of equations, we also need the equation of motion for these nondiagonal matrix elements. Using the Hamiltonian of (4.3) and (4.4), we obtain

$$i\hbar \frac{d}{dt}\rho_{ll} = J(\rho_{rl} - \rho_{lr}),$$

$$i\hbar \frac{d}{dt}\rho_{rr} = -J(\rho_{rl} - \rho_{lr}),$$

$$i\hbar \frac{d}{dt}\rho_{rl} = \delta\rho_{rl} + J(\rho_{ll} - \rho_{rr}) = \delta\rho_{rl} + J(1 - 2\rho_{rr}). \tag{4.29}$$

Instead of directly solving this set of coupled differential equations we first define new physical quantities: the "polarization" p, the "current" j, and the "inversion" I, by

$$p = \rho_{rl} + \rho_{lr},$$

$$j = i\rho_{rl} - i\rho_{lr},$$

$$I = 1 - 2\rho_{rr}. \tag{4.30}$$

The "inversion" I describes the dipole moment of the system. If the particle is in the left-hand site, $I = 1$, if it is in the right-hand site, $I = -1$. Its time derivative must therefore be given by the particle "current" j. This is indeed the case, since from (4.29) we obtain

$$\hbar \frac{d}{dt} I = -2Jj. \tag{4.31}$$

The nondiagonal elements of the density matrix, therefore, constitute the flow of the probability to be either in the left-hand or in the right-hand well. The "polarization" has, for the present system, a less obvious interpretation. Its equation of motion reads

$$\hbar \frac{d}{dt} p = -\delta j, \tag{4.32}$$

i.e., it is a constant for identical sites which correspond to $\delta = 0$. The "current" is governed by the equation of motion

$$\hbar \frac{d}{dt} j = \delta p + 2JI. \tag{4.33}$$

Defining the vectors

$$\boldsymbol{S} = (p, j, I),$$
$$\boldsymbol{\Omega} = (-2J, 0, \delta)/\hbar, \tag{4.34}$$

(4.31), (4.32), and (4.33) can be combined into a single equation

$$\frac{d}{dt} \boldsymbol{S} = \boldsymbol{\Omega} \times \boldsymbol{S}, \tag{4.35}$$

which is well-known to us from theoretical mechanics since it describes the precession of the vector $\boldsymbol{S}$ around the fixed vector $\boldsymbol{\Omega}$. In fact, the vector $\boldsymbol{S}$ has a fixed length. To prove this, we define

$$\bar{\rho}_{rl} = e^{i\delta t/\hbar} \rho_{rl},$$
$$\bar{J} = e^{i\delta t/\hbar} J, \tag{4.36}$$

and obtain

$$\hbar \frac{d}{dt} \bar{\rho}_{rl} = -i\bar{J}(1 - 2\rho_{rr}),$$
$$\hbar \frac{d}{dt} \rho_{rr} = -i\bar{J}(\bar{\rho}_{rl})^* + i\bar{J}^* \bar{\rho}_{rl}. \tag{4.37}$$

We now use (4.36) to construct the following identity:

$$\hbar(\bar{\rho}_{rl})^* \frac{d}{dt}\bar{\rho}_{rl} + \hbar\bar{\rho}_{rl}\frac{d}{dt}(\bar{\rho}_{rl})^* = -i\bar{J}(1 - 2\rho_{rr})(\bar{\rho}_{rl})^* + i\bar{J}^*(1 - 2\rho_{rr})\bar{\rho}_{rl}. \quad (4.38)$$

This can be rewritten as

$$\frac{d}{dt}[(\bar{\rho}_{rl})^*\bar{\rho}_{rl}] = -(2\rho_{rr} - 1)\frac{d}{dt}\rho_{rr} = -\frac{1}{4}\frac{d}{dt}(2\rho_{rr} - 1)^2, \quad (4.39)$$

which implies that

$$|\bar{\rho}_{rl}|^2 = |\rho_{rl}|^2 = -\frac{1}{4}(2\rho_{rr} - 1)^2 + c, \quad (4.40)$$

where c is a constant. Since we assumed as initial condition that the particle is at the left site, i.e., $\rho_{rr}(t = 0) = 0$ and $\rho_{rl}(t = 0) = 0$ the constant c must have the value $c = 1/4$. In terms of the p, j and I, we thus find indeed that

$$p^2 + j^2 + I^2 = 1. \quad (4.41)$$

Alternatively we can show, using (4.31)–(4.33), that

$$p\frac{dp}{dt} + j\frac{dj}{dt} + I\frac{dI}{dt} = 0, \quad (4.42)$$

from which, together with the initial condition $p(0) = j(0) = 0$ and $I(0) = 1$, (4.41) follows.

The solution of (4.31), (4.32), and (4.33) can now be obtained easily by taking the time derivative of (4.33) and using the two remaining equations:

$$\hbar^2 \frac{d^2}{dt^2}j + \delta^2 j + 4J^2 j = 0. \quad (4.43)$$

This is the equation of an undamped harmonic oscillator with frequency $\sqrt{\delta^2 + 4J^2}/\hbar = \Delta/\hbar$. With the initial value $I(t = 0) = 1$ and $j(t = 0) = p(t = 0) = 0$, we have

$$j(t) = \frac{2J}{\Delta}\sin\Delta(t/\hbar),$$

$$p(t) = \frac{2J}{\Delta}\frac{\delta}{\Delta}[\cos\Delta(t/\hbar) - 1],$$

$$I(t) = 1 - 2\frac{4J^2}{\Delta^2}\sin^2\left(\frac{\Delta}{2\hbar}t\right). \quad (4.44)$$

The result for I coincides with the result we have obtained above for ρ_{rr}, see (4.27) and (4.30). We see that for resonant tunneling, i.e., $\delta = 0$, the "polarization" $p = 0$ for all times. The vector S then rotates in the j-I plane, touching periodically $I = 1$ and $I = -1$, i.e., it rotates with a frequency $\Delta/\hbar = 2|J|/\hbar$ perpendicular to the vector $\Omega = (-2J/\hbar, 0, 0)$, which points into the p direction. We have optimal transfer of the particle between the sites and the amplitude of the "current" j is also maximal, i.e., unity.

For off-resonant tunneling, Ω has a component δ also in the I direction. The vector S, starting at $t = 0$ from $I = 1$, $p = j = 0$, therefore no longer touches the point $I = -1$. Physically, this means that the transfer of the particle between the wells is incomplete. Accordingly, the current no longer has amplitude unity, but $2J/\Delta < 1$, as already shown in Fig. 4.2.

According to Maxwell's equations, a charge that oscillates periodically between two spatial positions is the source of electromagnetic radiation. The source term in the wave equation is given by the time derivative of j. In semiconductor heterostructures, such periodic oscillations can be excited by pulsed optical excitation. The frequency of the motion can be adjusted by using suitably fabricated heterostructures. However, in real systems the periodic motion will always be damped by a number of interaction processes we have not discussed so far. Characteristic time scales for these interactions are in the pico- or even subpicosecond range. Therefore, the electromagnetic radiation has to be in the terahertz frequency range in order to be observable.

All of the formal derivations we have presented here will be recovered when we discuss optical excitation of two-level systems. In that context, the frequency $\Delta/\hbar$ will be called the Rabi frequency, and the notions "inversion" and "polarization" will become more transparent.

The discussion of coherent tunneling was included in this presentation in order to demonstrate that the dynamics of a quantum system can be treated by applying two different approaches. In the tunneling scenario the dynamical variables we are primarily interested in are the probabilities to find the particle either in the left–hand or in the right–hand state. In the first approach we have started to determine the eigenvalues and eigenstates $|1\rangle$ and $|2\rangle$ of the total system. The dynamical variables $\rho_{ll}(t)$ and $\rho_{rr}(t)$ have then been obtained by projecting the time-dependent mixed state $|t\rangle$, expressed by the eigenstates, onto $|l\rangle$ and $|r\rangle$, respectively. In the second approach we solved the equation of motion of the density matrix in the basis of $|l\rangle$ and $|r\rangle$. The respective matrix elements are the dynamical variables of interest. No diagonalization was necessary. Therefore, we will in the forthcoming chapters develop and apply the equation of motion method to treat optics. In systems that are characterized by both many–particle interaction and disorder the diagonalization of the system Hamiltonian is either rather tedious or even impossible. On the other hand, the equation of motion method allows us to compute the desired dynamical variables of the system without having to diagonalize the Hamiltonian.

4.4 Suggested Reading

1. A. Wurger, *From Coherent Tunnelling to Relaxation: Dissipative quantum dynamics of interacting defects*, Springer Tracts in Modern Physics, Vol. 135 (Springer, Berlin 1997)

5

The Semiconductor Model

In this chapter, we set up our one-dimensional semiconductor model which will be the basis of all the subsequent analysis. Although some features of a one-dimensional model seem to be artificial, there exist real systems which are effectively one-dimensional, like quantum wires, or semiconductor nanorings. Furthermore, this model is a good tutorial approach to introduce the relevant physical phenomena and their theoretical interpretation. In fact, a number of current research problems could so far only be solved for one-dimensional models because of the complexity of the relevant equations and the limited capabilities of even the most advanced supercomputers.

We start by introducing the model of noninteracting particles. The Coulomb interaction is added to the model later. Its inclusion leads to the fundamental semiconductor model used in this book. It is shown how ordered and disordered structures can be described and how the electron–phonon interaction can be included. Finally, the light–matter interaction is introduced.

5.1 Noninteracting Particles

5.1.1 The 1 × 1-Tight-Binding Model

We start from an ensemble of N two-level absorbers as shown in Fig. 5.1. They are positioned along a one-dimensional line at places ia, where $i = 1, \ldots, N$, N is the number of the two-level absorbers, and a is the lattice constant. One could also use a disordered spatial arrangement, where the nearest-neighbor distances are not constant but distributed around some mean value. Disorder, however, will be considered later in the form of energetic disorder, i.e., varying site energies.

The two states of any given two-level absorber i are denoted by $|ic\rangle$ and $|iv\rangle$. They have energies ϵ_i^c and ϵ_i^v, respectively. We assume that these states are all mutually orthonormal

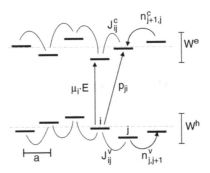

Fig. 5.1. The one-dimensional tight-binding model for the semiconductor. As described in the text, $J_{ij}^{c/v}$ are the couplings between the nearest-neighbor sites. Optical transitions are indicated by the *vertical arrow* from the valence to the conduction band. Diagonal disorder, i.e., site energies in the valence and the conduction band vary within widths W^h and W^e, respectively, is also added to the model. For other symbols see text

$$\langle ic|jc \rangle = \delta_{ij},$$
$$\langle iv|jv \rangle = \delta_{ij},$$
$$\langle ic|jv \rangle = 0. \tag{5.1}$$

However, there is tunneling from two-level absorber number i to j described by tunneling matrix elements J_{ij}^c and J_{ij}^v. We always assume that tunneling is only possible between nearest neighbors. This restriction is indicated by the notation $\langle ij \rangle$ in the summation indices. The material Hamiltonian $\hat{H}_0$ then reads

$$\hat{H}_0 = \sum_{i=1}^{N} \epsilon_i^c |ic\rangle\langle ic| + \sum_{\langle ij \rangle}^{N} J_{ij}^c |ic\rangle\langle jc|$$
$$+ \sum_{i=1}^{N} \epsilon_i^v |iv\rangle\langle iv| + \sum_{\langle ij \rangle}^{N} J_{ij}^v |iv\rangle\langle jv|. \tag{5.2}$$

Even though this two-band single-particle model is suitable as a basis for the study of many fundamental questions, in order to model real semiconductor heterostructures more closely, we have to add some features allowing us to include realistic optical selection rules.

5.1.2 The 2 × 2-Tight-Binding Model

Often, one is interested in optical excitations close to the fundamental band edge, i.e., only states close to the minimum of the conduction band and the

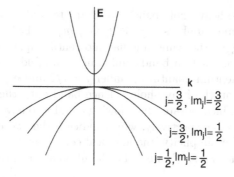

Fig. 5.2. Typical band structure around the fundamental gap at $k = 0$ of bulk III–V semiconductors

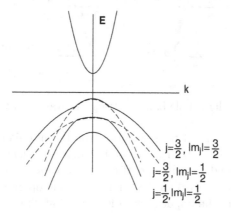

Fig. 5.3. Typical band structure around the fundamental gap at $k = 0$ of III–V semiconductor quantum wells. *Dashed lines* indicate heavy- and light-hole bands without their mutual interaction. They are shifted with respect to each other due to the reduced symmetry of the quantum well

maximum of the valence band are considered. In ordered bulk III–V semiconductors, e.g., the top of the valence band is three-fold degenerate if spin degeneracy and spin–orbit coupling are neglected. Including spin degeneracy, the top of the valence band is six-fold degenerate. The spin–orbit coupling leads to a splitting which results in a two-fold (Kramers) degenerate split-off band plus degenerate heavy- and light-hole bands which are still degenerate with each other in the center of the Brillouin zone, see Fig. 5.2.

In a semiconductor heterostructure, e.g., a quantum well, the reduced symmetry removes the heavy–light-hole degeneracy at the zone center, as shown in Fig. 5.3. This splitting is typically of the order of several meV depending on the material and structural parameters. For many purposes, it is sufficient to consider only the two-fold degenerate highest valence band, which is the heavy-hole band in unstrained heterostructures. At the zone center, the two degenerate heavy-hole bands are characterized by the total angular momentum quantum numbers $j = 3/2$, i.e., $m_j = \pm 3/2$, while the light-hole bands belong to the same angular momentum quantum number, but $m_j = \pm 1/2$. The conduction band minimum is two-fold degenerate and has total angular momentum quantum number $j = 1/2$ and $m_j = \pm 1/2$.

Concentrating on the heavy-hole valence band only, the resulting 2×2-band model can be incorporated by allowing c and v to have two values each, c_1 and c_2, v_1 and v_2, respectively, which indicate the two degenerate states per energy level. For simplicity, the identical couplings $J_{ij}^{c,v}$ and energies $\epsilon_i^{c,v}$ are assumed for the two subbands. The Hamiltonian then reads

$$
\hat{H}_0 = \sum_{i=1}^{N} \sum_c \epsilon_i^c |ic\rangle\langle ic| + \sum_{\langle ij \rangle} \sum_c J_{ij}^c |ic\rangle\langle jc|
$$
$$
+ \sum_{i=1}^{N} \sum_v \epsilon_i^v |iv\rangle\langle iv| + \sum_{\langle ij \rangle} \sum_v J_{ij}^v |iv\rangle\langle jv|. \tag{5.3}
$$

The inclusion of the light-hole band is straightforward and yields a 2×4-band model.

It is furthermore useful to write down the Hamiltonian in second quantization, which we need later to treat the many-particle interactions in a transparent way. We define single-particle operators a_i^c (a_i^v) which destroy an electron in the upper (lower) level c (v) at site i, and operators $a_i^{c\dagger}$($a_i^{v\dagger}$) which create an electron in the upper (lower) level c (v) at site i.

Sometimes, we introduce an extra "hat" to indicate the operator character of some quantities, e.g., $\hat{P}$ for the polarization operator. However, in order not to clutter the equations with redundant symbols we use this notation only if it is necessary to avoid misunderstandings.

We define the Hamiltonian matrices $\mathbf{T^c}$ and $\mathbf{T^v}$ with diagonal elements

$$
T_{ii}^c = \epsilon_i^c,
$$
$$
T_{ii}^v = \epsilon_i^v, \tag{5.4}
$$

and nondiagonal elements

$$
T_{ij}^c = J^c, \quad \text{for } i, j \text{ nearest neighbors,}
$$
$$
T_{ij}^v = J^v, \quad \text{for } i, j \text{ nearest neighbors,}
$$
$$
T_{ij}^c = T_{ij}^v = 0, \quad \text{else,} \tag{5.5}
$$

respectively, i.e., the nondiagonal elements are nonzero only if i, j denote nearest-neighbor sites. In this notation $\hat{H}_0$ reads

$$
\hat{H}_0 = \sum_{i,j=1}^{N} \sum_c T_{ij}^c a_i^{c\dagger} a_j^c + \sum_{i,j=1}^{N} \sum_v T_{ij}^v a_i^{v\dagger} a_j^v. \tag{5.6}
$$

5.1.3 The Electron–Hole Picture

Instead of discussing electrons in different states (bands), it is often more convenient to introduce the notion of electrons and holes, i.e., to work in the electron–hole picture. Here, one defines holes as missing electrons in the valence band. Generally, in equilibrium in semiconductors the upper band (conduction band) is not occupied by electrons and the lower band (valence band) is totally filled by N electrons (model with nondegenerate states) or $2N$ electrons (model with two-fold degenerate states). Hence, it is convenient to describe excited states, where electrons have been promoted from the valence into the conduction band, by counting the conduction-band electrons and the valence-band holes.

Formally, we introduce the new quasi-particles by defining (conduction-band) electron $(c_i^e, c_i^{e\dagger})$ and (valence-band) hole $(d_i^h, d_i^{h\dagger})$ operators via

$$c_i^e = a_i^c,$$
$$c_i^{e\dagger} = a_i^{c\dagger},$$
$$d_i^h = a_i^{v\dagger},$$
$$d_i^{h\dagger} = a_i^v. \tag{5.7}$$

The charge of the holes is that of the missing electrons, i.e., the hole charge is -1 times the electron charge. In this notation the Hamiltonian reads

$$\hat{H}_0 = \sum_{i,j=1}^{N} \sum_e T_{ij}^e c_i^{e\dagger} c_j^e + \sum_{i,j=1}^{N} \sum_h T_{ij}^h d_i^{h\dagger} d_j^h, \tag{5.8}$$

with

$$T_{ij}^e = T_{ij}^c,$$
$$T_{ij}^h = -T_{ij}^v. \tag{5.9}$$

Since we insist on "normal ordering", i.e., we always (anti-)commute creation operators to the left-hand side of destruction operators, the transformation to the electron–hole picture generates an additional constant term in the Hamiltonian. This term is omitted since it does not contribute to the system dynamics.

It is often useful to take the zero of the energy scale to coincide with the center of the gap between the valence and the conduction band, see Fig. 5.4. In the electron–hole picture, the electron energies are $(\epsilon_i^e = \epsilon_i^c$ and $J^e = J^c)$ and the hole energies are $(\epsilon_i^h = -\epsilon_i^v$ and $J^h = -J^v)$. The sums $\sum_e, \sum_h$ take care of the two degenerate levels (bands). They are absent if a model with nondegenerate states is considered. In this case, the superscripts e and h on the particle operators are redundant and can be omitted. In the remainder of this book, we always use the electron–hole picture in our semiconductor models.

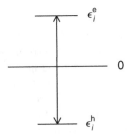

Fig. 5.4. The definition of electron ϵ_i^e and hole ϵ_i^h site energies. Both are positive, measured with respect to the zero of the energy scale, which coincides with the center of the gap between the valence and the conduction band

5.1.4 Periodic Boundary Conditions

The model introduced so far is a linear chain of N sites with the total length $L = Na$. In order to find the eigenstates and eigenvalues of this chain, we have to specify the boundary conditions at the ends. Following usual practice in solid state physics, we assume periodic boundary conditions, i.e., we consider a ring with circumference L, such that site $N + i$ is identical to site i. In particular, site $N + 1$ is identical to site 1, such that there are couplings J^e and J^h also between the now nearest neighbors N and 1.

For a model of semiconductor rings, these boundary conditions represent the correct physical situation. For a model of a large linear wire, they constitute a trick which leads to enumerable quantum numbers (states). It also enables us to introduce external electric fields without running into problems with boundary conditions at the ends or with ill-defined Hamiltonians. In this case, one has to take L large enough, such that, effectively, we are dealing with an infinite linear chain. We take these periodic boundary conditions as a pragmatic approach without going into the details of its justification which requires the discussion of open versus closed quantum systems and of dissipation.

5.1.5 Ordered and Disordered Systems

The Hamiltonian, (5.8), describes an ordered system if the upper level energies ϵ_i^e are all equal to ϵ^e, and the lower level energies ϵ_i^h are all equal to ϵ^h, as depicted in Fig. 5.5a. In this case, it is straightforward to diagonalize the Hamiltonian (see next chapter).

A disordered system is modeled by choosing the site energies $\epsilon_i^{e,h}$ randomly from given distributions $g_e(\epsilon)$ and $g_h(\epsilon)$ of width W^e and W^h, respectively. This kind of disorder is called diagonal disorder and the corresponding single-band Hamiltonian is known as the *Anderson Hamiltonian*, which is the generic model of the disorder-driven metal–insulator transition. It is the basis of most theories of disorder-driven Anderson localization and Anderson

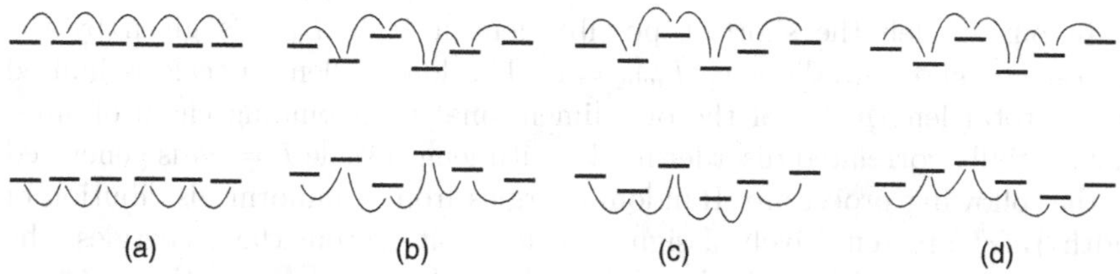

Fig. 5.5. The tight-binding model used in this presentation. (a) The ordered semiconductor. The (b) correlated, (c) the anticorrelated, and (d) the uncorrelated disordered semiconductor situation

metal–insulator transition. The same model has also been used in theories of optical absorption in disordered semiconductors.

If the upper and lower site energies are drawn independently from the two distribution functions, we have the situation of uncorrelated disorder which is illustrated in Fig. 5.5d. Correlated disorder means that the upper energies ϵ_i^e are drawn randomly from g_e, and the lower levels then follow from

$$\epsilon_i^h = \frac{|J^h|}{|J^e|}\epsilon_i^e, \tag{5.10}$$

while anticorrelated disorder is realized by

$$\Delta\epsilon_i^h = -\frac{|J^h|}{|J^e|}\Delta\epsilon_i^e, \tag{5.11}$$

see Figs. 5.5b and 5.5c, respectively, where $\Delta\epsilon_i^{h,e} = \epsilon_i^{h,e} - \epsilon_0^{h,e}$ are the deviations from the average values $\epsilon_0^{h,e} = \langle\epsilon_i^{h,e}\rangle$.

Note that in the case of correlated disorder the eigenstates $|\alpha e\rangle$ in the conduction band and $|\beta h\rangle$ in the valence band are pair wise identical, $|\alpha e\rangle = |\beta h\rangle$. The lowest eigenstate in the conduction band is identical to the uppermost state in the valence band, etc. Their energies scale like

$$E_\alpha^h = \frac{|J^h|}{|J^e|}E_\beta^e, \tag{5.12}$$

where α and β numerate the corresponding eigenstates in the conduction and valence band, respectively.

Our model for a disordered semiconductor allows us also to consider disorder potentials with various length scales. In some real systems, such as amorphous semiconductors, the disorder potentials vary strongly over distances comparable to interatomic separations. On the other hand, also extremely long-ranged disorder potentials are known, e.g., in weakly doped partially compensated semiconductors due to Coulomb fields originating from charged impurities, or in heterostructures with long-ranged interface roughness. Also, the simultaneous presence of both long-ranged and short-ranged disorder is quite common in semiconductor heterostructures.

In our model, the shortest possible length scale $L_{\min}$ is given by the nearest-neighbor site distance $L_{\min} = a$. The longest length scale is limited by the total length Na of the one-dimensional tight-binding chain of sites. The spatially correlated disorder model with length scale $L = na$ is generated by the following procedure: Random energies from a uniform distribution of width $W^{e,h}$ are tentatively assigned to each site. From these energies, the definitive energy of site i is determined from the average over the tentative energies of site i and that of its $n - 1$ neighbors. This sliding average results in the desired spatially correlated disorder potential.

5.2 Interacting Particles

The many-particle Coulomb interaction is described by a term

$$\hat{H}_C = \frac{1}{2} \sum_{ij} \left(\sum_{e'} c_i^{e'\dagger} c_i^{e'} - \sum_{h'} d_i^{h'\dagger} d_i^{h'} \right) V_{ij} \left(\sum_{e} c_j^{e\dagger} c_j^{e} - \sum_{h} d_j^{h\dagger} d_j^{h} \right),$$

$$(5.13)$$

which has the monopole–monopole form. When deriving the tight-binding representation of the interaction term from more realistic formulations of the crystal Hamiltonian, one realizes that there are actually also terms describing monopole–dipole and dipole–dipole interactions, which are omitted in our present model. These additional contributions result from the fact that, in reality, the solid is not built up from point-like objects (sites), but from spatially extended atoms or even groups of atoms. The interaction potential between these extended objects is then expanded in terms of multipoles. In our analysis, only the lowest-order contribution, i.e., the monopole–monopole term, is considered. As a consequence, in the present form our model is unable to describe, e.g., the propagation of Frenkel excitons which is due to a dipole–dipole interaction term.

The interaction term, (5.13), describes the repulsion of equally charged particles (electron–electron and hole–hole) and the attraction between opposite charges, i.e., between electrons and holes. This form is valid for both the ordered and the disordered cases.

For the case $e = e'$ (equal spin) and $i = j$, (5.13) contains a term

$$\frac{1}{2} \sum_i \sum_e c_i^{e\dagger} c_i^{e} V_{ii} c_i^{e\dagger} c_i^{e}.$$

$$(5.14)$$

Because of

$$c_i^{e\dagger} c_i^{e} c_i^{e\dagger} c_i^{e} = c_i^{e\dagger} c_i^{e},$$

$$(5.15)$$

this term adds to the single-particle energies $T_{ii}^e = \epsilon_i^e$ in (5.8) and is therefore assumed to be absorbed therein. The same holds for the corresponding hole contribution.

However, if $e \neq e'$ (different spin), but still $i = j$, we have a term

$$\frac{1}{2} \sum_i c_i^{e'\dagger} c_i^{e'} V_{ii} c_i^{e\dagger} c_i^e, \tag{5.16}$$

which describes the repulsion of two electrons with opposite spin that sit at the same site. There exist models for many-particle effects like magnetism and the Mott–Hubbard metal–insulator transitions which exclusively consider this on-site Coulomb repulsion term. (In materials which show these effects the single-particle eigenstates are well localized around certain atoms. This justifies the neglect of all interaction terms except the on-site term. In works dealing with the Mott–Hubbard transition a different notation is used. Instead of the nearest-neighbor coupling J the "hopping" parameter is denoted by t, while the on-site Coulomb repulsion V_{ii} is called the Hubbard correlation energy U.) In our treatment such effects for the electrons, and correspondingly for the holes, are included here but they are not considered to be dominant.

The matrix element V_{ij} is given by

$$V_{ij} = U_0 \frac{a}{a|i-j| + a_0}, \tag{5.17}$$

where U_0 quantifies the strength of the interaction. In one-dimensional tight-binding systems with the many-particle interaction given by (5.13) with $V_{ij} \propto |i-j|^{-1}$ the attractive electron–hole Coulomb interaction leads to a diverging ground-state energy. One way to overcome this model-dependent problem is to regularize the potential by adding a term a_0 in the denominator, which is of the order of the lattice constant a. This addition represents the fact that true physical systems are never one-dimensional in the mathematical sense, but have a finite small extension in the other two dimensions.

One of the applications of this model is related to semiconductor nanorings. In a small ring-like structure the distance between sites i and j on the circumference of the ring is $\frac{Na}{\pi} \sin\left[\frac{\pi}{N}|i-j|\right]$, such that the Coulomb interaction can be written as

$$V_{ij} = U_0 \frac{a}{\frac{Na}{\pi} \sin\left[\frac{\pi}{N}|i-j|\right] + a_0}. \tag{5.18}$$

5.3 Electron–Phonon Interaction

Although in this book we do not discuss details of the electron–phonon interaction, we want to mention the corresponding interaction Hamiltonian for completeness. The electron–phonon interaction can be included in our model by adding a term representing the phonons

$$\hat{H}_P = \sum_p \hbar\omega_p(b_p^\dagger b_p + \tfrac{1}{2}), \qquad (5.19)$$

where b_p ($b_p^\dagger$) destroys (creates) a phonon with quantum numbers (modes) p and energy $\hbar\omega_p$, and a term describing the interaction

$$\hat{H}_{EP} = \sum_{ijep} c_i^{e\dagger} c_j^e (A_{ijp}^e b_p^\dagger + A_{ijp}^{e*} b_p)$$

$$+ \sum_{ijhp} d_i^{h\dagger} d_j^h (A_{ijp}^h b_p^\dagger + A_{ijp}^{h*} b_p), \qquad (5.20)$$

where A_{ijp}^e and A_{ijp}^h are the coupling matrix elements for a phonon with quantum number p to electrons and holes, respectively. Equation (5.20) describes the transfer of electrons and holes from j to i due to the creation or annihilation of a phonon.

5.4 Electron–Light Interaction

As in the previous chapter, we consider only optical dipole transitions. The Hamiltonian describing the light–matter interaction is given by

$$\hat{H}_L = -\boldsymbol{E}(t) \cdot \hat{\boldsymbol{P}}. \qquad (5.21)$$

Here, $\boldsymbol{E}(t)$ is the two-dimensional electric vector field in the plane perpendicular to the propagation direction of the light beam. The interband polarization operator is given by (compare (3.9))

$$\hat{\boldsymbol{P}} = \sum_{ijeh} \left[\boldsymbol{\mu}_{ij}^{he} d_i^h c_j^e + H.C. \right]. \qquad (5.22)$$

The corresponding two-dimensional optical dipole moments $\boldsymbol{\mu}_{ij}^{he}$ are taken as

$$\boldsymbol{\mu}_{ij}^{h_1,e_1} = \delta_{ij} \frac{\mu_0}{\sqrt{2}}(1, i),$$

$$\boldsymbol{\mu}_{ij}^{h_1,e_2} = \boldsymbol{\mu}_{ij}^{h_2,e_1} = 0,$$

$$\boldsymbol{\mu}_{ij}^{h_2,e_2} = \delta_{ij} \frac{\mu_0}{\sqrt{2}}(1, -i), \qquad (5.23)$$

in order to model the optical selection rules in III–V semiconductor heterostructures. The prefactor μ_0 is the modulus of the matrix element for the heavy-hole transition. Consistent with the tight-binding model, the optical transitions are taken to be diagonal in the site index.

Note that without the Coulomb interaction the 2×2 Hamiltonian $\hat{H}_0 + \hat{H}_I$ for spin-degenerate heavy holes and electrons describes two separate subspaces with transitions $|-3/2h\rangle \rightarrow |-1/2e\rangle$ and $|3/2h\rangle \rightarrow |1/2e\rangle$, respectively. These two subspaces are, however, coupled by the many-body Coulomb interaction, (5.13), as can already be expected from Sect. 5.2 and is analyzed in more detail later.

6

Single-Particle Properties

In this chapter, we discuss the single-particle properties of the different models introduced so far. By single-particle properties, we mean features which are present if the many-particle Coulomb interaction is neglected, e.g., the single-particle states, the eigenvalues, and the density of states as a function of energy. Although we have not yet presented the equation of motion method that we use later extensively to calculate optical properties, we already discuss some features of the linear optical spectrum for the interaction-free case here. This can be done on the basis of Fermi's Golden Rule. We consider both the ordered and the disordered semiconductor model. The consequences of disorder are particularly interesting if particle propagation effects are studied. Such effects are investigated by analyzing the wave-packet dynamics in ordered and disordered systems.

6.1 Ordered Systems

6.1.1 States

An ordered system is characterized by the Hamiltonian (5.8) with $\epsilon_i^e = \epsilon_i^h = \epsilon_0/2$. Due to the perfect periodicity of this system, quantum numbers k exist, which define the crystal momentum $\hbar k$. The eigenstates are two-component Bloch states, i.e.,

$$|kh\rangle = \frac{1}{\sqrt{N}} \sum_l \exp\left(ikla\right)|lh\rangle \chi^h,$$

$$|ke\rangle = \frac{1}{\sqrt{N}} \sum_l \exp\left(ikla\right)|le\rangle \chi^e, \tag{6.1}$$

where $\chi^{e,h}$ are two-component spinors

$$\chi^e = (j = 1/2, m_j = \pm 1/2),$$
$$\chi^h = (j = 3/2, m_j = \pm 3/2). \tag{6.2}$$

Applying periodic boundary conditions leads to a discrete set of k values

$$k = 0, \pm\frac{2\pi}{Na}, \pm\frac{4\pi}{Na}, \ldots, \pm\frac{(N/2-1)2\pi}{Na}, \frac{\pi}{a}, \tag{6.3}$$

i.e., there are as many k values as we have sites (N) on the ring. The separation of the allowed k values is

$$\Delta k = \frac{2\pi}{Na} = \frac{1}{R}, \tag{6.4}$$

where R is the radius of the ring. As discussed in the previous chapter, we have a one-dimensional semiconductor model in the thermodynamic limit if both $N \to \infty$ and $R \to \infty$, such that $2\pi R/N = a$. A nanoring is represented by finite and small N and R.

Note that the following orthogonality and completeness relations hold:

$$\sum_{l=1}^{N} e^{\pm ikla} = N\delta_{k,0} \tag{6.5}$$

and

$$\sum_{|k|\leq\frac{\pi}{a}} e^{\pm ikla} = N\delta_{l,0}, \tag{6.6}$$

where the second sum extends over all discrete k values given in (6.3).

6.1.2 Eigenvalues

The noninteracting model is described by the Hamiltonian $\hat{H}_0$ of (5.8), which is the sum of the electronic $\hat{H}_e$ and the hole $\hat{H}_h$ parts. Both have the same structure, so we diagonalize as an example $\hat{H}_e$. We neglect the spin degrees of freedom, which can be reintroduced in our model after the Hamiltonian has been diagonalized. The system is periodic and, therefore, we Fourier-transform the single-particle states, i.e., we use the Bloch states given in (6.1)

$$|ke\rangle = \frac{1}{\sqrt{N}} \sum_{l=1}^{N} e^{ikla}|le\rangle \tag{6.7}$$

and using (6.5), the inverse transform is given by

$$|le\rangle = \frac{1}{\sqrt{N}} \sum_{|k|\leq\frac{\pi}{a}} e^{-ikla}|ke\rangle. \tag{6.8}$$

The electron Hamiltonian reads

$$\hat{H}_e = \sum_{l=1}^{N} \frac{\epsilon_0}{2} c_l^+ c_l + J^e \sum_{\langle lj \rangle} c_l^+ c_j. \tag{6.9}$$

In the noninteracting case, this Hamiltonian can just as well be written in the bra–ket notation as

$$\hat{H}_e = \frac{\epsilon_0}{2} \sum_{l=1}^{N} |le\rangle\langle le| + J^e \sum_{\langle lj \rangle} |le\rangle\langle je|. \tag{6.10}$$

We insert the transformation (6.8) and obtain

$$\hat{H}_e = \frac{1}{N} \frac{\epsilon_0}{2} \sum_{l=1}^{N} \sum_{|k|,|k'|\leq\frac{\pi}{a}} e^{i(k'-k)la} |ke\rangle\langle k'e|$$
$$+ \frac{1}{N} J^e \sum_{\langle lj \rangle} \sum_{|k|,|k'|\leq\frac{\pi}{a}} e^{ik'la-ikja} |ke\rangle\langle k'e|. \tag{6.11}$$

The sum over l can be performed using (6.5). This yields δ functions in k space

$$\hat{H}_e = \frac{1}{N}[N\frac{\epsilon_0}{2}\sum_{|k|\leq\frac{\pi}{a}}|ke\rangle\langle ke|$$
$$+ J^e \sum_{l}\sum_{|k|,|k'|\leq\frac{\pi}{a}}(e^{i(k'-k)la-ika}+e^{i(k'-k)la+ika})|ke\rangle\langle k'e|]$$
$$= \sum_{|k|\leq\frac{\pi}{a}}(\frac{\epsilon_0}{2}+2J^e\cos(ka))|ke\rangle\langle ke|. \tag{6.12}$$

In this representation, the Hamiltonian is diagonal. Its diagonal elements are the eigenvalues we are looking for.

Taking into account the spin degrees of freedom, we obtain the doubly degenerate cosine bands for electrons and holes

$$E^e(k) = \epsilon_0/2 + 2J^e\cos(ka),$$
$$E^h(k) = \epsilon_0/2 + 2J^h\cos(ka). \tag{6.13}$$

If we take both J^e and J^h to be negative, we have a model for a direct semiconductor with the gap at $k = 0$, as is illustrated in Fig. 6.1. The widths of the bands are given by

$$\Delta^\nu = |4J^\nu|, \tag{6.14}$$

where $\nu = e, h$.

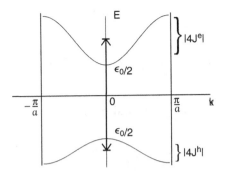

Fig. 6.1. The band structure $E(k)$ of the ordered semiconductor model. Here, the first Brillouin zone extending from $k = -\pi/a$ to $k = \pi/a$ is shown. The *vertical arrow* indicates the energy separation between the band centers which coincides with ϵ_0, i.e., the on-site energy separation

In the Bloch basis, using again second quantization, the Hamiltonian reads

$$\hat{H}_0 = \sum_k \left(\sum_e E^e(k) c_k^{e\dagger} c_k^e + \sum_h E^h(k) d_k^{h\dagger} d_k^h \right), \qquad (6.15)$$

where c_k^e ($c_k^{e\dagger}$) destroys (creates) an electron in state $|ke\rangle$ and d_k^h ($d_k^{h\dagger}$) destroys (creates) a hole in state $|kh\rangle$.

The effective electron and hole masses are obtained by expanding the cosine functions at $k = 0$ up to second order in k. This yields

$$E^\nu(k) \approx \epsilon_0/2 + |J^\nu| a^2 k^2. \qquad (6.16)$$

The effective mass m_ν in band ν close to the band extrema is defined by

$$E^\nu(k) = \text{const} + \frac{\hbar^2 k^2}{2m_\nu}, \qquad (6.17)$$

thus, m_ν is given by

$$m_\nu = \frac{\hbar^2}{2|J^\nu| a^2}. \qquad (6.18)$$

Rewriting this equation using $Na = 2\pi R$, yields

$$m_\nu R^2 \frac{|J^\nu|}{N^2} = \frac{\hbar^2}{2(2\pi)^2} = \text{const.} \qquad (6.19)$$

Here, we see that, once the physically relevant parameters m_ν and R are fixed, the ratio J^ν/N^2 is determined. Note that the separation of the k points is given by $1/R$. In order to model a one-dimensional semiconductor, R has to be taken large enough, such that the k points form a (quasi-)continuum. On the other hand, for the description of a semiconductor nanoring R is fixed and small, such that the k points remain discrete.

For numerical evaluations, it is often necessary to limit the number of sites N to relatively small values. Accordingly, also J^ν and thus the band widths have to be small. However, as long as we are interested in optical properties determined by states close to the fundamental gap, the small band width does not cause problems. In any case, for a given calculation, one has to perform a convergence check, i.e., the results are meaningful only if they do not depend critically on the ratio J^ν/N^2. Figure 6.2 shows band structures for given mass and radius, but various N (and hence J).

It should be noted that in the tight-binding description of a semiconductor the sites do not represent atoms. They rather can be viewed as building blocks from which the semiconductor is constructed. Thus, a is not a physical lattice constant of the atomic lattice. The tight-binding model serves as a parametrization of the electronic structure close to the band extrema. Here the physical relevant quantities for the electronic states are the effective masses. On the other hand, for semiconductor heterostructures like superlattices the sites can be identified with the individual wells which are coupled through the barriers by the $J^{e,h}$. Then a is the lattice constant of the superlattice. In this case the parameters of the model have a well-defined physical meaning.

The derivative

$$\frac{1}{\hbar}\frac{dE^\nu(k)}{dk} = v_k^\nu \tag{6.20}$$

is the velocity of a particle in band ν at k. Its maximum value at $k = |\pi/(2a)|$ is

$$|v_{\max}^\nu| = \frac{2|J^\nu|a}{\hbar}, \tag{6.21}$$

which, besides the physically relevant effective mass m_ν, also depends on the model parameter a, i.e., the lattice constant.

The light–matter interaction Hamiltonian in the k basis is

$$\hat{H}_L = -\boldsymbol{E}(t) \cdot \hat{\boldsymbol{P}}, \tag{6.22}$$

with the polarization operator

$$\hat{\boldsymbol{P}} = \sum_{keh} \left[\boldsymbol{\mu}^{eh} d_k^h c_k^e + H.C. \right], \tag{6.23}$$

where the k-independent optical dipole matrix element is given by (5.23), with the subscripts i, j omitted.

6.1.3 Density of States

The single-particle density of states $\rho^\nu(E)$ for one-dimensional cosine bands $\nu = e, h$ and for $R \to \infty$ but fixed J^ν and a (i.e., m_ν), has two integrable van

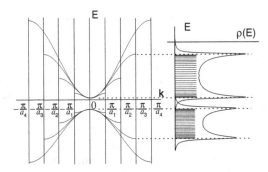

Fig. 6.2. For a given radius R and given effective masses m_ν, the band structure is shown for various numbers of sites N and lattice constants $a = 2\pi R/N$, where $a_1 : a_2 : a_3 : a_4 = 4 : 3 : 2 : 1$. With the number N of sites, also the lattice constant a and the band width, given by $4J^\nu$, are changed. Close to $k = 0$, however, the band structure remains nearly unchanged and is approximately determined by the effective mass. For one case, a_3, the allowed energy levels are given in the *right-hand figure* as *horizontal lines*. Broadening these levels slightly yields a continuous function, the single-particle density of states $\rho(E)$

Hove singularities at the band extrema. Actually, for finite N, the density of states consists of a series of δ peaks

$$\rho^\nu(E) = 2 \sum_k \delta(E - \epsilon_0/2 - 2J^\nu \cos(ka)), \qquad (6.24)$$

where the factor of 2 accounts for the two possible spin orientations per band. Using $\sum_k, \ldots, \rightarrow (Na/2\pi) \int, \ldots, dk$, we find in the continuum limit the result shown in Fig. 6.3. Close to the band edge at $k = 0$, where the effective mass approximation, (6.17), is valid, the singularities close to, e.g., the lower conduction band edge $E^e(k = 0) = \epsilon_0/2 - 2|J^e|$ can be approximated by

$$\rho^e(E) \approx \frac{N}{2\pi}(|J^e|)^{-1/2}[E - E(k = 0)]^{-1/2}. \qquad (6.25)$$

6.1.4 Optical Spectrum

Optical dipole transitions conserve the crystal momentum k if we neglect the small momentum of the photon. According to Fermi's Golden Rule, these "vertical" transitions yield an optical spectrum $\chi''(\omega)$ given by the joint density of states times the squared modulus of the optical dipole matrix element

$$\chi''(\omega) = 2\pi|\mu_0|^2 \sum_k \delta(\hbar\omega - E^e(k) - E^h(k)). \qquad (6.26)$$

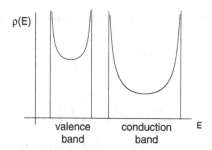

Fig. 6.3. The density of single-particle states $\rho(E)$ for the one-dimensional semi-conductor model for $R \to \infty$ and fixed J^ν and a (i.e., m_ν)

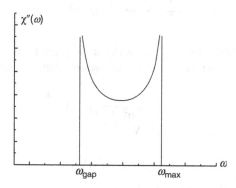

Fig. 6.4. The absorption spectrum of the ordered one-dimensional two-band semi-conductor model. Here, $\hbar\omega_{\text{gap}}$ and $\hbar\omega_{\text{max}}$ are the energies corresponding to the fundamental single-particle gap and the maximum optical absorption energy

In the continuum limit, i.e., for $R \to \infty$, this spectrum has the same overall appearance as the single-particle density of states with the two van Hove singularities at the fundamental gap and at the upper limit of the spectrum, see Fig. 6.4. Again, for a nanoring, i.e., for finite N and R, the spectrum consists of peaks at photon energies $\hbar\omega = E^e(k) - E^h(k)$ weighted by $2|\mu_0|^2$, where the factor of 2 again accounts for the two spin directions.

6.1.5 Wave Packets and Propagation

Since we later want to study the propagation of locally generated particles, it is useful to present the general idea already at this point. We assume that a spinless electron is generated at the central site, denoted by $l = 0$, at time $t = 0$ in the conduction band. The corresponding state $|0e\rangle$ is not an eigenstate of the Hamiltonian, but a wave packet composed of the Bloch eigenstates $|ke\rangle$ of the conduction band. It is given by the projections of $|0e\rangle$ onto all Bloch eigenstates via

$$|0e\rangle = \sum_k |ke\rangle \langle ke|0e\rangle. \tag{6.27}$$

Let us assume nondegenerate bands for the moment. Inserting (6.1), we find

$$\langle ke|0e \rangle = \frac{1}{\sqrt{N}}, \tag{6.28}$$

which gives

$$|0e\rangle = \frac{1}{\sqrt{N}} \sum_k |ke\rangle. \tag{6.29}$$

At later times, the Bloch functions are multiplied by the corresponding phase factors and the original wave packet develops in time according to

$$|\psi(t)\rangle = \frac{1}{\sqrt{N}} \sum_k \exp\left[-iE^e(k)t/\hbar\right]|ke\rangle. \tag{6.30}$$

To quantify the temporal extension of the wave packet, there are at least two possibilities. We can introduce the mean square displacement R_{msd}, which is given by

$$R_{\text{msd}}(t) = \left(a^2 \sum_l l^2 n_{ll}^e \right)^{1/2}, \tag{6.31}$$

where

$$n_{ll}^e = \langle c_l^\dagger c_l \rangle = |\langle \psi(t)|le\rangle|^2. \tag{6.32}$$

This gives, using (6.30), (6.1), and (6.21)

$$n_{ll}^e = \left| \frac{1}{N} \sum_k \exp\left[ikla + i\frac{v_{\text{max}}^e}{a}t\cos ka\right] \right|^2. \tag{6.33}$$

In the continuum limit, we obtain

$$n_{ll}^e = \left| \frac{a}{2\pi} \int_{-\pi/a}^{\pi/a} dk \exp\left[ikla + i\frac{v_{\text{max}}^e}{a}t\cos ka\right] \right|^2$$
$$= J_l^2\left(\frac{v_{\text{max}}^e}{a}t\right), \tag{6.34}$$

where $J_l(x)$ is the lth-order Bessel function. The relation

$$\sum_{l=-\infty}^{\infty} J_l^2\left(\frac{v_{\text{max}}^e}{a}t\right) = 1 \tag{6.35}$$

reflects particle conservation and

$$\sum_{l=-\infty}^{\infty} lJ_l^2\left(\frac{v_{\text{max}}^e}{a}t\right) = 0 \tag{6.36}$$

indicates that the center of mass of the wave packet remains at site $l = 0$. However,

$$R_{\mathrm{msd}}^2(t) = a^2 \sum_{l=-\infty}^{\infty} l^2 J_l^2 \left(\frac{v_{\mathrm{max}}^e}{a}t\right) = \frac{1}{2}(v_{\mathrm{max}}^e t)^2 \qquad (6.37)$$

shows that $R_{\mathrm{msd}}(t)$ follows a ballistic behavior, i.e., $R_{\mathrm{msd}}(t) \propto t$.

On the other hand, one can quantify the contribution of a given site j to the wave packet at time $t > 0$ by defining the participation number

$$\Lambda(t) = \frac{[\sum_i n_{ii}(t)]^2}{\sum_i n_{ii}^2}. \qquad (6.38)$$

For a packet localized at a single site, the participation number is unity, while for a packet uniformly distributed over all sites, it is $\Lambda \propto N$. For our present case of local excitation in an ordered system, we find

$$\Lambda(t) = \left[\sum_{l=-N}^{N} J_l^4 \left(\frac{v_{\mathrm{max}}^e}{a}t\right)\right]^{-1}, \qquad (6.39)$$

which is plotted in Fig. 6.5 for $N = 1000$. We see that the participation number increases nearly linearly in time, which is characteristic for ballistic propagation. The small wiggles reflect the discrete structure of the one-dimensional lattice. The quantity defined by $v_\Lambda \approx a\Lambda/t$ can also be taken as a measure of the spreading of the wave packet. It is slightly larger than the maximum velocity v_{max}^e.

6.1.6 Dimerized Systems

Dimerized systems can be taken as a first step towards a spatially inhomogeneous system. The sites are now pairwise identical, i.e., we have perfect periodicity with a unit cell containing two sites. The Hamiltonian is that of (5.8) and dimerization is introduced by defining a dimerization parameter δ by

$$\epsilon_i^{e,h} = [\epsilon_0 + (-1)^i \delta_{e,h}]/2, \qquad (6.40)$$

which results in the Hamiltonian presented pictorially in Fig. 6.6.

6.2 Disordered Systems

6.2.1 States

The stationary single-particle Schrödinger equation with the Hamiltonian (5.8) and diagonal disorder can be solved numerically only for rather small

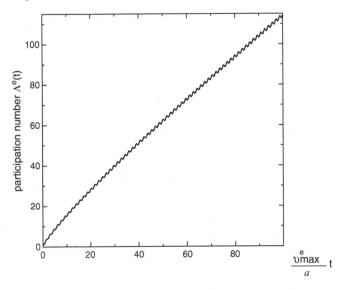

Fig. 6.5. The participation number Λ^e for a system of $N = 1000$ sites as a function of time for an ordered system

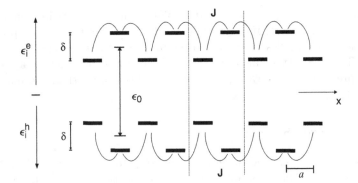

Fig. 6.6. The dimerized semiconductor model for $\delta_e = \delta_h = \delta$. The *vertical dashed lines* indicate a unit cell

numbers of sites N. As we consider the influence of the many-particle interaction later in this book, we do not go into details of these calculations here, since, with interactions included, the direct diagonalization of the Hamiltonian, (5.8), is restricted to even smaller values of N. Therefore, we do not use the diagonalization method in these cases. Instead, we apply the equation of motion method and solve the coupled dynamic equations in the time domain. This procedure is also well adapted to describe real optical experiments, which are performed in the time domain and do not directly yield information about eigenstates. Only in the limit $t \to \infty$, can one obtain such information. However, in real situations phase-breaking interactions restrict the experimentally relevant time scale for the investigation of coherent effects to rather small values of the order of picoseconds.

A number of facts are known for a noninteracting electron moving in a disordered environment. Since there is no translational symmetry in the presence of disorder, we do not have any quantum numbers except spin quantum numbers. In particular, there is no momentum index k that classifies the states. In contrast to the Bloch states of ordered systems, which are extended over all sites with constant moduli of their amplitudes, with disorder all eigenstates are localized. The length scale of these localized eigenstates is called the localization length ξ_{loc}. The envelope displays an exponential behavior

$$\propto \exp(-(n - n_0)a/\xi_{loc}), \tag{6.41}$$

where n_0 is the central site of that particular state.

The localization length is largest in the center of the band and, for small disorder, depends on energy as

$$\xi_{loc} \approx 24\frac{4J^2 - E^2}{W^2}a. \tag{6.42}$$

Here, $E = 0$ refers to the band center and we assume a distribution function of the site energies given by a box of width W.

6.2.2 Eigenvalues

A band structure in the sense of (6.13) does not exist for disordered systems due to the absence of translational symmetry. However, if the disorder is weak, the site-energy distribution functions do not have extensive tails and, therefore, a gap between the conduction band and valence band can still be expected. In the following, we always assume that this is the case, which basically means that the energy scale of the disorder is much smaller than the band gap of the corresponding ordered system. In particular, for a distribution function with maximum (minimum) deviation of ϵ_i given by ϵ_{max} (ϵ_{min}), the width of the respective band is always smaller than $4J + \epsilon_{max} - \epsilon_{min}$.

6.2.3 Density of States

The density of states function $\rho(E)$ for the band of the disordered system has a certain finite width and decays away from the center showing more or less developed tails that depend on the distribution function. In particular, in the extreme tail region the density of states follows the distribution function. This limit is called the atomic limit. As mentioned above, we always assume that the density of states functions of the valence band and the conduction band do not overlap, i.e., there exists an energy gap.

6.2.4 Optical Spectrum

The form of the optical spectrum for noninteracting particles depends strongly on the underlying disorder. For uncorrelated disorder, quite similar to the single-particle density of states, the optical spectrum looks like a broadened version of the corresponding spectrum for the ordered case. Instead of the two singularities at the band edges, broadened peaks appear with a tail extending into the gap region. While for uncorrelated disorder no selection rule exists, for correlated disorder a given hole state with energy ϵ_λ^h is optically coupled to just one electron state. The corresponding energies fulfill the relation $\epsilon_\lambda^e/\epsilon_\lambda^h = |J^e/J^h|$ and the spectrum resembles the single-particle density of states function.

For perfectly anticorrelated disorder, see (5.11), and $J^h = J^e$, on the other hand, the optical spectrum is dominated by a peak in the center which results from transitions between states in the tails of the single-particle spectra. These states are strongly localized and appear in pairs with fixed energetic separation. Their weight is therefore large and they all contribute to the optical spectrum at a photon energy that is equal to the site-energy separation ϵ_0.

For more general cases, numerical calculations have to be performed. One either determines the eigenvectors and eigenvalues of the Hamiltonian and uses Fermi's Golden Rule to obtain the spectrum or, alternatively, one uses the equation of motion approach explained below. Examples of such calculations are shown later after we have discussed how to calculate optical spectra for the various kinds of disorder, i.e., for uncorrelated, correlated and anticorrelated disorder.

6.2.5 Wave Packets and Propagation

We have already studied the propagation of wave packets in an ordered model. Due to localization we expect for a disordered model that the wave packet is unable to extend towards infinity in the course of time in the thermodynamic limit, i.e., N and $R \rightarrow \infty$. Let us assume again that the packet is initially generated at a given site, denoted by $l = 0$ in the conduction band. Its state is then $|l = 0, e, t = 0\rangle$. The subsequent dynamics depends on the energy of that particular site and its environment.

Imagine first that the site has an energy $\epsilon_{l=0}$ in the wings of the distribution function. It is then extremely hard for the electron to tunnel via many intermediate sites with strongly deviating energies to another site having an energy close to $\epsilon_{l=0}$. In other words, to a very good approximation, the state of the isolated site is already an eigenfunction of the total system. This case we have called the atomic limit. Consequently, the particle is confined to that site, the mean square displacement is zero and the participation number is unity for all times.

On the other hand, for an energy $\epsilon_{l=0}$ of the site $l = 0$ well in the center of the distribution, the probability to find similar energies in its vicinity is large. Thus, the initial state $|l = 0, e, t = 0\rangle$ describes an electron confined to site $l = 0$, which is not an eigenstate of the system. At $t = 0$, it has the representation in terms of the eigenstates $|e, \lambda\rangle$, where $\lambda = 1, \ldots, N$ denotes the corresponding eigenvalues

$$|l = 0, e, t = 0\rangle = \sum_\lambda |e, \lambda\rangle\langle e, \lambda|l = 0, e, t = 0\rangle, \qquad (6.43)$$

with (in principle) known expansion coefficients $\langle e, \lambda|l = 0, e, t = 0\rangle$. At later times, we have

$$|l = 0, e, t\rangle = \sum_\lambda e^{iE_\lambda^e t/\hbar}|e, \lambda\rangle\langle e, \lambda|l = 0, e, t = 0\rangle. \qquad (6.44)$$

While at $t = 0$, all amplitudes $\langle le|l = 0, e, t = 0\rangle$ interfere destructively for all l other than 0, finite amplitudes will develop in the course of time at $l \neq 0$. This indicates coherent propagation in a certain region of space, which is limited by the largest localization length of the eigenstates $|e, \lambda\rangle$ contributing to the sum (6.43).

Since in the disordered situation the eigenstates are not known analytically, we calculate the solution of the equation of motion numerically. In Fig. 6.7, the temporal increase of the participation number $\Lambda_0(t)$ is shown for various widths W of the box-shaped site energy distribution function. In this particular calculation, the particle has been generated optically in the center of the band at site $l = 0$ and a configuration average over the disorder realizations has been applied. Further details are provided in Chap. 17.

After a rapid increase on the time scale of the optical excitation pulse, the participation number $\Lambda_0(t)$ saturates, indicating the localization of the states that contribute to the initial electronic wave packet. For weaker disorder, the

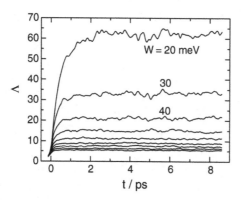

Fig. 6.7. Time dependence of the participation number Λ_0 for the noninteracting case for various disorder parameters W. Note that after a very rapid increase the participation number, i.e., the extension of the wave packet, saturates. This indicates Anderson localization

saturation value is larger, indicating the disorder dependence of the localization lengths involved.

In Chap. 17, the influence of the many-particle interaction on wave-packet propagation is studied. This provides information about the simultaneous influence of disorder and interactions on the coherent dynamics of correlated electron–hole pairs.

At this point, a side remark is in order. If there is a coupling of the electron to a bath of other excitations, e.g., phonons, the coherent propagation of the particle is limited by a typical inelastic scattering time T_{inel}. For longer times, the phases of the contributing eigenstates and also their weights are altered in some random way, which depends on the details of the scattering process. After the scattering event, the particle again starts a coherent propagation, however, from a different site. It is therefore conceivable that after a succession of such phase destroying interactions, the particle is able to propagate over distances larger than in the purely coherent situation. We come back to this phenomenon of interaction-assisted delocalization later.

From the present discussion it can be learned that coherence is required if localization is to be observed. In fact, Anderson localization is a coherent phenomenon.

The Equation of Motion Approach

In the previous chapters, we have already discussed optical properties of the ordered and disordered semiconductor model using Fermi's Golden Rule, which is applicable as long as linear optical spectra are considered. Most of the following discussions focus on nonlinear and time-dependent optical phenomena induced by excitations with short pulses. For the theoretical analysis, it is therefore most appropriate to apply the equation of motion method, which in principle yields the full information about the dynamics in the linear and the nonlinear optical regime.

In this chapter, we focus on the description of the material excitations resulting from irradiation with classical light fields. We do not include the possible feedback of these excitations on the light field, which would require the simultaneous self-consistent solution of the material equations together with Maxwell's equations. Since we avoid the self-consistent treatment at this stage, we cannot discuss phenomena such as radiative decay or propagation effects of optical excitations through the medium.

We start our discussion with the simplest model system, i.e., an ensemble of two-level absorbers. We proceed by considering few-level systems and Fano situations.

7.1 Observables

Before setting up the equations of motion, we have to specify the relevant observables. Let us consider as an example the two-level absorber with Hamiltonian $\hat{H}_0$ given by (3.4). Without an external light field, its eigenstates are the ground state $|1v\rangle$ and the excited state $|1c\rangle$. Once the light field acts on the system, the states change into a superposition of the ground state and the excited state. We have already encountered the polarization operator $\hat{P}$, which describes the transition from $|1v\rangle$ to $|1c\rangle$, in (3.9)

$$\hat{P} = \mu_{vc}|1v\rangle\langle 1c| + H.C. \tag{7.1}$$

In addition to a change in the state of the system, also the populations in the ground state n_{vv} (assumed to be unity before the light field arrived) and in the excited state n_{cc} (initially zero) change in the course of time. The corresponding operators are

$$\hat{n}_{vv} = |1v\rangle\langle 1v|,$$
$$\hat{n}_{cc} = |1c\rangle\langle 1c|. \tag{7.2}$$

The relevant observables are the expectation values of these operators, which in the Schrödinger representation read

$$A(t) = Tr(\hat{\rho}(t)\hat{A}), \tag{7.3}$$

where $\hat{A}$ stands for $\hat{P}$ or $\hat{n}$. The density matrix $\hat{\rho}(t)$ has to be calculated by solving its equation of motion (i.e., the Liouville-von Neumann equation)

$$\hbar\frac{d\hat{\rho}}{dt} = -i[\hat{H}(t), \hat{\rho}], \tag{7.4}$$

where

$$\hat{H}(t) = \hat{H}_0 + \hat{H}_L(t) \tag{7.5}$$

and $\hat{H}_L(t)$ is given by (3.10).

As an alternative, one can use the Heisenberg picture and calculate the temporal development of the operators $\hat{A}(t)$ according to the Heisenberg equation

$$\hbar\frac{d\hat{A}(t)}{dt} = i[\hat{H}(t), \hat{A}(t)], \tag{7.6}$$

and then take its expectation value

$$A(t) = Tr(\hat{\rho}_0\hat{A}(t)), \tag{7.7}$$

using the ground-state (equilibrium) density matrix $\hat{\rho}_0$. In the following, we are going to use both approaches.

More generally, the polarization operator for an M-M' system, see (3.15), is given by

$$\hat{P} = \sum_{\nu}^{M}\sum_{\nu'}^{M'} \mu_{\nu\nu'}|\nu v\rangle\langle\nu'c| + H.C., \tag{7.8}$$

while the population operators read

$$\hat{n}_{\nu,\nu} = |\nu v\rangle\langle\nu v|,$$
$$\hat{n}_{\nu',\nu'} = |\nu'c\rangle\langle\nu'c|. \tag{7.9}$$

In addition, we also have to take into account nondiagonal operators for pairs of states within the upper and the lower set of levels. These are called intraband coherences and are defined by

$$\hat{n}_{\nu,\mu} = |\nu v\rangle\langle\mu v|,$$
$$\hat{n}_{\nu',\mu'} = |\nu' c\rangle\langle\mu' c|. \tag{7.10}$$

7.2 Noninteracting Particles

In this section, we neglect the vector character of both the dipole matrix element and the electric field, i.e., we assume that both vectors are parallel.

7.2.1 The Two-Level Bloch Equations

As explained above, in order to obtain the polarization $P(t)$ in the Schrödinger picture, we have to solve (7.4) for $\hat{\rho}(t)$ and insert the result into (7.3). We evaluate the trace in the complete set of the two eigenstates of the two-level absorber, $|1v\rangle$ and $|1c\rangle$, i.e.,

$$P(t) = \langle 1v|\hat{\rho}(t)|1c\rangle\langle 1c|\hat{P}|1v\rangle$$
$$+\langle 1c|\hat{\rho}(t)|1v\rangle\langle 1v|\hat{P}|1c\rangle. \tag{7.11}$$

Note that the diagonal elements $\langle 1v|\hat{P}|1v\rangle$ and $\langle 1c|\hat{P}|1c\rangle$ of $\hat{P}$ are zero because of (7.1). We thus obtain

$$P(t) = \mu_{vc}^*\rho(t)_{vc} + \mu_{vc}(\rho(t)_{vc})^*, \tag{7.12}$$

where

$$\rho(t)_{vc} = \langle 1v|\hat{\rho}(t)|1c\rangle \tag{7.13}$$

and

$$\rho(t)_{cv} = \langle 1c|\hat{\rho}(t)|1v\rangle = (\rho(t)_{vc})^*. \tag{7.14}$$

We now take matrix elements of the equation of motion (7.4), i.e.,

$$\left\langle 1v \left| \hbar\frac{d\hat{\rho}}{dt} \right| 1c \right\rangle = -i\langle 1v|[\hat{H}(t),\hat{\rho}]|1c\rangle$$

$$= -i\langle 1v|\hat{H}(t)\hat{\rho}|1c\rangle + i\langle 1v|\hat{\rho}\hat{H}(t)|1c\rangle$$

$$= -i\langle 1v|\hat{H}(t)|1v\rangle\langle 1v|\hat{\rho}|1c\rangle - i\langle 1v|\hat{H}(t)|1c\rangle\langle 1c|\hat{\rho}|1c\rangle$$

$$+i\langle 1v|\hat{\rho}|1v\rangle\langle 1v|\hat{H}(t)|1c\rangle + i\langle 1v|\hat{\rho}|1c\rangle\langle 1c|\hat{H}(t)|1c\rangle$$

$$= -i\langle 1v|\hat{H}_0|1v\rangle\langle 1v|\hat{\rho}|1c\rangle + i\langle 1v|\hat{\rho}|1v\rangle\langle 1v|\hat{H}_0|1c\rangle$$

$$-i\langle 1v|\hat{H}_0|1c\rangle\langle 1c|\hat{\rho}|1c\rangle + i\langle 1v|\hat{\rho}|1c\rangle\langle 1c|\hat{H}_0|1c\rangle$$

$$-i\langle 1v|\hat{H}_L(t)|1v\rangle\langle 1v|\hat{\rho}|1c\rangle + i\langle 1v|\hat{\rho}|1v\rangle\langle 1v|\hat{H}_L(t)|1c\rangle$$

$$-i\langle 1v|\hat{H}_L(t)|1c\rangle\langle 1c|\hat{\rho}|1c\rangle + i\langle 1v|\hat{\rho}|1c\rangle\langle 1c|\hat{H}_L(t)|1c\rangle$$

$$= -i\epsilon_1^v \rho_{vc} + i\rho_{vc}\epsilon_1^c$$

$$-iE(t)\rho_{vv}\langle 1v|\hat{P}(t)|1c\rangle + iE(t)\langle 1v|\hat{P}(t)|1c\rangle\rho_{cc}$$

$$= +i\hbar\omega_{cv}\rho_{vc} - iE(t)\rho_{vv}\mu_{vc} + iE(t)\mu_{vc}\rho_{cc}, \qquad (7.15)$$

with

$$\hbar\omega_{cv} = \epsilon_1^c - \epsilon_1^v. \qquad (7.16)$$

In this calculation, we used the fact that within our present model the non-diagonal elements of $\hat{H}_0$ and the diagonal elements of $\hat{H}_L$ vanish. We thus obtain the equation of motion

$$i\hbar\frac{d\rho_{vc}}{dt} = -\hbar\omega_{cv}\rho_{vc} + E(t)\mu_{vc}(\rho_{vv} - \rho_{cc}). \qquad (7.17)$$

The first term on the right-hand side describes the free rotation of the interband coherence ρ_{vc} with frequency ω_{cv}. The second term is the source term, containing a product of the exciting field, the dipole matrix element, and the difference between the diagonal (intraband) elements, $(\rho_{vv} - \rho_{cc})$, which are defined by

$$\rho_{vv} = \langle 1v|\hat{\rho}(t)|1v\rangle,$$
$$\rho_{cc} = \langle 1c|\hat{\rho}(t)|1c\rangle. \qquad (7.18)$$

These diagonal elements quantify the populations in the lower (n_{vv}) and upper (n_{cc}) state, respectively. Inserting the operators from (7.9) into (7.3) we see that

$$n_{vv} = \rho_{vv},$$
$$n_{cc} = \rho_{cc}. \qquad (7.19)$$

In equilibrium, i.e., before the excitation has started, $\rho_{vv} - \rho_{cc} = 1$, since the lower state $|1v\rangle$ is occupied and the upper state $|1c\rangle$ is empty. When the external field $E(t)$ is switched on, the term $E(t)\mu_{vc}$ is thus initially multiplied by the factor $\rho_{vv} - \rho_{cc} = 1$. This factor decreases in the course of time, since, in the ensemble of two-level absorbers, the probability for the upper state to be occupied increases, while that of the lower state decreases. The source term therefore decreases in time. This phenomenon is called pump saturation. It originates from the fact that due to the Pauli exclusion principle any given quantum state can at most be occupied by one fermion. Hence, the pump saturation is a consequence of Pauli blocking, i.e., phase-space filling.

In order to gain more quantitative insight into the population dynamics, one has to set up the equations of motion for these quantities. Following the steps leading to (7.17) we obtain the equations

$$i\hbar\frac{d\rho_{vv}}{dt} = -E(t)\mu_{vc}(\rho_{vc})^* + E(t)\mu_{vc}^*\rho_{vc}, \tag{7.20}$$

and

$$i\hbar\frac{d\rho_{cc}}{dt} = -E(t)\mu_{vc}^*\rho_{vc} + E(t)\mu_{vc}(\rho_{vc})^*. \tag{7.21}$$

Summing these equations, we find

$$\frac{d}{dt}(\rho_{vv} + \rho_{cc}) = 0. \tag{7.22}$$

Due to the initial condition $\rho_{vv}(t=0) = 1$ and $\rho_{cc}(t=0) = 0$, we have

$$\rho_{vv} = 1 - \rho_{cc}, \tag{7.23}$$

which reflects particle number conservation.

Equations (7.17), (7.20), and (7.21) form a closed set, which enables us to study optical excitation problems for an ensemble of two-level absorbers. This set of equations is called optical Bloch equations, in analogy to the Bloch equations for spin-1/2 systems.

For further analysis, it is advantageous to define a reduced complex valued polarization function $\bar{P}(t)$ by

$$P(t) = e^{i\omega_L t}\bar{P}(t) + c.c. \tag{7.24}$$

and accordingly

$$\rho(t)_{vc} = e^{i\omega_L t}\bar{\rho}(t)_{vc}, \tag{7.25}$$

such that, from (7.12),

$$\bar{P}(t) = \mu_{vc}^*\bar{\rho}(t)_{vc}. \tag{7.26}$$

Here, ω_L is the central frequency of the laser excitation field.

Taking the time derivative of (7.25), we arrive after some simple algebra at the closed set of equations "in the rotating frame"

$$\hbar\frac{d\bar{\rho}_{vc}}{dt} - i\hbar(\omega_{cv} - \omega_L)\bar{\rho}_{vc} = -iE(t)e^{-i\omega_L t}\mu_{vc}(\rho_{vv} - \rho_{cc}),$$

$$\hbar\frac{d\rho_{vv}}{dt} = iE(t)e^{-i\omega_L t}\mu_{vc}(\bar{\rho}_{vc})^* - iE(t)e^{i\omega_L t}(\mu_{vc})^*\bar{\rho}_{vc},$$

$$\hbar\frac{d\rho_{cc}}{dt} = iE(t)e^{i\omega_L t}(\mu_{vc})^*\bar{\rho}_{vc} - iE(t)e^{-i\omega_L t}\mu_{vc}(\bar{\rho}_{vc})^*. \tag{7.27}$$

A considerable simplification is possible if the exciting light field is nearly resonant with ω_{cv}, or more precisely, if the relative detuning is small, i.e.,

$$\left| \frac{\omega_L - \omega_{cv}}{\omega_{cv}} \right| \ll 1. \tag{7.28}$$

Inserting the field in the form of (3.11), we see that some terms are proportional to the slowly varying factor $\exp\left[-i(\omega_L - \omega_{cv})t\right]$ and some terms rotate with about twice the light frequency ω_L, i.e., they are proportional to $\exp\left[-i(\omega_L + \omega_{cv})t\right]$. For not too short pulses, for which a central reference frequency is still well defined, one may often neglect those terms which rotate with about $2\omega_L$ since their weight as a source term is negligible compared to the other terms which do not contain such rapidly oscillating exponential factors (Riemann–Lebesgue Lemma).

Summarizing this procedure, we have first transformed the equation of motion into the rotating frame introducing the polarization component $\bar{P}(t)$. Then we have eliminated all rapidly varying source terms, i.e., the terms that are spectrally far from resonance. This approximation is called the "rotating-wave approximation". Clearly, the validity of the rotating-wave approximation has to be critically examined in each case. For example, it becomes invalid for large relative detunings, since then $\bar{\rho}(t)$ is no longer a slowly varying function due to the large frequency $\omega_{cv} - \omega_L$ appearing in the equation of motion for $\bar{\rho}(t)$. The rotating-wave approximation is also not applicable for light pulses that are so short that a considerable part of the pulse spectrum is far from resonance with the material system.

Applying the rotating-wave approximation, we finally arrive at

$$\hbar \frac{d\bar{\rho}_{vc}}{dt} - i\hbar(\omega_{cv} - \omega_L)\bar{\rho}_{vc} = -iE_l^*(t)e^{-i\mathbf{k}_l \cdot \mathbf{r}}\mu_{vc}(\rho_{vv} - \rho_{cc})$$

$$\hbar \frac{d\rho_{vv}}{dt} = iE_l^*(t)e^{-i\mathbf{k}_l \cdot \mathbf{r}}\mu_{vc}(\bar{\rho}_{vc})^* - iE_l(t)e^{i\mathbf{k}_l \cdot \mathbf{r}}(\mu_{vc})^*\bar{\rho}_{vc}$$

$$\hbar \frac{d\rho_{cc}}{dt} = iE_l(t)e^{i\mathbf{k}_l \cdot \mathbf{r}}(\mu_{vc})^*\bar{\rho}_{vc} - iE_l^*(t)e^{-i\mathbf{k}_l \cdot \mathbf{r}}\mu_{vc}(\bar{\rho}_{vc})^*$$

$$\tag{7.29}$$

where we have concentrated on the action of pulse l with envelope $E_l(t)$ having the spatial phase factor $\exp\left(i\mathbf{k}_l \cdot \mathbf{r}\right)$. This closed set of equations is called the optical Bloch equations in the rotating-wave approximation. It describes the coupled dynamics of two-level absorbers for excitations with small detunings.

The optical Bloch equations have been derived here for systems without any relaxation mechanisms. One can, phenomenologically, introduce damping terms into these equations by adding

$$\hbar \frac{\bar{\rho}_{vc}}{T_2} \tag{7.30}$$

to the left-hand side of the first Bloch equation, and

$$\hbar\frac{\rho_{cc}}{T_1} \tag{7.31}$$

and

$$\hbar\frac{\rho_{vv}}{T_1} \tag{7.32}$$

to the left-hand side of the second and third equation.

Here, a word of caution is in order. A proper description of damping mechanisms requires a fully microscopic theory that explicitly treats the relevant interaction mechanisms. Following such approaches, one finds that the damping mechanisms of polarization and population are usually not independent, and typically not even simply exponential (as assumed by introducing $T_{1,2}$ times). Moreover, strong cancellations occur between contributions from different scattering processes. Consequently, the phenomenological introduction of T_1 times and T_2 times is often not well justified and only a meaningful procedure for very schematic studies.

As another side remark, we mention here that the damping of the polarization is often referred to as "dephasing", suggesting that phase-breaking interactions lead to a decay of the polarization without changing the populations. However, also this notion has to be taken with great care. One has to define precisely what is meant by the word "dephasing" and which process one refers to.

If we have a microscopic process that leads to an exponential decay of the population, such as radiative decay, then also the polarization is destroyed by the same process. In this limit, the relation

$$T_2 = 2T_1, \tag{7.33}$$

is fulfilled, provided no other interactions are relevant. If this relation is assumed to be valid, one speaks of the "coherent limit".

7.2.2 An Analogy

At this point, it is instructive to make contact with the description of coherent tunneling, which is presented in Chap. 4. We ignore the spatial phase factors (which are not relevant in the present situation) and write down the three equations, (7.29), for stationary excitation, i.e., $E_l(t) = E_0$ (assumed to be real), together with the corresponding three equations, (4.29),

$$i\hbar\frac{\bar{\rho}_{vc}}{dt} + \hbar(\omega_{cv} - \omega_L)\bar{\rho}_{vc} = E_0\mu_{vc}(\rho_{vv} - \rho_{cc}),$$

$$i\hbar\frac{\rho_{rl}}{dt} - \delta\rho_{rl} = J(\rho_{ll} - \rho_{rr}), \tag{7.34}$$

$$ i\hbar \frac{\rho_{vv}}{dt} = -E_0 \mu_{vc}(\bar{\rho}_{vc})^* + E_0(\mu_{vc})^* \bar{\rho}_{vc}, $$

$$ i\hbar \frac{\rho_{ll}}{dt} = -J(\rho_{rl})^* + J\rho_{rl}, \tag{7.35} $$

$$ i\hbar \frac{\rho_{cc}}{dt} = -E_0(\mu_{vc})^* \bar{\rho}_{vc} + E_0 \mu_{vc}(\bar{\rho}_{vc})^*, $$

$$ i\hbar \frac{\rho_{rr}}{dt} = -J\rho_{rl} + J(\rho_{rl})^*. \tag{7.36} $$

By identifying

- $\bar{\rho}_{vc}$ with ρ_{rl},
- ρ_{vv} with ρ_{ll},
- ρ_{cc} with ρ_{rr},
- $\hbar(\omega_L - \omega_{cv})$ with δ,
- $\mu_{vc}E_0$ with J,

we see that the equations which govern the coherent tunneling are identical to the optical Bloch equations. Physically, the detuning of the two levels involved in the tunneling process corresponds to the detuning in the optical excitation of the two-level system. The coupling J between the two levels governing the frequency of the tunnel transition corresponds to μE_0, which is up to $\hbar$ half of the so-called Rabi frequency Ω, as is discussed in detail below.

Consequently, in analogy with (4.40) we have the conservation law

$$ |\rho_{vc}|^2 = -\frac{1}{4}(2\rho_{cc} - 1)^2 + \frac{1}{4}. \tag{7.37} $$

Furthermore, in analogy with (4.30) we define polarization p, current j, and inversion I by

$$ p = \bar{\rho}_{vc} + (\bar{\rho}_{vc})^*, $$

$$ j = i\bar{\rho}_{vc} - i(\bar{\rho}_{vc})^*, $$

$$ I = 1 - 2\rho_{cc}. \tag{7.38} $$

The notion of polarization p in the context of optical absorption makes obvious physical sense, see (7.12), in contrast to the scenario of coherent tunneling, where we have already formally introduced this quantity. On the other hand, the notion of current j is less obvious in the optical situation as compared to the case of resonant tunneling.

In analogy to the tunneling analysis, we also have in the optical case the conservation law

$$ p^2 + j^2 + I^2 = 1, \tag{7.39} $$

which tells us that the Bloch vector $\boldsymbol{S} = (p, j, I)$ is always normalized to unity. The dynamics of optical excitations can, therefore, be visualized as the dynamics of a Bloch vector with length unity, defining the Bloch sphere. In the ground state (equilibrium), this vector points to the north pole, while the totally inverted system is described by the vector pointing to the south pole. The azimuthal plane is spanned by the polarization p and the current j.

The picture of the Bloch sphere is often helpful in the discussion of echo dynamics of spin-1/2 systems, of optical excitations of atomic systems, and in problems related to nuclear magnetic resonance. In our context of optical excitation of semiconductors, the relevant equations turn out to be much more complex, and the Bloch vector and Bloch sphere picture is no longer adequate. In this book, we therefore do not dwell on this concept any longer.

Finally, as an example for the analogy between coherent tunneling and optical two-level dynamics, we consider stationary and resonant excitation. We then have $\delta = 0$, i.e., $\Delta = 2J$, see (4.11), and obtain in analogy to (4.44)

$$I(t) = 1 - 2\rho_{cc}$$
$$= 1 - 2\sin^2 \frac{E_0 \mu_{vc} t}{\hbar}, \tag{7.40}$$

i.e.,

$$\rho_{cc} = \sin^2 \frac{2E_0 \mu_{vc} t}{2\hbar}$$
$$= \frac{1}{2}\left(1 - \cos \frac{2E_0 \mu_{vc} t}{\hbar}\right). \tag{7.41}$$

Of course, for finite detuning ($\delta \neq 0$) we also arrive at an expression analogous to (4.44).

In summary, we see that the tunneling dynamics is directly analogous to the population dynamics of a two-level absorber. Left and right in the tunneling scenario correspond to down and up in the optical system. Current has a more direct meaning in the tunneling case while polarization is a more obvious concept in optics.

7.2.3 Dynamics of Optical Excitations

In the tunneling situation, J determines the frequency of the coherent spatial motion from site l to site r, while for optical excitation the product $2\mu_{vc}E_0$ determines the frequency of the population moving from the lower level to the upper one.

We define the Rabi frequency as

$$\Omega = 2E_0 \mu_{vc}/\hbar \tag{7.42}$$

and assume E_0 and μ_{vc} to be real. In the case of resonant stationary excitation, the optical Bloch equations

$$\frac{d\bar{\rho}_{vc}}{dt} = -i\frac{1}{2}\Omega(\rho_{vv} - \rho_{cc}),$$
$$\frac{d\rho_{vv}}{dt} = i\frac{1}{2}\Omega(\bar{\rho}_{cv} - \bar{\rho}_{vc}),$$
$$\frac{d\rho_{cc}}{dt} = i\frac{1}{2}\Omega(\bar{\rho}_{vc} - \bar{\rho}_{cv}), \tag{7.43}$$

are solved (note that $\rho_{cc}+\rho_{vv} = 1$) by the manifestly nonlinear result, already encountered in (7.41)

$$\rho_{cc} = \frac{1}{2}(1 - \cos \Omega t). \tag{7.44}$$

For resonant stationary excitation, the population oscillates between the ground state and the excited state with the Rabi frequency Ω. The evolution of the population from the ground state via the excited state back to the ground state is called a Rabi flop. At the time when the population of the system has moved completely into the upper state, the system is inverted, i.e., $I = -1$.

Equations (7.29) contain the important conservation law which we already encountered in (7.37) when we discussed the analogy between coherent tunneling and optical excitation of two-level absorbers. Here, we repeat the derivation of this conservation law in the framework of the optical Bloch equations. We apply the rotating-wave approximation and assume excitation by light, which may or may not be pulsed. Therefore, our present derivation is somewhat more general than the previous one.

The light field is given by

$$E(t) = E_0(t) \exp\left(-i\omega_L t\right) + c.c. \tag{7.45}$$

To obtain a compact notation, we define

$$\rho = \bar{\rho}_{vc},$$
$$\mu = \mu_{vc},$$
$$n = \rho_{cc},$$
$$K = E_0(t)\mu e^{-i(\omega_{cv}-\omega_L)t}/\hbar = \Omega(t)e^{-i(\omega_{cv}-\omega_L)t}/2. \tag{7.46}$$

We introduce the new variable

$$q = \rho e^{-i(\omega_{cv}-\omega_L)t} \tag{7.47}$$

and obtain its equation of motion

$$\frac{dq}{dt} = -iK(1 - 2n). \tag{7.48}$$

We also need to calculate $n(t)$ from

$$\frac{dn}{dt} = -iKq^* + iK^*q. \tag{7.49}$$

Using these equations, we find that

$$\frac{d|q|^2}{dt} = \left(\frac{dq}{dt}q^* + q\frac{dq^*}{dt}\right)$$
$$= -iK(1 - 2n)q^* + iK^*(1 - 2n)q$$
$$= (1 - 2n)\frac{dn}{dt}$$
$$= -\frac{1}{4}\frac{d}{dt}(1 - 2n)^2 \tag{7.50}$$

and thus,

$$\frac{d}{dt}\left[qq^* + \frac{1}{4}(2n-1)^2\right] = 0. \tag{7.51}$$

The expression in square brackets must therefore be a constant, a^2. Since without excitation $n = 0$ and $q = 0$, this constant must be $a^2 = 1/4$, thus we have

$$\rho\rho^* + \frac{1}{4}(2n-1)^2 = \frac{1}{4}, \tag{7.52}$$

which is identical to (7.37). Solving for n yields the conservation law

$$n = \frac{1}{2} \pm \frac{1}{2}\sqrt{1 - 4\rho\rho^*}. \tag{7.53}$$

In the low-excitation limit ($\rho^*\rho \ll 1/4$ and $n \ll 1$), we obtain

$$n = \rho^*\rho, \tag{7.54}$$

i.e.,

$$\rho_{cc} = \rho_{vc}^*\rho_{vc}. \tag{7.55}$$

For low excitation levels, we may therefore replace the population (or more generally, intraband) quantities by products of polarization (i.e., interband) quantities. If we decide to treat dephasing and population relaxation by introducing phenomenological damping terms defined by T_1 and T_2, we have to make sure that the "coherent limit" relation $T_2 = 2T_1$ is fulfilled in order to be consistent with (7.55). Up to third order in the light field and in the coherent limit, the nonlinear optical response is, according to the optical Bloch equations, described by a single equation for the polarization, i.e.,

$$\hbar\frac{d\bar{\rho}_{vc}}{dt} - i\hbar\left(\omega_{cv} - \omega_L + \frac{i}{T_2}\right)\bar{\rho}_{vc} = -iE_0(t)\mu_{vc}(1 - 2|\bar{\rho}_{vc}|^2). \tag{7.56}$$

Let us now study the response of the ensemble of two-level absorbers to a short light pulse. The excitation initially increases the population in the upper state $|1c\rangle$ up to a certain value. Let us assume that the excitation is resonant, i.e., $\omega_{cv} = \omega_L$ and, that for simplicity, $K(t)$ is real. Then ρ_{vc} is purely imaginary since the populations ρ_{cc} and ρ_{vv} are real and thus the imaginary part is a constant, which is zero due to the initial condition $\bar{\rho}_{vc}(t = 0) = 0$. We then have

$$\bar{\rho}_{vc} = \rho = -ip, \tag{7.57}$$

with real p.

As above, see (7.38), we define the inversion I by (note that in order to be consistent with the previous section, we define the inversion with an opposite sign with respect to the usual definition)

$$I = 1 - 2n = 1 - 2\rho_{cc}. \tag{7.58}$$

The optical Bloch equations read

$$\frac{dp}{dt} = KI,$$

$$\frac{dI}{dt} = -4Kp. \tag{7.59}$$

Exploiting the conservation law (7.52), we obtain

$$\frac{dI}{dt} = \pm 2K\sqrt{1 - I^2} \tag{7.60}$$

or

$$\frac{\frac{dI}{dt}}{\sqrt{1 - I^2}} = \pm 2K. \tag{7.61}$$

Integrating this equation yields

$$\int \frac{dI}{\sqrt{1 - I^2}} = \pm 2 \int K\,dt + C = \pm\Theta + C, \tag{7.62}$$

where we define the Rabi angle

$$\Theta = 2 \int K(t)dt = \int \Omega(t)dt. \tag{7.63}$$

Using the initial condition that before the arrival of the light pulse $\rho_{cc} = 0$, i.e., $I = 1$, we find $C = \pi/2$, such that

$$I = \cos\Theta \tag{7.64}$$

or

$$\rho_{cc} = \frac{1}{2}(1 - \cos\Theta). \tag{7.65}$$

Total inversion ($\rho_{cc} = 1$ and $\rho_{vv} = 0$) of the two-level system (actually the ensemble of identical two-level systems) is therefore achieved by applying a pulse that leads to $\Theta = \pi$. This is called a π pulse. A $\pi/2$ pulse would lead to an ensemble of systems with equal populations in the upper and lower states. A total Rabi flop is generated by applying a 2π pulse.

Here, we have considered the simplest case of resonant excitation. For nonresonant excitation the same procedure can be followed, leading to slightly more complicated, but interesting, results. As this case is treated in many textbooks dealing with spin echoes and nuclear magnetic resonance, we do not repeat this more general situation here.

7.2.4 The Few-Level Bloch Equations

After applications of the optical Bloch equations to the simplest model of an ensemble of two-level absorbers, we now proceed to analyze the optical Bloch equations for more general few-level systems. The underlying Hamiltonian is that of (3.13), and the light–matter interaction is given by (3.14). As before, we assume that the light field and the optical dipole matrix elements μ are parallel.

The relevant observables are now the total polarization

$$P(t) = \sum_{\nu=1}^{M} \sum_{\nu'=1}^{M'} (\mu_{\nu\nu'})^* \rho_{\nu\nu'} + c.c. \tag{7.66}$$

and the intraband coherences

$$n_{\nu\mu}(t) = \rho_{\nu\mu}(t) \tag{7.67}$$

and

$$n_{\nu'\mu'}(t) = \rho_{\nu'\mu'}(t), \tag{7.68}$$

where the undashed indices refer to the lower lying (valence) states $|\nu v\rangle$ while the dashed indices refer to the upper (conduction) states $|\nu' c\rangle$. It is assumed that the two sets of states are well separated in energy and that optical dipole matrix elements only connect states in the two different sets. This means that we ignore generally possible intraband dipole elements here. The diagonal elements

$$n_{\nu\nu}(t) = \rho_{\nu\nu}(t) \tag{7.69}$$

and

$$n_{\nu'\nu'}(t) = \rho_{\nu'\nu'}(t) \tag{7.70}$$

are the populations in the states $|\nu v\rangle$ and $|\nu' c\rangle$, respectively.

Following the same reasoning as before, we derive the optical Bloch equations in the rotating-wave approximation for the ensemble of M-M' systems and obtain

$$\hbar \frac{d\bar{\rho}_{\nu\nu'}}{dt} - i\hbar(\omega_{\nu'\nu} - \omega_L)\bar{\rho}_{\nu\nu'} = -iE_l^*(t)e^{-i\boldsymbol{k}_l\cdot\boldsymbol{r}} \left(\sum_{\mu=1}^{M} \mu_{\mu\nu'}\rho_{\nu\mu} - \sum_{\mu'=1}^{M'} \mu_{\nu\mu'}\rho_{\mu'\nu'} \right),$$

$$\hbar \frac{d\rho_{\nu\mu}}{dt} - i(\epsilon_\mu - \epsilon_\nu)\rho_{\nu\mu} = iE_l^*(t)e^{-i\boldsymbol{k}_l\cdot\boldsymbol{r}} \sum_{\kappa'=1'}^{M'} \mu_{\nu\kappa'}(\bar{\rho}_{\mu\kappa'})^*$$

$$- iE_l(t)e^{i\boldsymbol{k}_l\cdot\boldsymbol{r}} \sum_{\kappa'=1'}^{M'} (\mu_{\mu\kappa'})^* \bar{\rho}_{\nu\kappa'}, \tag{7.71}$$

$$\hbar \frac{d\rho_{\nu'\mu'}}{dt} - i(\epsilon_{\mu'} - \epsilon_{\nu'})\rho_{\nu'\mu'} = iE_l(t)e^{i\boldsymbol{k}_l \cdot \boldsymbol{r}} \sum_{\kappa=1}^{M} (\mu_{\kappa\nu'})^* \bar{\rho}_{\kappa\mu'}$$

$$-iE_l^*(t)e^{-i\boldsymbol{k}_l \cdot \boldsymbol{r}} \sum_{\kappa=1}^{M} \mu_{\kappa\mu'}(\bar{\rho}_{\kappa\nu'})^*.$$

$$(7.72)$$

These equations constitute an obvious generalization of the two-level optical Bloch equations. We see that, although we did not introduce intraband dipole matrix elements into our model, intraband coherences $\rho_{\nu\mu}$ and $\rho_{\nu'\mu'}$ are generated.

7.2.5 Fano Situations

Fano situations are a special case of the model discussed in the previous section. We consider only one single ground state $|v\rangle$, while the excited states $|\kappa\rangle$ result from the diagonalization discussed in Sect. 3.4 leading to the matrix elements indicated in (3.21). Therefore, the optical Bloch equations (7.71) and (7.72) can be used to describe Fano situations, provided that the sum over the lower set of states is absent.

Dynamical Equations for Semiconductors

In this chapter, we present and discuss theoretical approaches which allow one to analyze the coherent optical properties of semiconductors. As an introduction, we first discuss the free-carrier case where we ignore the many-particle interaction. Even though this is not very realistic for semiconductors, it is a useful exercise since the free-carrier model yields a closed set of material equations.

In the next step, we include the many-body interactions in the analysis. Considering the Coulomb interaction has the consequence that the set of equations for the correlation functions yields an infinite hierarchy due to the coupling to increasingly more complicated correlations of higher order. We present an approximate, but systematic treatment of the set of coupled equations in the coherent limit. The lowest-order treatment is the time-dependent Hartree–Fock approximation in which all so-called many-body correlations are neglected. The Hartree–Fock equations are exact in the linear regime including the predominant coherent Coulomb coupling effects. However, higher-order correlations have to be included into the semiconductor Bloch equations once we study nonlinear phenomena like four-wave-mixing or pump–probe experiments.

8.1 Noninteracting Particles

For the discussion of the semiconductor model, it is advantageous to work in the electron–hole picture and to formulate the theory using second quantization. In the following, we explicitly consider the vector character of both the light field and the optical dipole matrix element, see (5.23), i.e., the Hamiltonian is, cf. (5.8) and (5.21),

$$\hat{H} = \hat{H}_0 + \hat{H}_L, \tag{8.1}$$

$$\hat{H}_0 = \sum_{i,j} \sum_e T_{ij}^e c_i^{e\dagger} c_j^e + \sum_{i,j} \sum_h T_{ij}^h d_i^{h\dagger} d_j^h, \tag{8.2}$$

$$\hat{H}_L = -\boldsymbol{E}(t) \cdot \hat{\boldsymbol{P}}, \tag{8.3}$$

with

$$\hat{\boldsymbol{P}} = \sum_{ijeh} \left[\boldsymbol{\mu}_{ij}^{he} d_i^h c_j^e + H.C. \right]. \tag{8.4}$$

The polarization $\boldsymbol{P}(t)$ is given by the expectation value of the polarization operator, see (8.4), i.e., we have to determine the expectation values

$$p_{ij}^{he} = \langle d_i^h c_j^e \rangle. \tag{8.5}$$

All equations are now understood to be written in the rotating-wave approximation, i.e., interband energy differences are measured relative to the laser frequency, and the exciting light field is

$$\boldsymbol{E}(t) = \sum_{l=1}^n \boldsymbol{E}_l(t) \exp(i\boldsymbol{k}_l \cdot \boldsymbol{r}), \tag{8.6}$$

where $\boldsymbol{E}_l(t)$ is assumed to contain a real envelope function of the laser pulse. The total interband optical polarization is

$$\boldsymbol{P}(t) = 2\Re \sum_{ijhe} \boldsymbol{\mu}_{ij}^{he} p_{ij}^{he} \exp\left(-i\omega_L t\right). \tag{8.7}$$

We work in the Heisenberg picture and calculate the temporal evolution of $p_{ij}^{he}(t)$ according to

$$\hbar \frac{d}{dt} p_{ij}^{he} = \hbar \left\langle \frac{d}{dt} (d_i^h c_j^e) \right\rangle = i \langle [\hat{H}(t), d_i^h c_j^e] \rangle. \tag{8.8}$$

As we have to sum over the two conduction (e_1, e_2) and valence (h_1, h_2) bands when evaluating the commutator, we encounter the typical notational problem that more and more indices condense around the physical symbols like, e.g., $p_{ij}^{h_1 e_2}$.

The equation of motion for $p_{12}^{h_1 e_2}$, where $1, 2$ are two specific site indices, reads

$$\hbar \frac{d}{dt} p_{12}^{h_1 e_2} = -i \sum_j T_{2j}^e p_{1j}^{h_1 e_2} - i \sum_i T_{i1}^h p_{i2}^{h_1 e_2}$$

$$+ i\boldsymbol{E}(t) \cdot \left[(\mu_{12}^{h_1 e_2})^* - \sum_{je} (\mu_{1j}^{h_1 e})^* n_{j2}^{ee_2} \right.$$

$$\left. - \sum_{ih} (\mu_{i2}^{he_2})^* n_{i1}^{hh_1} \right].$$

(8.9)

Again, we encounter intraband coherences like, e.g.,

$$n_{ij}^{e_1 e_2} = \langle c_i^{e_1\dagger} c_j^{e_2} \rangle,$$
$$n_{ij}^{h_1 h_2} = \langle d_i^{h_1\dagger} d_j^{h_2} \rangle.$$

(8.10)

Their dynamics is determined via

$$\hbar \frac{d}{dt} n_{ij}^{e_1 e_2} = \hbar \left\langle \frac{d}{dt} (c_i^{e_1\dagger} c_j^{e_2}) \right\rangle = i\langle [\hat{H}(t), c_i^{e_1\dagger} c_j^{e_2}] \rangle$$

(8.11)

and the resulting equations are

$$\hbar \frac{d}{dt} n_{12}^{e_1 e_2} = -i \sum_j T_{2j}^e n_{1j}^{e_1 e_2} + i \sum_i T_{i1}^e n_{i2}^{e_1 e_2}$$

$$-i \left[\boldsymbol{E}^*(t) \cdot \sum_{jh} \mu_{j1}^{he_1} p_{j2}^{he_2} - \boldsymbol{E}(t) \cdot \sum_{ih} (\mu_{i2}^{he_2})^* (p_{i1}^{he_1})^* \right], \quad (8.12)$$

$$\hbar \frac{d}{dt} n_{12}^{h_1 h_2} = -i \sum_j T_{2j}^h n_{1j}^{h_1 h_2} + i \sum_i T_{i1}^h n_{i2}^{h_1 h_2}$$

$$-i \left[\boldsymbol{E}^*(t) \cdot \sum_{je} \mu_{1j}^{h_1 e} p_{2j}^{h_2 e} - \boldsymbol{E}(t) \cdot \sum_{ie} (\mu_{2i}^{h_2 e})^* (p_{1i}^{h_1 e})^* \right]. \quad (8.13)$$

Equations (8.9), (8.12), and (8.13) are a closed set of equations which can be solved numerically in any desired order of the external light field $\boldsymbol{E}(t)$.

In this book, we restrict our discussion to coherent optical phenomena in semiconductors. Moreover, in most of the following discussions, we consider relatively weak optical excitations, such that the optical response is often calculated up to the third order in the light field. Under these conditions, we can employ a straightforward generalization of the conservation law encountered in (7.55) in the low excitation limit:

$$n_{12}^{e_1 e_2} = \sum_{ah_a} p_{a2}^{h_a e_2} (p_{a1}^{h_a e_1})^*,$$

(8.14)

and

$$n_{12}^{h_1 h_2} = \sum_{ae_a} p_{1a}^{h_1 e_a} (p_{2a}^{h_2 e_a})^*.$$

(8.15)

Equations (8.14) and (8.15) are valid to second order in the optical field. These relations can be proven by first limiting (8.12) and (8.13) to second order, i.e.,

$$\hbar\frac{d}{dt}n_{12}^{e_1e_2(2)} = -i\sum_j T_{2j}^e n_{1j}^{e_1e_2(2)} + i\sum_i T_{i1}^e n_{i2}^{e_1e_2(2)}$$
$$-i\left[\boldsymbol{E}^*(t)\cdot\sum_{jh}\boldsymbol{\mu}_{j1}^{he_1}p_{j2}^{he_2(1)} - \boldsymbol{E}(t)\cdot\sum_{ih}(\boldsymbol{\mu}_{i2}^{he_2})^*(p_{i1}^{he_1(1)})^*\right], \quad (8.16)$$

$$\hbar\frac{d}{dt}n_{12}^{h_1h_2(2)} = -i\sum_j T_{2j}^h n_{1j}^{h_1h_2(2)} + i\sum_i T_{i1}^h n_{i2}^{h_1h_2(2)}$$
$$-i\left[\boldsymbol{E}^*(t)\cdot\sum_{je}\boldsymbol{\mu}_{1j}^{h_1e}p_{2j}^{h_2e(1)} - \boldsymbol{E}(t)\cdot\sum_{ie}(\boldsymbol{\mu}_{2i}^{h_2e})^*(p_{1i}^{h_1e(1)})^*\right], \quad (8.17)$$

where the superscripts (i) denote that the respective quantities are of ith order. The linear polarization appearing in (8.16) and (8.17) is the solution of

$$\hbar\frac{d}{dt}p_{12}^{h_1e_2(1)} = -i\sum_j T_{2j}^e p_{1j}^{h_1e_2(1)} - i\sum_i T_{i1}^h p_{i2}^{h_1e_2(1)} + i\boldsymbol{E}(t)\cdot(\boldsymbol{\mu}_{12}^{h_1e_2})^*. \quad (8.18)$$

Now, one can argue that: (i) (8.14) and (8.15) are valid before the optical excitation, since in this case both sides vanish; and (ii) use (8.16)–(8.18) to verify that the time derivatives of both sides of (8.14) and (8.15) are identical.

Applying these conservation laws, we obtain instead of (8.9)

$$-i\hbar\frac{d}{dt}p_{12}^{h_1e_2} = -\sum_j T_{2j}^e p_{1j}^{h_1e_2} - \sum_i T_{i1}^h p_{i2}^{h_1e_2}$$
$$+\boldsymbol{E}(t)\cdot\left[(\boldsymbol{\mu}_{12}^{h_1e_2})^* - \sum_{abh'e'}((\boldsymbol{\mu}_{1b}^{h_1e'})^*(p_{ab}^{h'e'})^*p_{a2}^{h'e_2}\right.$$
$$\left.+(\boldsymbol{\mu}_{b2}^{h'e_2})^*(p_{ba}^{h'e'})^*p_{1a}^{h_1e'})\right], \quad (8.19)$$

which only contains interband quantities. The intraband quantities follow from (8.14) and (8.15) and, therefore, (8.12) and (8.13) need not be solved.

By omitting all the super- and subscripts, we see that (8.19) has the structure

$$-i\hbar\frac{d}{dt}p = -\hbar\omega p + \mu^* E - \mu^* E p^* p. \quad (8.20)$$

We realize the close correspondence to the optical Bloch equation for the two-level system (7.29). The first term on the right-hand side is the free rotation of the polarization given by the single-particle energies; the second and third terms are the source terms including the phase-space filling $1 - p^*p$.

8.2 Interacting Particles

We now turn to a more realistic semiconductor model, which includes the Coulomb many-particle interaction in the form of the monopole–monopole interaction given by (5.13). We find that in this case the equations of motion are no longer closed. Instead, we encounter an infinite hierarchy of equations, which cannot be treated exactly. Hence, any solution can only be approximate and thus we have to look for systematic approximation schemes. In this book, we are primarily interested in coherent optical phenomena. In the purely co-herent limit, i.e., if all phase-breaking interactions with baths of other degrees of freedom like, e.g., phonons or incoherently generated populations of elec-trons and/or holes are ignored, there exists a systematic way of constructing successive approximations.

8.2.1 Hierarchy of Equations

Extending the semiconductor Hamiltonian of (8.1) by the many-body Coulomb interaction gives

$$\hat{H} = \hat{H}_0 + \hat{H}_C + \hat{H}_L, \tag{8.21}$$

with

$$\hat{H}_C = \frac{1}{2} \sum_{ij} \left(\sum_{e'} c_i^{e'\dagger} c_i^{e'} - \sum_{h'} d_i^{h'\dagger} d_i^{h'} \right) V_{ij} \left(\sum_{e} c_j^{e\dagger} c_j^{e} - \sum_{h} d_j^{h\dagger} d_j^{h} \right). \tag{8.22}$$

Evaluating the commutators with the Hamiltonian $\hat{H}$, we obtain the following equation of motion for $p_{12}^{h_1 e_2}$

$$-i\hbar \frac{d}{dt} p_{12}^{h_1 e_2} = -\sum_j T_{2j}^e p_{1j}^{h_1 e_2} - \sum_i T_{i1}^h p_{i2}^{h_1 e_2} + V_{12} p_{12}^{h_1 e_2}$$

$$+ \boldsymbol{E}(t) \cdot \left[(\boldsymbol{\mu}_{12}^{h_1 e_2})^* - \sum_{je} (\boldsymbol{\mu}_{1j}^{h_1 e})^* n_{j2}^{ee_2} - \sum_{ih} (\boldsymbol{\mu}_{i2}^{he_2})^* n_{i1}^{hh_1} \right]$$

$$+ \sum_{ah_a} V_{a1} \langle d_a^{h_a} d_1^{h_1} d_a^{h_a \dagger} c_2^{e_2} \rangle - \sum_{ah_a} V_{2a} \langle d_1^{h_1} d_a^{h_a} c_2^{e_2} d_a^{h_a \dagger} \rangle$$

$$+ \sum_{ae_a} V_{a1} \langle c_a^{e_a \dagger} d_1^{h_1} c_a^{e_a} c_2^{e_2} \rangle - \sum_{ae_a} V_{2a} \langle d_1^{h_1} c_a^{e_a \dagger} c_2^{e_2} c_a^{e_a} \rangle, \tag{8.23}$$

which differs from (8.9) by the terms containing the interaction V times the four-point correlation functions.

Next, we could derive the equations of motion for the intraband coherences $n_{ij}^{e_1 e_2}$ and $n_{ij}^{e_1 e_2}$. Evaluating the commutators with $\hat{H}$ gives equations of the same level of complexity as (8.23). All these equations contain four-point correlation functions. Writing down equations of motions for these four-point correlation functions, one finds that they couple to six-point correlation functions, etc., i.e., (8.23) is the beginning of an infinite hierarchy of coupled equations of motion.

8.2.2 Coherent Dynamics-Controlled Truncation Scheme

Equations (8.14) and (8.15) are special cases of more general rules which hold in the coherent limit up to a certain finite order in the optical field even in the presence of the many-body Coulomb interaction, as has been shown by Axt and Stahl (1994) and Lindberg et al. (1994). These approaches rely on basically two assumptions:

(i) The nonlinear optical response is written as an expansion in powers of the optical field. Limiting this expansion to a certain finite order introduces a systematic truncation for the hierarchy of many-body Coulomb correlations. This is due to the fact that, simply speaking, up to the nth order in the field, the light–matter coupling can at most generate n electron–hole pairs and thus, all higher-order correlation functions do not contribute to the nth-order response.

(ii) It is assumed that before the optical excitation the semiconductor is in its ground state, i.e., no electrons and holes are present, and that the dynamics of the system is fully coherent. In this case, correlation functions containing mixtures of creation and annihilation operators can be factorized into products of expectation values which contain only one type of operators, i.e., either only creation or only annihilation operators.

Combining (i) and (ii), it is possible to describe the nth-order response by $[(n + 1)/2]$ correlation functions which represent single- and multi-exciton transitions, see Victor et al. (1995). This means that the first-order response is just described by the single-exciton coherence p, in the third-order, the biexciton coherence (B) also has to be calculated, in the fifth-order additionally the triexciton coherence (W) has to be calculated, etc. By excitonic transitions we denote Coulomb-correlated electron and hole complexes. A single-excitonic transition consists of an electron and a hole. A biexcitonic one is composed from two electrons and two holes, etc. We will discuss excitonic transitions in more detail later.

In the following, it is shown how one can derive generalizations of the relations given by (8.14) and (8.15) which allow one to formulate the nonlinear optical response solely in terms of single- and multi-exciton coherences. Since the required derivations are rather lengthy but straightforward, we outline in the following the basic steps that are involved in this procedure by following the approach of Lindberg et al. (1994). More details and further information can be found in the publications that are listed in the section "Suggested Reading" at the end of this chapter.

Let us start from a general multiband electron–hole Hamiltonian in an arbitrary representation

$$
\begin{aligned}
\hat{H} &= \hat{H}_0 + \hat{H}_C + \hat{H}_L \\
&= \sum_{\nu_1,\nu_2} \epsilon(\nu_1,\nu_2)a^\dagger(\nu_1)a(\nu_2) \\
&\quad + \frac{1}{2}\sum_{\nu_1,\nu_2,\nu_3,\nu_4} V(\nu_1,\nu_2,\nu_3,\nu_4)a^\dagger(\nu_1)a^\dagger(\nu_2)a(\nu_3)a(\nu_4) \\
&\quad + \frac{1}{2}\sum_{\nu_1,\nu_2} \left((\boldsymbol{\mu}_{\nu_1,\nu_2}\cdot\boldsymbol{E})a^\dagger(\nu_1)a^\dagger(\nu_2) + (\boldsymbol{\mu}_{\nu_1,\nu_2}\cdot\boldsymbol{E})^*a(\nu_2)a(\nu_1) \right). \quad (8.24)
\end{aligned}
$$

The ν_i are multi-indices which refer to a specific state in a specific band. Thus, depending on whether ν_i denotes a conduction or a valence band, the corresponding operators $a^\dagger(\nu_i)$ and $a(\nu_i)$ act as electron or hole field operators in the respective band. The diagonal part of $\epsilon(\nu_1,\nu_2)$ denotes the single-particle energies whereas its off-diagonal part describes the coupling between the single-particle states. Due to the hermiticity of the Hamiltonian, the relation $\epsilon(\nu_1,\nu_2) = \epsilon(\nu_2,\nu_1)^*$ holds. $\boldsymbol{\mu}_{\nu_1,\nu_2}$ is the antisymmetrized dipole matrix element for interband transitions, i.e., $\boldsymbol{\mu}_{\nu_1,\nu_2} = \langle\nu_2|e\boldsymbol{r}|\nu_1\rangle$ if ν_2 refers to a conduction band and ν_1 to a valence band and $\boldsymbol{\mu}_{\nu_1,\nu_2} = -\langle\nu_2|e\boldsymbol{r}|\nu_1\rangle$ in the opposite case. The fact that optical intraband and intersubband excitations are neglected means that $\boldsymbol{\mu}_{\nu_1,\nu_2}$ vanishes if ν_1 and ν_2 both refer either to a conduction band or to a valence band. Therefore, the light–matter interaction in (8.24) either creates or destroys electron–hole pairs through the operators $a^\dagger(\nu_1)a^\dagger(\nu_2)$ and $a(\nu_2)a(\nu_1)$, respectively. In $\hat{H}_C$ of (8.24), as in (8.22), only those terms are considered which change neither the number of electrons nor the number of holes. Therefore, the light–matter interaction $\hat{H}_L$ is the only term in the Hamiltonian $\hat{H}$ which changes the number of particles. Because of the hermiticity of the Hamiltonian and the anticommutation of the field operators $a(\nu)$ and $a^\dagger(\nu)$, the Coulomb matrix elements can be chosen to fulfill

$$
\begin{aligned}
V(\nu_1,\nu_2,\nu_3,\nu_4) &= V(\nu_4,\nu_3,\nu_2,\nu_1)^*, \\
V(\nu_1,\nu_2,\nu_3,\nu_4) &= -V(\nu_2,\nu_1,\nu_3,\nu_4) = -V(\nu_1,\nu_2,\nu_4,\nu_3). \quad (8.25)
\end{aligned}
$$

The antisymmetrized matrix elements are used to compress the notation in the following equations of motion. The Heisenberg equations for the field operators $a(\nu)$ and $a^\dagger(\nu)$ are obtained by evaluating the commutators with the Hamiltonian as

$$i\hbar \frac{d}{dt} a(\nu) = \sum_{\nu_1} \epsilon(\nu, \nu_1) a(\nu_1)$$

$$+ \sum_{\nu_1, \nu_2, \nu_3} V(\nu, \nu_1, \nu_2, \nu_3) a^\dagger(\nu_1) a(\nu_2) a(\nu_3)$$

$$+ \sum_{\nu_1} (\boldsymbol{\mu}_{\nu, \nu_1} \cdot \boldsymbol{E}) a^\dagger(\nu_1), \qquad (8.26)$$

$$i\hbar \frac{d}{dt} a^\dagger(\nu) = - \sum_{\nu_1} \epsilon(\nu, \nu_1)^* a^\dagger(\nu_1)$$

$$- \sum_{\nu_1, \nu_2, \nu_3} V(\nu_1, \nu_2, \nu_3, \nu) a^\dagger(\nu_1) a^\dagger(\nu_2) a(\nu_3)$$

$$- \sum_{\nu_1} (\boldsymbol{\mu}_{\nu, \nu_1} \cdot \boldsymbol{E})^* a(\nu_1). \qquad (8.27)$$

Let us now define a normally ordered operator product of creation and annihilation operators via

$$\{N, M\} \equiv a^\dagger(\nu_N) a^\dagger(\nu_{N-1}), \ldots, a^\dagger(\nu_1) a(\eta_1), \ldots, a(\eta_{M-1}) a(\eta_M), \quad (8.28)$$

where again, depending on ν_i and η_j, the operators $a^\dagger$ and a are electron or hole operators for certain states, respectively. The quantities $\{N, M\}$ contain the full information about the temporal evolution of the photoexcited many-body system. For example, the microscopic polarization p^{he} corresponds to $\{0, 2\}$, its complex conjugate $(p^{he})^*$ to $\{2, 0\}$, and the carrier occupations and intraband coherences $n^{e_1 e_2}$ and $n^{h_1 h_2}$ to $\{1, 1\}$.

The time derivative of the normally ordered operator product is given by

$$\frac{d}{dt} \{N + 1, M + 1\} = \left(\frac{da^\dagger(\nu_{N+1})}{dt} \right) \{N, M\} a(\eta_{M+1})$$

$$+ a^\dagger(\nu_{N+1}) \left(\frac{d\{N, M\}}{dt} \right) a(\eta_{M+1})$$

$$+ a^\dagger(\nu_{N+1}) \{N, M\} \left(\frac{da(\eta_{M+1})}{dt} \right). \qquad (8.29)$$

In order to determine to which type of correlations functions $\{N, M\}$ is coupled, one just needs to analyze (8.29) together with the equations of motion for the individual field operators, i.e. (8.26) and (8.27). The latter equations show that $\{0, 1\}$ is coupled to itself by the single-particle part, coupled to $\{1, 0\}$ by the light–matter interaction, and coupled to $\{1, 2\}$ by the many-body Coulomb interaction, whereas $\{1, 0\}$ is coupled to itself, to $\{0, 1\}$ and to $\{2, 1\}$, respectively. Combining these results with (8.29), one finds, after restoring normal order of the operators on the right-hand side, that the light–matter interaction couples $\{N, M\}$ to $\{N-1, M+1\}$, $\{N+1, M-1\}$, $\{N-2, M\}$, and $\{N, M-2\}$. The many-body Coulomb interaction, however, couples $\{N, M\}$ to itself and to $\{N+1, M+1\}$. This coupling scheme and its important consequences was first noticed by Axt and Stahl (1994). These findings verify once again that

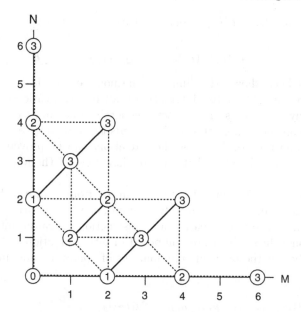

Fig. 8.1. Schematic drawing of relevant dynamic variables $\langle\{N, M\}\rangle$ up to third order in the light field. The *numbers in the symbols* correspond to the minimum order in the external field in which $\langle\{N, M\}\rangle$ is finite. The *dotted lines* denote the couplings due to the light–matter interaction, whereas the *solid lines* indicate the couplings via the Coulomb interaction. [After M. Lindberg et al., Phys. Rev. **B50**, 18060 (1994)]

it is the four-operator Coulomb part of the Hamiltonian which generates the coupling to products that contain more operators, i.e., the many-body hierarchy. All couplings to a smaller number of operators are solely due to the light–matter interaction.

Clearly, operators with $N - M$ being odd are only coupled to other operators where $N - M$ is odd and operators with $N - M$ being even are only coupled to other operators where $N - M$ is even. Thus, if we start from the ground state as the initial condition, i.e., $\langle\{0, 0\}\rangle = 1$ and all other expectation values vanish before the photoexcitation, the operators $\{N, M\}$ with $N - M$ being odd vanish at all times, since their equations of motion contain no sources.

The reasoning presented above is visualized in Fig. 8.1. This figure clearly shows that in any finite order in the external field, only a finite number of expectation values contribute to the optical response. To find the lowest order in the light field at which any term $\{N, M\}$ is nonvanishing, one only needs to evaluate the minimum number of dotted lines in Fig. 8.1, which are needed to connect it to $\{0, 0\}$. As one can see from Fig. 8.1 and as is proven more rigorously below, the minimum order in the field in which $\langle\{N, M\}\rangle$ is finite is $(N + M)/2$ if N and M are both even and $(N + M)/2 + 1$ if N and M are both odd, i.e.,

$$\langle a^\dagger(\nu_{2N}),\ldots,a^\dagger(\nu_1)\,a(\eta_1),\ldots,a(\eta_{2M})\rangle \; = \; O(|\boldsymbol{E}|^{N+M}) \qquad (8.30)$$

and

$$\langle a^\dagger(\nu_{2N+1}),\ldots,a^\dagger(\nu_1)\,a(\eta_1),\ldots,a(\eta_{2M+1})\rangle \; = \; O(|\boldsymbol{E}|^{N+M+2}). \;\; (8.31)$$

So far, it has been shown that limiting the nonlinear optical response to a finite order in the light–matter interaction leads to a truncation of the many-body hierarchy. Now, we start to discuss generalizations of (8.14) and (8.15). In order to point out where this analysis leads to, we first give the results to lowest order and then outline how these relations can be proven, which also shows how they can be extended if expressions beyond the lowest order are required.

For the case where one initially starts from the ground state of the semiconductor and where the dynamics is fully coherent, one can factorize expectation values containing a mix of creation and destruction operators into products of expectation values which are composed of either creation or destruction operators only. For the relevant combinations of operators, one finds

$$\langle a^\dagger(\nu_{2N}),\ldots,a^\dagger(\nu_1)\,a(\eta_1),\ldots,a(\eta_{2M})\rangle$$
$$= \langle a^\dagger(\nu_{2N}),\ldots,a^\dagger(\nu_1)\rangle\langle a(\eta_1),\ldots,a(\eta_{2M})\rangle + O(|\boldsymbol{E}|^{N+M+2}) \qquad (8.32)$$

and

$$\langle a^\dagger(\nu_{2N+1}),\ldots,a^\dagger(\nu_1)\,a(\eta_1),\ldots,a(\eta_{2M+1})\rangle$$
$$= \sum_{\delta_1}\langle a^\dagger(\nu_{2N+1}),\ldots,a^\dagger(\nu_1)a^\dagger(\delta_1)\rangle\langle a(\delta_1)a(\eta_1),\ldots,a(\eta_{2M+1})\rangle$$
$$+ O(|\boldsymbol{E}|^{N+M+4}). \qquad\qquad\qquad (8.33)$$

Clearly, (8.14) and (8.15) are special cases of (8.33) if one considers the factorization of the carrier occupations and intraband coherences $n^{e_1 e_2}$ and $n^{h_1 h_2}$ in second order. If one limits the analysis to the third order and applies (8.32) and (8.33) for all quantities which appear in Fig. 8.1, one finds that only the single-exciton coherences $\langle\{2,0\}\rangle$ and $\langle\{0,2\}\rangle$, the biexciton coherences $\langle\{4,0\}\rangle$ and $\langle\{0,4\}\rangle$, and the triexciton coherences $\langle\{6,0\}\rangle$ and $\langle\{0,6\}\rangle$ are needed to fully determine the response. Since the optical polarization is determined by $\langle\{2,0\}\rangle$ and $\langle\{0,2\}\rangle$ and the coupling of the triexciton coherences $\langle\{6,0\}\rangle$ and $\langle\{0,6\}\rangle$ to the polarization is at least of fifth order, see Fig. 8.1, one finds that only the single-exciton and biexciton coherences are required in the so-called coherent $\chi^{(3)}$ limit. In the coherent $\chi^{(5)}$ limit, additionally the triexciton coherence has to be computed. This means that the coherent third-order response is fully described by two quantities, whereas the fifth order requires the calculation of three excitonic correlation functions. In general, the coherent nth-order response can be described by $[(n+1)/2]$ correlation functions which represent single- and multi-exciton transitions.

To prove (8.32) and (8.33) and generalizations which include higher-order terms, it is important to note that in the Heisenberg picture (H) the operators are time-dependent and all expectation values are taken with respect to the initial state, which is the ground state $|0\rangle$ of the semiconductor, i.e.,

$$\langle a^\dagger(\nu_N) a^\dagger(\nu_{N-1}), \ldots, a^\dagger(\nu_1) a(\eta_1), \ldots, a(\eta_M) \rangle =$$
$$\langle 0 | a_H^\dagger(\nu_N, t), \ldots, a_H^\dagger(\nu_1, t) a_H(\eta_1, t), \ldots, a_H(\eta_M, t) | 0 \rangle . \quad (8.34)$$

Now, we can define an interaction picture with respect to the light–matter interaction by decomposing the Hamiltonian $\hat{H} = \hat{H}_0 + \hat{H}_C + \hat{H}_L$ into two parts, via $\hat{H} = \hat{H}_1 + \hat{H}'$, with $\hat{H}_1 = \hat{H}_0 + \hat{H}_C$ and $\hat{H}' = \hat{H}_L$. In the interaction picture (I), the time dependence of an operator $\mathcal{A}$ is only due to $\hat{H}_1$ and given by

$$\mathcal{A}_I(t) = e^{i\hat{H}_1 t} \mathcal{A} e^{-i\hat{H}_1 t}, \quad (8.35)$$

where $\mathcal{A}$ is the operator in the Schrödinger picture. The time evolution of the wave functions is solely due to $\hat{H}'$ according to

$$|\Psi_I(t)\rangle = S_I(t)|\Psi_H\rangle, \quad (8.36)$$

where the time evolution operator in the interaction representation $S_I(t)$ is the solution of

$$i\hbar \frac{d}{dt} S_I(t) = \hat{H}'(t) S_I(t), \quad (8.37)$$

with the initial condition $S_I(0) = 1$. Equation (8.37) has a formal solution given by

$$S_I(t) = \sum_{k=0}^{\infty} S_{I,k}(t) = 1 + \sum_{k=1}^{\infty} \left(-i \int_0^t dt' \hat{H}'(t') S_{I,k-1}(t') \right). \quad (8.38)$$

with $S_{I,0}(t) = 1$. Equation (8.38) shows that the terms $S_{I,k}$ with $k > 0$, which are induced by the light–matter interaction, can be obtained iteratively, i.e., $S_{I,1}(t) = -i \int_0^t dt' \hat{H}'(t') S_{I,0}(t')$, $S_{I,2}(t) = -i \int_0^t dt' \hat{H}'(t') S_{I,1}(t')$, etc. Since $\hat{H}' = \hat{H}_L$ is of the order E, $S_{I,k}$ is of the order $|E|^k$.

Any expectation value of an operator $\langle \{N, M\} \rangle$ can be written as

$$\langle \{N, M\} \rangle = \langle \Psi | S_I^\dagger(t) \{N, M\}_I S_I(t) | \Psi \rangle$$
$$= \left\langle \Psi \left| \sum_{k_1=0}^{\infty} S_{I,k_1}^\dagger(t) \{N, M\}_I \sum_{k_2=0}^{\infty} S_{I,k_2}(t) \right| \Psi \right\rangle$$
$$= \sum_{k=0}^{\infty} \sum_{l=0}^{k} \langle \Psi | S_{I,k-l}^\dagger(t) \{N, M\}_I S_{I,l}(t) | \Psi \rangle$$
$$\equiv \sum_{k=0}^{\infty} \{N, M\}_k(t), \quad (8.39)$$

where $\{N, M\}_k$ is obviously of the order $|E|^k$.

Let us now determine the lowest order k in which $\langle\{N, M\}\rangle$ is not zero. We note that $S_{I,k}$ can be expressed as a linear combination of operators which create or annihilate up to k pairs of carriers, i.e.,

$$S_{I,k}(t) = \sum_{l=-k}^{k} S_{2l}^{k}(t),\tag{8.40}$$

where S_{2l}^{k} creates or annihilates $2|l|$ carriers, respectively, depending on whether l is positive or negative. When $S_{I,k}$ acts on the ground state, which contains no carriers, the annihilation terms vanish and we have

$$S_{I,k}|0\rangle = \sum_{l=0}^{k} S_{2l}^{k}(t)|0\rangle.\tag{8.41}$$

Applying the normally ordered operator $\{N, M\}_I$ leads to

$$\{N, M\}_I S_{I,l}(t)|0\rangle = 0,\tag{8.42}$$

if $M > 2l$. Similarly, by letting the operators act on $\langle 0|$, it can also straightforwardly be shown that

$$\langle 0|S_{I,k-l}^{\dagger}(t)\{N, M\}_I S_{I,l}(t)|0\rangle = 0,\tag{8.43}$$

if $N > 2k - 2l$. This means that $\langle\{N, M\}\rangle$ is finite if $2l \geq M$ and $2k - 2l \geq N$, i.e., $k \geq (N + M)/2$. Therefore, an expectation value $\langle\{N, M\}\rangle$ is at least of order $|\boldsymbol{E}|^{(N+M)/2}$.

By inserting the interaction representation of the identity operator in between the creation and destruction operators of a ground-state expectation value of Heisenberg operators, i.e., in between $a_H^{\dagger}(\nu_1, t)$ and $a_H(\eta_1, t)$ of (8.34), we obtain

$$\begin{aligned}
&\langle 0| a_H^{\dagger}(\nu_N, t), \ldots, a_H^{\dagger}(\nu_1, t)\, a_H(\eta_1, t), \ldots, a_H(\eta_M, t)|0\rangle \\
&= \langle 0| a_H^{\dagger}(\nu_N, t), \ldots, a_H^{\dagger}(\nu_1, t)|0\rangle \langle 0| a_H(\eta_1, t), \ldots, a_H(\eta_M, t)|0\rangle \\
&\quad + \sum_{\delta_1} \langle 0| a_H^{\dagger}(\nu_N, t), \ldots, a_H^{\dagger}(\nu_1, t)\, a_I^{\dagger}(\delta_1, t)|0\rangle \\
&\qquad \times \langle 0| a_I(\delta_1, t)\, a_H(\eta_1, t), \ldots, a_H(\eta_M, t)|0\rangle \\
&\quad + \frac{1}{2} \sum_{\delta_1, \delta_2} \langle 0| a_H^{\dagger}(\nu_N, t), \ldots, a_H^{\dagger}(\nu_1, t)\, a_I^{\dagger}(\delta_2, t)\, a_I^{\dagger}(\delta_1, t)|0\rangle \\
&\qquad \times \langle 0| a_I(\delta_1, t)\, a_I(\delta_2, t)\, a_H(\eta_1, t), \ldots, a_H(\eta_M, t)|0\rangle \\
&\quad + \cdots.
\end{aligned}\tag{8.44}$$

Here, the interaction picture for the identity operator has been used since the vacuum state is an eigenstate of the interaction picture operators, which evolve in time according to $\hat{H}_1$, see (8.35) and do not change the number of particles. In contrast, the vacuum state is not an eigenstate of the Heisenberg operators. In the following, we show that if N and M are both even, the first term on the right-hand side of (8.44) is the lowest-order factorized contribution and the further terms are of higher order in $\boldsymbol{E}$. If N and M are both odd, the first term on the right-hand side of (8.44) vanishes since, as explained above, in this case quantities like $\langle\langle\{N,0\}\rangle\rangle$ and $\langle\langle\{0,M\}\rangle\rangle$ are not generated by the coherent dynamical evolution of the system. Therefore, the second term provides the lowest-order factorization if N and M are both odd.

The transformation between the Heisenberg and the interaction operators is

$$a_H(\delta,t) = S_I^\dagger(t)a_I(\delta,t)S_I(t). \tag{8.45}$$

Using the expansion of $S_I(t)$, i.e. (8.38) and (8.40), we can write

$$a_H(\delta,t) = \sum_{n=-\infty}^{\infty} A_{H,2n-1}(\delta,t), \tag{8.46}$$

where $A_{H,2n-1}(\delta,t)$ creates or annihilates $2n-1$ carriers, respectively, depending on whether $2n-1$ is positive or negative, and is at least of the order $|\boldsymbol{E}|^{|n|}$. From this relation it follows that

$$\langle 0\,|\,a_I(\delta_1),\ldots,a_I(\delta_{2p})\,a_H(\eta_1),\ldots,a_H(\eta_{2M})|0\rangle = O(|\boldsymbol{E}|^{M+p}). \tag{8.47}$$

Inserting (8.47) into (8.44) immediately yields

$$\langle 0\,|\,a_H^\dagger(\nu_{2N},t),\ldots,a_H^\dagger(\nu_1,t)\,a_H(\eta_1,t),\ldots,a_H(\eta_{2M},t)\,|\,0\rangle$$
$$= \langle 0\,|\,a_H^\dagger(\nu_{2N},t),\ldots,a_H^\dagger(\nu_1,t)\,|\,0\rangle\langle 0\,|\,a_H(\eta_1,t),\ldots,a_H(\eta_{2M},t)\,|\,0\rangle$$
$$+O(|\boldsymbol{E}|^{N+M+2}). \tag{8.48}$$

If the number of creation and annihilation operators on the left-hand side of (8.44) is odd, the first term on the right-hand side vanishes. The second term contains interaction operators and is thus not directly an expectation value. However, since $A_{H,-1}(\delta,t) = a_I(\delta,t) + O(|\boldsymbol{E}|^2)$, we have

$$\langle 0\,|\,a_H(\delta_1)\,a_H(\eta_1),\ldots,a_H(\eta_{2M+1})|0\rangle$$
$$= \langle 0\,|\,a_I(\delta_1)\,a_H(\eta_1),\ldots,a_H(\eta_{2M+1})|0\rangle + O(|\boldsymbol{E}|^{M+3}). \tag{8.49}$$

Therefore, we obtain the lowest-order factorization for odd N and M as

$$\langle 0\,|\,a_H^\dagger(\nu_{2N+1},t),\ldots,a_H^\dagger(\nu_1,t)\,a_H(\eta_1,t),\ldots,a_H(\eta_{2M+1},t)\,|\,0\rangle$$
$$= \sum_{\delta_1}\langle 0\,|\,a_H^\dagger(\nu_{2N+1},t),\ldots,a_H^\dagger(\nu_1,t)\,a_I^\dagger(\delta_1,t)\,|\,0\rangle$$
$$\times\langle 0\,|\,a_I(\delta_1,t)\,a_H(\eta_1,t),\ldots,a_H(\eta_{2M+1},t)\,|\,0\rangle$$
$$+O(|\boldsymbol{E}|^{N+M+4}). \tag{8.50}$$

It has been shown that the lowest order expansions in powers of $|\boldsymbol{E}|$ of the first two terms of (8.44) indeed coincide with the factorizations given in (8.32) and (8.33). This decoupling scheme can be used to obtain the relevant equations of motion in the lowest nontrivial order, i.e., in the coherent $\chi^{(3)}$ limit, see next section. The derivations presented here can also be extended to higher orders in the field. As an example, we present in Sect. 8.2.5 the equations of motion for the coherent $\chi^{(5)}$ limit.

8.2.3 Dynamical Truncation in the Coherent $\chi^{(3)}$ Limit

As has been shown above, in the coherent limit, the two-point intraband correlation functions and the four-point correlation functions appearing in (8.23) can be written as summations over products of single- and multi-exciton transitions. We can now truncate the equations of motion by explicitly imposing the condition that the resulting polarization has only driving terms up to a certain order in $|\boldsymbol{E}|$. The lowest order is the linear response, i.e., $p \propto |\boldsymbol{E}|$, and the intraband quantities n vanish. The latter are at least second order in $|\boldsymbol{E}|$ and (8.33) shows that the two-point intraband quantities $n^{e_1 e_2}$ and $n^{h_1 h_2}$ are in this order given by (8.14) and (8.15). The third order is of particular interest, since it is the lowest order in which pump–probe and four-wave-mixing experiments can be described. Using (8.32), we can write

$$\langle d_a^{h_a\dagger} d_1^{h_1} d_a^{h_a} c_2^{e_2} \rangle = \sum_{be_b} \langle d_1^{h_1} c_b^{e_b} d_a^{h_a} c_2^{e_2} \rangle (p_{ab}^{h_a e_b})^* + O(|\boldsymbol{E}|^5), \qquad (8.51)$$

where the left-hand side is one of the four-point terms appearing in (8.23).

Applying this decoupling scheme to all terms in (8.23), the optical response up to the third order in the light field can be expressed using two transition-type quantities. These are the interband coherences $p_{12}^{h_1 e_2}$, i.e., single-exciton excitations, and

$$B_{1324}^{h_1 e h e_2} = -\langle d_1^{h_1} d_2^{h} c_3^{e} c_4^{e_2} \rangle, \qquad (8.52)$$

which describes biexciton excitations.

In order to be able to analyze pure correlation effects, it is advantageous to remove the uncorrelated (Hartree–Fock factorized) parts from the four-point quantities, i.e., to define

$$\bar{B}_{1324}^{h_1 e h e_2} = B_{1324}^{h_1 e h e_2} + p_{14}^{h_1 e_2} p_{23}^{he} - p_{13}^{h_1 e} p_{24}^{he_2}. \qquad (8.53)$$

This procedure results in closed equations of motion for the single-exciton amplitude p_{12}^{he} and the biexciton amplitude $\bar{B}_{1324}^{h_1 e h e_2}$, which completely determine the optical response within the coherent $\chi^{(3)}$ limit. No other correlation function contributes to the polarization up to third order.

Following the procedure outlined above, the equation of motion for p is obtained as

$$-i\hbar\frac{d}{dt}p_{12}^{he} = -\sum_j T_{2j}^e p_{1j}^{he} - \sum_i T_{i1}^h p_{i2}^{he} + V_{12}p_{12}^{he}$$

$$+ \sum_{abh'e'} (V_{a2} - V_{a1} - V_{b2} + V_{b1}) \left[(p_{ba}^{h'e'})^* p_{b2}^{h'e} p_{1a}^{he'} \right.$$

$$- (p_{ba}^{h'e'})^* p_{ba}^{h'e'} p_{12}^{he} - (p_{ba}^{h'e'})^* \bar{B}_{ba12}^{h'e'he} \right]$$

$$+ \boldsymbol{E}(t) \cdot \left[(\mu_{12}^{he})^* - \sum_{abh'e'} ((\mu_{1b}^{he'})^* (p_{ab}^{h'e'})^* p_{a2}^{h'e} \right.$$

$$+ (\mu_{b2}^{h'e})^* (p_{ba}^{h'e'})^* p_{1a}^{he'}) \right]. \tag{8.54}$$

Again, it is useful to repeat this equation in an abbreviated form, i.e., without the super- and subscripts,

$$-i\hbar\frac{d}{dt}p = -\hbar\omega_x p + V p^* pp + V p^* \bar{B} + \mu^* E - \mu^* E p^* p. \tag{8.55}$$

In (8.54), the first three terms on the right-hand side define the homogeneous part of the equation of motion, which includes the single-particle electronic energies and couplings (T) and the electron–hole Coulomb attraction (V_{12}). This part can be diagonalized yielding the so-called exciton states. Therefore, the eigenenergies are excitonic energies $\hbar\omega_x$, see (8.55). The following lines of (8.54) describe different types of inhomogeneities. In addition to the linear source term given by the external field times the dipole transition matrix element, $\boldsymbol{E} \cdot \boldsymbol{\mu}^*$, in the coherent $\chi^{(3)}$ limit, there are optical nonlinearities arising from Pauli blocking (also called phase-space filling) $(\mu^* E p^* p)$ and from the many-body Coulomb interaction. The latter include the first-order Coulomb contribution $(V p^* pp)$ and the correlation contribution $(V p^* \bar{B})$. The nonlinear response can be written as a sum over the contributions from these three types of optical nonlinearities, which makes it possible to investigate them separately.

The term $\propto p_{ba}^{h'e'*} p_{ba}^{h'e'} p_{12}^{he}$ in line three of (8.54) looks like an energy renormalization. Using the typical selection rules for heavy-hole to conduction electron $(hh\text{-}e)$ transitions, this term, which is of first order in the Coulomb interaction, introduces a coupling between the spin subspaces. This term vanishes only in a homogeneous system. It is, however, finite and contributes to the nonlinear response in inhomogeneous, for example, disordered systems.

The equation of motion for $\bar{B}$ is obtained as

$$
\begin{aligned}
-i\hbar\frac{d}{dt}\bar{B}_{ba12}^{h'e'he} = &-\sum_i (T_{2i}^e \bar{B}_{ba1i}^{h'e'he} + T_{i1}^h \bar{B}_{bai2}^{h'e'he} \\
&+ T_{ai}^e \bar{B}_{bi12}^{h'e'he} + T_{ib}^h \bar{B}_{ia12}^{h'e'he}) \\
&+ (V_{ba} + V_{b2} + V_{1a} + V_{12} - V_{b1} - V_{a2})B_{ba12}^{h'e'he} \\
&- (V_{ba} + V_{12} - V_{b1} - V_{a2})p_{1a}^{he'} p_{b2}^{h'e} \\
&+ (V_{1a} + V_{b2} - V_{b1} - V_{a2})p_{ba}^{h'e'} p_{12}^{he}.
\end{aligned}
\tag{8.56}
$$

In abbreviated form, this equation has the structure

$$
-i\hbar\frac{d}{dt}\bar{B} = -\hbar\omega_{2x}\bar{B} + Vpp.
\tag{8.57}
$$

The first three lines on the right-hand side of (8.56) constitute the homogeneous part of the equation of motion, which includes the electronic energies and couplings (T) as well as the six possible Coulomb interactions between the two electrons and the two holes and, therefore, corresponds to the biexciton energy $\hbar\omega_{2x}$. The further terms represent the inhomogeneity. Since the uncorrelated first-order Coulomb contributions have been removed from $\bar{B}$, it is purely driven by sources which include the many-body interaction V, i.e., by terms proportional to Vpp.

The semiconductor Bloch equations, (8.54) and (8.56), fully determine the interband polarization $\mathbf{P}$ within the coherent $\chi^{(3)}$ limit. For a spatially homogeneous system, the complexity of solving these equations can be reduced, since in this case, the center-of-mass motion is irrelevant. When p and B depend only on the relative motion of the particles, one can easily transform the equations of motion in k space. However, in spatially inhomogeneous, e.g., disordered systems, the complete equations (8.54) and (8.56) have to be solved. Furthermore, these expressions can be easily extended to include propagation effects.

Instead of removing the uncorrelated parts from the four-point functions, one can also derive equations which describe the optical response up to $\chi^{(3)}$ using the original four-point functions $B_{1324}^{h_1 e h e_2}$. This results in the following equations for p and B

$$
\begin{aligned}
-i\hbar\frac{d}{dt}p_{12}^{he} = &-\sum_j T_{2j}^e p_{1j}^{he} - \sum_i T_{i1}^h p_{i2}^{he} + V_{12}p_{12}^{he} \\
&- \sum_{abh'e'} (V_{a2} - V_{a1} - V_{b2} + V_{b1})(p_{ba}^{h'e'})^* B_{ba12}^{h'e'he}
\end{aligned}
$$

$$+E(t) \cdot \left[(\boldsymbol{\mu}_{12}^{he})^* - \sum_{abh'e'} \left((\boldsymbol{\mu}_{1b}^{he'})^* (p_{ab}^{h'e'})^* p_{a2}^{h'e} \right. \right.$$

$$\left. \left. + (\boldsymbol{\mu}_{b2}^{h'e})^* (p_{ba}^{h'e'})^* p_{1a}^{he'} \right) \right], \tag{8.58}$$

and

$$-i\hbar \frac{d}{dt} B_{ba12}^{h'e'he} = -\sum_i (T_{2i}^e B_{ba1i}^{h'e'he} + T_{i1}^h B_{bai2}^{h'e'he}$$

$$+ T_{ai}^e B_{bi12}^{h'e'he} + T_{ib}^h B_{ia12}^{h'e'he})$$

$$+ (V_{ba} + V_{b2} + V_{1a} + V_{12} - V_{b1} - V_{a2}) B_{ba12}^{h'e'he}$$

$$- E(t) \cdot \left[(\boldsymbol{\mu}_{12}^{he})^* p_{ba}^{h'e'} + (\boldsymbol{\mu}_{ba}^{h'e'})^* p_{12}^{he} \right.$$

$$\left. - (\boldsymbol{\mu}_{1a}^{he'})^* p_{b2}^{h'e} - (\boldsymbol{\mu}_{b2}^{h'e})^* p_{1a}^{he'} \right]. \tag{8.59}$$

Symbolically, these equations read

$$-i\hbar \frac{d}{dt} p = -\hbar\omega_x p - V p^* B + \mu^* E - \mu^* E p^* p, \tag{8.60}$$

and

$$-i\hbar \frac{d}{dt} B = -\hbar\omega_{2x} B - \mu^* E p. \tag{8.61}$$

Equations (8.58) and (8.59) are equivalent to (8.54) and (8.56), i.e., solving either one of these equation pairs gives exactly the same result. What is, however, different is the appearance of the Coulomb-induced optical nonlinearities in the equation of p. Whereas in (8.54) and (8.56) the Coulomb terms are split into a first-order (Vp^*pp) and a correlation part ($Vp^*\bar{B}$), no such distinction appears in (8.58) and (8.59), but all many-body nonlinearities are proportional to Vp^*B.

The comparison of (8.54) and (8.56) and (8.58) and (8.59) indicates that strong compensations between the first-order and the higher-order Coulomb correlations can be expected. This is confirmed by the numerical solutions presented in later chapters. As another difference between (8.56) and (8.59) we note that $\bar{B}$ is driven by inhomogeneous terms proportional to Vpp whereas B is driven by μEp.

8.2.4 Hartree–Fock Approach in the Coherent $\chi^{(3)}$ Limit

The equations derived above contain all terms driven by the light field up to third order. However, due to the numerical complexity, one may sometimes not want to solve equations containing biexcitons. Also, situations exist in which biexcitons are not of large importance. The equations relevant in such cases are obtained by neglecting the correlation function $\bar{B}$ in (8.54),

$$
\begin{aligned}
-i\hbar\frac{d}{dt}p_{12}^{he} = &-\sum_j T_{2j}^e p_{1j}^{he} - \sum_i T_{i1}^h p_{i2}^{he} + V_{12}p_{12}^{he}\\
&+ \sum_{abh'e'} (V_{a2} - V_{a1} - V_{b2} + V_{b1})[(p_{ba}^{h'e'})^* p_{b2}^{h'e} p_{1a}^{he'}\\
&- (p_{ba}^{h'e'})^* p_{ba}^{h'e'} p_{12}^{he}]\\
&+ E(t) \cdot \left[(\mu_{12}^{he})^* - \sum_{abh'e'} \left((\mu_{1b}^{he'})^* (p_{ab}^{h'e'})^* p_{a2}^{h'e}\right.\right.\\
&\left.\left. + (\mu_{b2}^{h'e})^* (p_{ba}^{h'e'})^* p_{1a}^{he'}\right)\right].
\end{aligned} \tag{8.62}
$$

This is the semiconductor Bloch equation in the time-dependent Hartree–Fock approximation for the interband coherence p. $\bar{B}$ has been neglected and thus, according to (8.53), B is approximated by its factorized contributions, i.e.,

$$
B_{1324}^{h_1 e h e_2} \approx -p_{14}^{h_1 e_2} p_{23}^{he} + p_{13}^{h_1 e} p_{24}^{he_2}. \tag{8.63}
$$

In the linear response limit, one neglects all terms proportional to powers of p higher than unity and thus has

$$
-i\hbar\frac{d}{dt}p_{12}^{he} = -\sum_j T_{2j}^e p_{1j}^{he} - \sum_i T_{i1}^h p_{i2}^{he} + V_{12}p_{12}^{he} + E(t) \cdot (\mu_{12}^{he})^*, \tag{8.64}
$$

which is nothing but the Wannier equation for the exciton. It complies with our previous notation that the observable p describes excitonic excitations.

In second order in the field, one obtains the intraband coherences n by inserting the linear response result for p into the conservation law equations (8.14) and (8.15).

Although the semiconductor Bloch equations in the time-dependent Hartree–Fock approximation contain terms which describe third-order processes $\propto p^*pp$, they are not a consistent approximation to $\chi^{(3)}$ processes, like four-wave-mixing and pump–probe experiments, except in cases where biexcitonic resonances are unimportant. Only the equations containing the four-point correlation functions (8.54) and (8.56) or alternatively (8.58) and (8.59) consistently contain all terms driven by the light field up to third order.

8.2.5 Dynamical Truncation in the Coherent $\chi^{(5)}$ Limit

As shown in Sect. 8.2.2, in the coherent limit, intraband quantities can be expressed via combinations of interband single- and multi-exciton transitions. These rules have been applied in Sect. 8.2.3 to the lowest nontrivial order. There, it has been shown that in the coherent $\chi^{(3)}$ limit the nonlinear optical response is completely determined by single and biexciton coherences, i.e., p and B. In the following, we present the equations of motion which describe the coherent $\chi^{(5)}$ limit. To obtain these equations, one thus has to analyze the expressions derived in Sect. 8.2.2 beyond the lowest order. Since these derivations are rather elaborate, we demonstrate them for one term. Following these lines, also all other quantities which are relevant in the coherent $\chi^{(5)}$ limit can be obtained.

Let us consider the electron population and interband coherence $n_{ab} = \langle c_a^\dagger c_b \rangle$, where in order to reduce the notation a and b are multi indices. Starting from the ground state, the coherent dynamics generates an equal number of electrons N_e and holes N_h, which allows us to write

$$
\begin{aligned}
0 &= \langle c_a^\dagger c_b N_e \rangle - \langle c_a^\dagger c_b N_h \rangle \\
&= \sum_k \langle c_a^\dagger c_b c_k^\dagger c_k \rangle - \sum_j \langle c_a^\dagger c_b d_j^\dagger d_j \rangle \\
&= \sum_k \langle c_a^\dagger c_k^\dagger c_k c_b \rangle + \langle c_a^\dagger c_b \rangle - \sum_j \langle c_a^\dagger d_j^\dagger d_j c_b \rangle. \qquad (8.65)
\end{aligned}
$$

We thus get

$$
\langle c_a^\dagger c_b \rangle = \sum_j \langle c_a^\dagger d_j^\dagger d_j c_b \rangle - \sum_k \langle c_a^\dagger c_k^\dagger c_k c_b \rangle. \qquad (8.66)
$$

The first term on the right-hand side of (8.66) corresponds to an exciton occupation. Since it describes the generation and the annihilation of one electron–hole pair, it is finite to second order in the field. Factorizing it using only the first term of (8.44) gives

$$
\sum_j \langle c_a^\dagger d_j^\dagger d_j c_b \rangle = \sum_j \langle c_a^\dagger d_j^\dagger \rangle \langle d_j c_b \rangle = \sum_j (p_{ja})^* p_{jb} + O(|E|^4), \qquad (8.67)
$$

which reproduces (8.14).

The second term on the right-hand side of (8.66) is an electron density–density correlation. Since four electron and no hole operators appear, this term is at least of fourth order in the field. It can be factorized by using the third term of (8.44) according to

$$
\begin{aligned}
-\sum_k \langle c_a^\dagger c_k^\dagger c_k c_b \rangle = &-\frac{1}{2} \sum_{k,l,j} \langle 0| c_{H,a}^\dagger c_{H,k}^\dagger c_{I,l}^\dagger c_{I,j}^\dagger |0\rangle \, \langle 0| c_{I,j} c_{I,l} c_{H,k} c_{H,b} |0\rangle \\
&-\frac{1}{2} \sum_{k,l,j} \langle 0| c_{H,a}^\dagger c_{H,k}^\dagger d_{I,l}^\dagger d_{I,j}^\dagger |0\rangle \, \langle 0| d_{I,j} d_{I,l} c_{H,k} c_{H,b} |0\rangle
\end{aligned}
$$

$$-\frac{1}{2}\sum_{k,l,j}\langle 0|c_{H,a}^\dagger c_{H,k}^\dagger d_{I,l}^\dagger c_{I,j}^\dagger|0\rangle\,\langle 0|c_{I,j}d_{I,l}c_{H,k}c_{H,b}|0\rangle$$

$$-\frac{1}{2}\sum_{k,l,j}\langle 0|c_{H,a}^\dagger c_{H,k}^\dagger c_{I,l}^\dagger d_{I,j}^\dagger|0\rangle\,\langle 0|d_{I,j}c_{I,l}c_{H,k}c_{H,b}|0\rangle$$

$$+O(|\boldsymbol{E}|^6). \tag{8.68}$$

Here, obviously only the second term is of fourth order whereas the others are of sixth and higher order. Thus, (8.68) simplifies to

$$-\sum_k\langle c_a^\dagger c_k^\dagger c_k c_b\rangle = -\frac{1}{2}\sum_{k,l,j}\langle c_a^\dagger c_k^\dagger d_l^\dagger d_j^\dagger\rangle\,\langle d_j d_l c_k c_b\rangle + O(|\boldsymbol{E}|^6)$$

$$= -\frac{1}{2}\sum_{k,l,j}B_{jalk}^* B_{lkjb} + O(|\boldsymbol{E}|^6), \tag{8.69}$$

where $B_{lkjb} = \langle d_l c_k d_j c_b\rangle$ and we have commuted some operators to arrive at the last line.

What now remains is to obtain the fourth-order factorized contribution arising from the first term in (8.66), i.e., $\sum_j\langle c_a^\dagger d_j^\dagger d_j c_b\rangle$. To calculate it, we use the first and third term of (8.44) together with (8.67) and get

$$\sum_j\langle c_a^\dagger d_j^\dagger d_j c_b\rangle - \sum_j(p_{ja})^* p_{jb}$$

$$=\frac{1}{2}\sum_{k,l,j}\langle 0|c_{H,a}^\dagger d_{H,j}^\dagger c_{I,l}^\dagger c_{I,k}^\dagger|0\rangle\,\langle 0|c_{I,k}c_{I,l}d_{H,j}c_{H,b}|0\rangle$$

$$+\frac{1}{2}\sum_{k,l,j}\langle 0|c_{H,a}^\dagger d_{H,j}^\dagger d_{I,l}^\dagger d_{I,k}^\dagger|0\rangle\,\langle 0|d_{I,k}d_{I,l}d_{H,j}c_{H,b}|0\rangle$$

$$+\frac{1}{2}\sum_{k,l,j}\langle 0|c_{H,a}^\dagger d_{H,j}^\dagger d_{I,l}^\dagger c_{I,k}^\dagger|0\rangle\,\langle 0|c_{I,k}d_{I,l}d_{H,j}c_{H,b}|0\rangle$$

$$+\frac{1}{2}\sum_{k,l,j}\langle 0|c_{H,a}^\dagger d_{H,j}^\dagger c_{I,k}^\dagger d_{I,l}^\dagger|0\rangle\,\langle 0|d_{I,l}c_{I,k}d_{H,j}c_{H,b}|0\rangle + O(|\boldsymbol{E}|^6).$$

$$=\sum_{k,l,j}\langle 0|c_{H,a}^\dagger d_{H,j}^\dagger c_{I,k}^\dagger d_{I,l}^\dagger|0\rangle\,\langle 0|d_{I,l}c_{I,k}d_{H,j}c_{H,b}|0\rangle + O(|\boldsymbol{E}|^6), \tag{8.70}$$

where the two identical fourth-order terms have been combined and the higher-order terms are neglected.

It is important to note that, as is shown below, the factorized result of the right-hand side of (8.70) is up to fourth order *not* just given by $\sum_{k,l,j}B_{lkja}^* B_{lkjb}$. For further analysis, it is advantageous to transform (8.70) into the interaction picture

$$\sum_{k,l,j}\langle 0|c^\dagger_{H,a}d^\dagger_{H,j}c^\dagger_{I,k}d^\dagger_{I,l}|0\rangle\,\langle 0|d_{I,l}c_{I,k}d_{H,j}c_{H,b}|0\rangle + O(|\boldsymbol{E}|^6) \tag{8.71}$$

$$=\sum_{k,l,j}\langle 0|S^\dagger_I(t)c^\dagger_{I,a}d^\dagger_{I,j}S_I(t)c^\dagger_{I,k}d^\dagger_{I,l}|0\rangle\,\langle 0|d_{I,l}c_{I,k}S^\dagger_I(t)d_{I,j}c_{I,b}S_I(t)|0\rangle + O(|\boldsymbol{E}|^6).$$

Here, each expectation value is at least of second order, i.e., by expanding the time evolution operator in orders of the field, see (8.38), we therefore get

$$\sum_{k,l,j}\langle 0|S^\dagger_I(t)c^\dagger_{I,a}d^\dagger_{I,j}S_I(t)c^\dagger_{I,k}d^\dagger_{I,l}|0\rangle\,\langle 0|d_{I,l}c_{I,k}S^\dagger_I(t)d_{I,j}c_{I,b}S_I(t)|0\rangle + O(|\boldsymbol{E}|^6)$$

$$=\sum_{k,l,j}\langle 0|S^\dagger_{I,2}(t)c^\dagger_{I,a}d^\dagger_{I,j}c^\dagger_{I,k}d^\dagger_{I,l}|0\rangle\,\langle 0|d_{I,l}c_{I,k}d_{I,j}c_{I,b}S_{I,2}(t)|0\rangle$$

$$+\sum_{k,l,j}\langle 0|S^\dagger_{I,1}(t)c^\dagger_{I,a}d^\dagger_{I,j}S_{I,1}(t)c^\dagger_{I,k}d^\dagger_{I,l}|0\rangle\,\langle 0|d_{I,l}c_{I,k}S^\dagger_{I,1}(t)d_{I,j}c_{I,b}S_{I,1}(t)|0\rangle$$

$$+\sum_{k,l,j}\langle 0|S^\dagger_{I,2}(t)c^\dagger_{I,a}d^\dagger_{I,j}c^\dagger_{I,k}d^\dagger_{I,l}|0\rangle\,\langle 0|d_{I,l}c_{I,k}S^\dagger_{I,1}(t)d_{I,j}c_{I,b}S_{I,1}(t)|0\rangle$$

$$+\sum_{k,l,j}\langle 0|S^\dagger_{I,1}(t)c^\dagger_{I,a}d^\dagger_{I,j}S_{I,1}(t)c^\dagger_{I,k}d^\dagger_{I,l}|0\rangle\,\langle 0|d_{I,l}c_{I,k}d_{I,j}c_{I,b}S_{I,2}(t)|0\rangle$$

$$+O(|\boldsymbol{E}|^6). \tag{8.72}$$

The first term of (8.72) simply gives

$$\sum_{k,l,j}\langle 0|S^\dagger_{I,2}(t)c^\dagger_{I,a}d^\dagger_{I,j}c^\dagger_{I,k}d^\dagger_{I,l}|0\rangle\,\langle 0|d_{I,l}c_{I,k}d_{I,j}c_{I,b}S_{I,2}(t)|0\rangle$$

$$=\sum_{k,l,j}\langle 0|S^\dagger_{I,2}(t)c^\dagger_{I,a}d^\dagger_{I,j}c^\dagger_{I,k}d^\dagger_{I,l}S_{I,0}|0\rangle\,\langle 0|S^\dagger_{I,0}d_{I,l}c_{I,k}d_{I,j}c_{I,b}S_{I,2}(t)|0\rangle$$

$$=\sum_{k,l,j}B^*_{jalk}B_{lkjb}, \tag{8.73}$$

where $S_{I,0}=S^\dagger_{I,0}=1$, see (8.38). Similarly, the expectation values of the third and fourth terms of (8.72) which contain the time-evolution operator in second order are B^*_{jalk} and B_{lkjb}, respectively.

To determine the results of the expectation values which contain two first-order time-evolution operators, we, e.g., manipulate the second term of (8.72)

$$\sum_{k,l,j}\langle 0|S^\dagger_{I,1}(t)c^\dagger_{I,a}d^\dagger_{I,j}S_{I,1}(t)c^\dagger_{I,k}d^\dagger_{I,l}|0\rangle\,\langle 0|d_{I,l}c_{I,k}S^\dagger_{I,1}(t)d_{I,j}c_{I,b}S_{I,1}(t)|0\rangle$$

$$=\sum_{k,l,j}\langle 0|S^\dagger_{I,1}(t)c^\dagger_{I,a}d^\dagger_{I,j}S_{I,0}S_{I,1}(t)c^\dagger_{I,k}d^\dagger_{I,l}S_{I,0}|0\rangle$$

$$\langle 0|S^\dagger_{I,0}d_{I,l}c_{I,k}S^\dagger_{I,1}(t)S^\dagger_{I,0}d_{I,j}c_{I,b}S_{I,1}(t)|0\rangle$$

$$=\sum_{k,l,j}\langle 0|S^\dagger_{I,1}(t)c^\dagger_{I,a}d^\dagger_{I,j}S_{I,0}|0\rangle\langle 0|S_{I,1}(t)c^\dagger_{I,k}d^\dagger_{I,l}S_{I,0}|0\rangle$$

$$\langle 0|S^\dagger_{I,0}d_{I,l}c_{I,k}S^\dagger_{I,1}(t)|0\rangle\langle 0|S^\dagger_{I,0}d_{I,j}c_{I,b}S_{I,1}(t)|0\rangle$$

$$= \sum_{k,l,j} \langle 0|S_{I,1}^{\dagger}(t)c_{I,a}^{\dagger}d_{I,j}^{\dagger}S_{I,0}|0\rangle(-1)\langle 0|S_{I,1}^{\dagger}(t)c_{I,k}^{\dagger}d_{I,l}^{\dagger}S_{I,0}|0\rangle$$

$$(-1)\langle 0|S_{I,0}^{\dagger}d_{I,l}c_{I,k}S_{I,1}(t)|0\rangle\langle 0|S_{I,0}^{\dagger}d_{I,j}c_{I,b}S_{I,1}(t)|0\rangle$$

$$= \sum_{k,l,j}(-(p_{ja})^{*}(p_{lk})^{*})(-p_{lk}p_{jb}), \tag{8.74}$$

where we have used that $S_{I,1}(t) = -S_{I,1}^{\dagger}(t)$, see (8.38).

Combining all these results, we have

$$\sum_{j}\langle c_{a}^{\dagger}d_{j}^{\dagger}d_{j}c_{b}\rangle = \sum_{j}(p_{ja})^{*}p_{jb} \tag{8.75}$$

$$+ \sum_{k,l,j}(B_{jalk}^{*} - (p_{ja})^{*}(p_{lk})^{*})(B_{lkjb} - p_{lk}p_{jb}) + O(|\boldsymbol{E}|^{6}).$$

Inserting (8.69) and (8.76) into (8.66) yields

$$n_{ab} = \langle c_{a}^{\dagger}c_{b}\rangle = \sum_{j}(p_{ja})^{*}p_{jb}$$

$$+ \sum_{k,l,j}\left[\frac{1}{2}B_{jalk}^{*}B_{lkjb} - (p_{ja})^{*}(p_{lk})^{*}B_{lkjb}\right.$$

$$\left. - p_{lk}p_{jb}B_{jalk}^{*} + (p_{ja})^{*}(p_{lk})^{*}p_{lk}p_{jb}\right] + O(|\boldsymbol{E}|^{6}). \tag{8.76}$$

Thus, the electronic intraband coherences and populations can be expressed as summations over products of single- and biexciton transitions up to fourth order in the field.

To describe all $\chi^{(5)}$ processes, such conservation laws have to be applied to all other relevant mixed expectation values which contain creation and annihilation operators. As another example let us consider $\langle d_{a}^{\dagger}d_{b}d_{c}c_{d}\rangle$, which appears in the equation of motion for p, (8.23). This term can be written as

$$\langle c_{a}^{\dagger}c_{b}d_{c}c_{d}\rangle = \sum_{j}(p_{ja})^{*}B_{cdjb}$$

$$+ \sum_{jkl}(\tfrac{1}{2}B_{jalk}^{*} - (p_{ja})^{*}(p_{lk})^{*})(W_{lkcdjb} - p_{lk}B_{cdjb})$$

$$+ \mathcal{O}(|\boldsymbol{E}|^{7}). \tag{8.77}$$

Here, the first line contains the expressions arising in third order and the second line gives the fifth-order contributions. The latter include the triexciton transition which is defined as $W = \langle d_{l}c_{k}d_{c}c_{d}d_{j}c_{b}\rangle$.

Now, we can proceed and set up the equations of motion which contain all coherent $\chi^{(5)}$ processes. The equations presented below are, however, not complete, since the terms W describing triexcitons have been omitted completely. In many situations, these terms do not seem to play a crucial role, probably since typically no new bound states arise as a consequence of W. Furthermore, solving equations including W is a formidable numerical task which can presently be performed only for very small systems. For example, a study of $\chi^{(5)}$ processes including triexcitons in small semiconductor rings is given in Meier et al. (2003). Alternatively, instead of neglecting the triexciton coherence W altogether, one could have considered the factorized parts of W, i.e., $W - \bar{W}$, and only ignore $\bar{W}$. However, results obtained on the $\chi^{(3)}$ level already show that the complete neglect of terms may often yield better results than approximating them by their factorized version.

The resulting coupled equations of motion for p and B describing coherent $\chi^{(5)}$ processes are

$$
\begin{aligned}
-i\hbar\frac{d}{dt}p_{12}^{he} =\; & -\sum_j T_{2j}^e p_{1j}^{he} - \sum_i T_{i1}^h p_{i2}^{he} + V_{12}p_{12}^{he} \\
& -\sum_{abh'e'}(V_{a2}-V_{a1}-V_{b2}+V_{b1})(p_{ba}^{h'e'})^* B_{ba12}^{h'e'he} \\
& +\boldsymbol{E}\cdot\left[(\mu_{12}^{he})^* - \sum_{abh'e'}(\mu_{1b}^{he'})^*(p_{ab}^{h'e'})^* p_{a2}^{h'e} + (\mu_{b2}^{h'e})^*(p^{h'e'})^*_{ba} p_{1a}^{he'}\right] \\
& -\frac{1}{2}\boldsymbol{E}\cdot\sum_{\substack{ijkl\\h'e'h''e''}}[(\mu_{1i}^{he'})^*(B_{lkji}^{h''e''h'e'})^* B_{lkj2}^{h''e''h'e} \\
& \hspace{4cm} +(\mu_{j2}^{h'e})^*(B_{lkji}^{h''e''h'e'})^* B_{lk1i}^{h''e''he'}] \\
& +\boldsymbol{E}\cdot\sum_{\substack{ijkl\\h'e'h''e''}}[(\mu_{1i}^{he'})^*(p_{lk}^{h''e''})^*(p_{ji}^{h'e'})^* B_{lkj2}^{h''e''h'e} \\
& \hspace{4cm} +(\mu_{j2}^{h'e})^*(p_{lk}^{h''e''})^*(p_{ji}^{h'e'})^* B_{lk1i}^{h''e''he'}] \\
& +\boldsymbol{E}\cdot\sum_{\substack{ijkl\\h'e'h''e''}}[(\mu_{1i}^{he'})^*(B_{lkji}^{h''e''h'e'})^* p_{lk}^{h''e''} p_{j2}^{h'e} \\
& \hspace{4cm} +(\mu_{j2}^{h'e})^*(B_{lkji}^{h''e''h'e'})^* p_{lk}^{h''e''} p_{1i}^{he'}] \\
& -\boldsymbol{E}\cdot\sum_{\substack{ijkl\\h'e'h''e''}}[(\mu_{1i}^{he'})^*(p_{lk}^{h''e''})^*(p_{ji}^{h'e'})^* p_{lk}^{h''e''} p_{j2}^{h'e} \\
& \hspace{4cm} +(\mu_{j2}^{h'e})^*(p_{lk}^{h''e''})^*(p_{ji}^{h'i'})^* p_{lk}^{h''e''} p_{1i}^{he'}] \\
& -\sum_{\substack{ijkl\\h'e'h''e''}}(V_{i1}-V_{i2}-V_{j1}+V_{j2})(B_{jilk}^{h'e'h''e''})^* p_{lk}^{h''e''} B_{12ji}^{heh'e'} \\
& +\sum_{\substack{ijkl\\h'e'h''e''}}(V_{i1}-V_{i2}-V_{j1}+V_{j2})(p_{ji}^{h'e'})^*(p_{lk}^{h''e''})^* p_{lk}^{h''e''} B_{12ji}^{heh'e'}, \quad (8.78)
\end{aligned}
$$

and for the four-point correlations we obtain

$$
-i\hbar\frac{d}{dt}B_{abcd}^{heh'e'} = -\sum_i (T_{di}^e B_{abci}^{heh'e'} + T_{bi}^e B_{aicd}^{heh'e'} + T_{ic}^h B_{abid}^{heh'e'} + T_{ia}^h B_{ibcd}^{heh'e'})
$$
$$
+(V_{ab} + V_{ad} + V_{cb} + V_{cd} - V_{ac} - V_{bd})B_{abcd}^{heh'e'}
$$
$$
+\mathbf{E}\cdot[(\mu_{cd}^{h'e'})^* p_{ab}^{he} + (\mu_{ab}^{he})^* p_{cd}^{h'e'} - (\mu_{cb}^{h'e})^* p_{ad}^{he'} - (\mu_{ad}^{he'})^* p_{cb}^{h'e}]
$$
$$
-\mathbf{E}\cdot\sum_{ije''h''} (p^{h''e''})_{ji}^* [(\mu_{ai}^{he''})^* B_{jbcd}^{h''eh'h'} + (\mu_{ci}^{h'e''})^* B_{abjd}^{heh''e'}
$$
$$
+(\mu_{jb}^{h''e})^* B_{aicd}^{he''h'e'} + (\mu_{jd}^{h''e'})^* B_{abci}^{heh'e''}]. \tag{8.79}
$$

The semiconductor Bloch equations, (8.78) and (8.79), fully determine the interband polarization $\mathbf{P}$ within the coherent $\chi^{(5)}$ limit if the triexciton transitions W are neglected. The first three lines in (8.78) and (8.79) are identical to the third-order equations (8.58) and (8.59), respectively. The following lines additionally contain the terms that contribute to coherent $\chi^{(5)}$ processes.

8.3 Suggested Reading

1. V.M. Axt and A. Stahl, "A dynamics-controlled truncation scheme for the hierarchy of density matrices in semiconductor optics", Z. Phys. B **93**, 195 (1994)
2. V.M. Axt and A. Stahl, "The role of the biexciton in a dynamic density matrix theory of the semiconductor bend edge", Z. Phys. B **93**, 205 (1994)
3. H. Haug and S.W. Koch, *Quantum Theory of the Optical and Electronic Properties of Semiconductors*, 4th edn. (World Scientific, Singapore 2004)
4. M. Lindberg, Y.Z. Hu, R. Binder, and S.W. Koch, "$\chi^{(3)}$ formalism in optically excited semiconductors and its applications in four-wave-mixing spectroscopy", Phys. Rev. **B50**, 18060 (1994)
5. T. Meier and S.W. Koch, "Coulomb correlation signatures in the excitonic optical nonlinearities of semiconductors", in *Ultrafast Physical Processes in Semiconductors*, Vol. **67** in Series *Semiconductors and Semimetals*, ed. K.T. Tsen (Academic Press, New York 2001), p. 231
6. T. Meier, C. Sieh, E. Finger, W. Stolz, W.W. Rühle, P. Thomas, and S.W. Koch, "Signatures of biexcitons and triexcitons in coherent nondegenerate semiconductor optics", Phys. Stat. Sol. B **238**, 537 (2003)
7. W. Schäfer and M. Wegener, *Semiconductor Optics and Transport Phenomena* (Springer, Berlin 2002)
8. K. Victor, V.M. Axt, and A. Stahl, "Hierarchy of density matrices in coherent semiconductor optics", Phys. Rev. **B51**, 14164 (1995)
9. K. Victor, V.M. Axt, G. Bartels, A. Stahl, K. Bott, and P. Thomas, "Microscopic foundation of the phenomenological few-level approach to coherent semiconductor optics", Z. Phys. B. **99**, 197 (1996)

Part II

Applications I

9

Linear Optical Response

In this and the following chapters, we apply the theory developed above to discuss a variety of optical phenomena. We begin by summarizing some of the concepts known from the introductory sections, such as linear response theory. The response function, i.e., the linear optical susceptibility χ, relates the optical polarization $P(t)$ to the exciting light field $E(t)$

$$P(t) = \int_{-\infty}^{\infty} \chi(t - t')E(t')dt'. \tag{9.1}$$

In general, $E(t)$ and $P(t)$ are vectors, such that χ is a tensorial quantity. For isotropic media, however, the susceptibility can be taken as a scalar function.

Due to causality, $\chi(t-t') = 0$ holds if $t < t'$, i.e., there can be no response at time t if the excitation arrives at a later time t'. The Laplace transform $\chi(z)$ of the susceptibility $\chi(t)$ as a function of the complex frequency z is

$$\chi(z) = \int_0^{\infty} \chi(t)e^{izt}dt, \tag{9.2}$$

with $\Im z > 0$. Causality requires that the real and imaginary parts of $\chi(z)$ for $z \to \omega + i\varepsilon$, with real ω and an infinitesimal $\varepsilon > 0$, obey the Kramers–Kronig relations. The imaginary part

$$\chi''(\omega) = \lim_{\epsilon \to 0} \Im \chi(\omega + i\epsilon) \tag{9.3}$$

is related to the absorption in homogeneous materials. We refer to this quantity as the *linear optical spectrum*.

To derive the linear optical spectrum, one has to solve the microscopic equation of motion for the optical polarization, i.e., for the interband coherence, in lowest order in the light field. For all models discussed so far, the corresponding equations of motion have the same structure: There is the free rotation given by a frequency term, there are source terms containing the phase-space-filling nonlinearities (called Pauli blocking), and there are

Coulomb-induced linear and nonlinear terms (for systems with interacting particles). Clearly, the linear response is obtained by ignoring all possible nonlinear terms. We are left with an equation that contains the free rotation, a driving term given by the Rabi frequency $\Omega = 2\boldsymbol{E}(t) \cdot \boldsymbol{\mu}/\hbar$ and, for interacting systems, the electron–hole Coulomb attraction.

The restriction to the low excitation limit excludes excitation by a stationary light field, as this excitation would drive the system into a state where the nonlinear terms are no longer negligible. However, by introducing a phenomenological damping term into the free rotation part of the equation, see the discussion after (7.29), one can restrict the excitation to low values even for stationary excitation. This trick is also inherently used in the treatment of linear response using Fermi's Golden Rule. On the other hand, if short light pulses are considered as excitation, no such damping terms need to be included. One only has to make sure that the Rabi angle Θ, see (7.63), remains sufficiently small.

A rather elegant way to obtain the linear optical spectrum is to use a very short weak pulse. In the calculation, one then has to compute the temporal evolution of the interband coherence, keeping only terms linear in the excitation field and neither populations nor intraband coherences. As an idealization of a very short light pulse, we use

$$E(t) = \eta \delta(t). \tag{9.4}$$

Note that η has the dimension of electric field times time, such that $2\eta\mu_0/\hbar$ is the Rabi angle Θ, see (7.42) and (7.63). Inserting $E(t)$ into (9.1) yields

$$P(t) = \eta \chi(t) \tag{9.5}$$

with

$$\chi(t) = 0, \quad \text{for} \quad t < 0. \tag{9.6}$$

As mentioned above, the linear absorption spectrum is given by

$$\chi''(\omega) = \lim_{\epsilon \to 0} \Im \int_0^\infty \chi(t) e^{i(\omega + i\epsilon)t} dt, \tag{9.7}$$

with positive ϵ. This procedure is valid for homogeneous and isotropic systems and is illustrated below first for the simplest system, i.e., the ensemble of two-level absorbers, and then for more complicated systems.

9.1 Linear Response for Two-Level Systems

The equation of motion for the interband coherence $\bar{\rho}_{vc}$ neglecting nonlinear terms is given by, see (7.29),

$$\hbar\frac{d\bar{\rho}_{vc}}{dt} = i\hbar\left(\omega_0 - \omega_L + \frac{i}{T_2}\right)\bar{\rho}_{vc} - iE(t)\mu_0,\tag{9.8}$$

where we have ignored the space dependence. In (9.8), we have added a damping term $-\frac{\hbar}{T_2}\bar{\rho}_{cv}$. The dephasing time T_2 can be be taken as $T_2 \to \infty$ if the undamped case is to be considered. This term provides the physical motivation for the imaginary part $\varepsilon > 0$ of ω in (9.7).

For an excitation according to (9.4), the solution of (9.8) is

$$\bar{\rho}_{vc} = -i\frac{\eta}{\hbar}\mu_0\theta(t)e^{i(\omega_0-\omega_L)t}e^{-t/T_2},\tag{9.9}$$

where $\theta(t)$ is the unit step function. Transforming back into the stationary frame, we have

$$\rho_{vc} = -i\frac{\eta}{\hbar}\mu_0\theta(t)e^{i\omega_0 t}e^{-t/T_2}.\tag{9.10}$$

Note that we have actually violated the necessary condition for the application of the rotating-wave approximation, since the $\delta(t)$ pulse does not have a well-defined central frequency, but contains all frequencies with equal weight. So the condition of small relative detuning is not fulfilled. However, using a $\delta(t)$ excitation and the equation of motion for ρ_{vc} gives exactly the same solution (9.10).

The linear optical polarization is given by

$$\begin{aligned}P(t) &= \eta\chi(t)\\ &= \mu_0(\rho_{vc})^* + (\mu_0)^*\rho_{vc}\\ &= i\frac{\eta}{\hbar}|\mu_0|^2\theta(t)(e^{-i\omega_0 t} - e^{i\omega_0 t})e^{-t/T_2},\end{aligned}\tag{9.11}$$

with

$$\chi(t) = \frac{2}{\hbar}|\mu_0|^2\theta(t)\sin\omega_0 t\,e^{-t/T_2}.\tag{9.12}$$

In the limit $T_2 \to \infty$, the imaginary part of $\chi(\omega + i\epsilon)$, see (9.7), for positive frequencies ω is

$$\chi''(\omega) = \frac{\pi}{\hbar}|\mu_0|^2\delta(\omega - \omega_0).\tag{9.13}$$

As expected, the linear optical spectrum consists of a single line at the resonance frequency ω_0. Note that for all frequencies, we have $\omega\chi''(\omega) \geq 0$, which is the signature of a dissipative system.

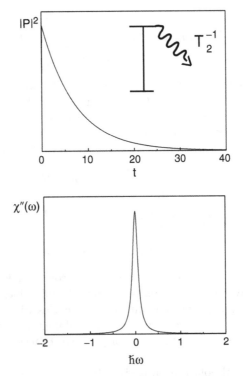

Fig. 9.1. Temporal dynamics and spectrum of the linear optical polarization of a homogeneously broadened ensemble of two-level absorbers. The rate of decay of the polarization and the width of the spectrum are determined by the dephasing rate $\gamma = T_2^{-1}$. Units are arbitrary and we take $\omega_0 = 0$

For finite T_2, the spectrum is

$$\chi''(\omega) = \frac{\pi}{\hbar}|\mu_0|^2 \frac{(\pi T_2)^{-1}}{(\omega - \omega_0)^2 + T_2^{-2}}, \tag{9.14}$$

i.e., a Lorentzian line with a width that is proportional to the inverse of the dephasing time T_2. Such a resonance is called a *homogeneously* broadened line, see Fig. 9.1.

An inhomogeneous ensemble corresponds to an ensemble of independent two-level systems with resonance frequencies ω_i distributed according to a certain distribution function $G(\omega_i)$. For $T_2 \to \infty$, the linear spectrum of this ensemble is *inhomogeneously* broadened, i.e., it coincides with $G(\omega)$ weighted by the optical matrix elements for the excitation energy ω squared.

If the individual lines of an inhomogeneous ensemble are homogeneously broadened, the total spectrum is both homogeneously and inhomogeneously broadened, see Fig. 9.2 for an example.

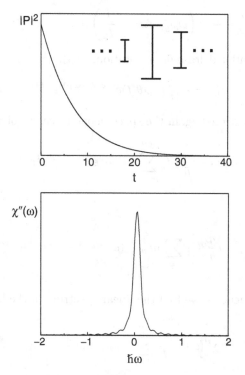

Fig. 9.2. Temporal dynamics and spectrum of the linear optical polarization of a dominantly inhomogeneously broadened ensemble of two-level absorbers. In order to obtain smooth curves, a dephasing rate of $\gamma = 0.06$ has been assumed. Units are arbitrary and we take $\omega_0 = 0$

At this point, a technical remark is in order. One could, of course, have taken the Fourier transform of the equation of motion in frequency space at the very beginning. The pulsed excitation is then represented by a constant term, and the resulting algebraic equation is easily solved. This procedure becomes, however, increasingly cumbersome for more general situations. On the other hand, as is shown below, the solution in the time domain is, at least numerically, possible even for quite general systems which, e.g., include both disorder and interactions.

9.2 Linear Response for Few-Level Systems

The calculation of the linear optical response of systems with more than two levels is a straightforward extension of that for two-level systems. We start from (7.71) and (7.72) but, for simplicity, assume instead of the M states $|\nu v\rangle$ just a single ground state $|g\rangle$. The equation of motion in linear response then is

$$\hbar \frac{d\bar{\rho}_{g\nu'}}{dt} - i\hbar \left(\omega_{\nu'g} - \omega_L + \frac{i}{T_2} \right) \bar{\rho}_{g\nu'} = -iE(t)\mu_{g\nu'}. \tag{9.15}$$

With the solution for delta-pulse excitation, (9.4),

$$\bar{\rho}_{g\nu'} = -i\frac{\eta}{\hbar}\mu_{g\nu'}\theta(t)e^{i(\omega_{\nu'g}-\omega_L)t}e^{-t/T_2}, \tag{9.16}$$

where we have neglected again the space dependence, we obtain

$$\bar{P}(t) = -i\frac{\eta}{\hbar}\theta(t)\sum_{\nu'=1}^{M'}|\mu_{g\nu'}|^2 e^{i(\omega_{\nu'g}-\omega_L)t}e^{-t/T_2} \tag{9.17}$$

and

$$P(t) = i\frac{\eta}{\hbar}\theta(t)\sum_{\nu'=1}^{M'}|\mu_{g\nu'}|^2(e^{-i\omega_{\nu'g}t} - e^{i\omega_{\nu'g}t})e^{-t/T_2}. \tag{9.18}$$

For positive frequencies, we find the linear spectrum in the limit $T_2 \to \infty$

$$\chi''(\omega) = \frac{\pi}{\hbar}\sum_{\nu'=1}^{M'}|\mu_{g\nu'}|^2\delta(\omega - \omega_{\nu'g}), \tag{9.19}$$

which is simply a superposition of the transitions from the ground state to all excited states. Note that we would have obtained exactly the same spectrum if we had considered an ensemble of isolated two-level absorbers, see previous section, with individual frequencies $\omega_{\nu'g}$ and corresponding optical dipole matrix elements $\mu_{g\nu'}$. We have thus shown that the linear spectrum is not able to distinguish between these two cases. Also, the insertion of a phenomenological finite dephasing time $T_2 < \infty$ would have led to identical results as obtained above, i.e., a superposition of Lorentzian lines.

9.2.1 Continua

On the basis of the above $1+M'$-level system, one can derive the optical spectrum for transitions from a ground state $|g\rangle$ into a continuum of excited states, once one has specified the density of the excited levels $g(E)$ as a function of energy. The optical dipole matrix elements $\mu(E)$ determine the linear polarization according to (9.17). To proceed from the discrete to the continuous case, we only have to replace

$$\sum_{\nu'=1}^{\infty}\cdots \to \int g(E)\cdots dE. \tag{9.20}$$

We therefore have

$$\bar{P}(t) = -i\frac{\eta}{\hbar}\theta(t)e^{-i\omega_L t}e^{-t/T_2}\int g(E)|\mu(E)|^2 e^{iEt/\hbar}dE, \qquad (9.21)$$

from which we obtain the optical spectrum for positive ω and $T_2 \to \infty$ as

$$\chi''(\omega) = \pi g(\omega)|\mu(\omega)|^2. \qquad (9.22)$$

Let us consider two simple cases. As a first example, we mention a constant optical spectrum representing an unstructured continuum. Due to the integral appearing in (9.21), this corresponds to a polarization in the time domain that is proportional to $\delta(t)$.

The second example consists of an ensemble of identical systems with a single ground state and a continuum which happens to have a Lorentzian spectrum, i.e.,

$$g(\omega)|\mu(\omega)|^2 \propto \frac{\gamma}{(\omega - \omega_0)^2 + \gamma^2}. \qquad (9.23)$$

In this case, the corresponding polarization decays exponentially in time, i.e.,

$$\begin{aligned}\bar{P}(t) &\propto \theta(t)e^{-i\omega_L t}\int \frac{\gamma}{(\omega - \omega_0)^2 + \gamma^2}e^{i\omega t}d\omega \\ &= \pi\theta(t)e^{i(\omega_0 - \omega_L)t}e^{-\gamma t}.\end{aligned} \qquad (9.24)$$

Note that this decay is a property of the particular continuous spectrum, and is not due to a dephasing time introduced into the equation of motion in an *ad hoc* way.

9.2.2 Homogeneous versus Inhomogeneous Ensembles

In Sect. 9.1, we discussed an ensemble of identical two-level absorbers which all have the same excitation energy $\hbar\omega_0$ and are coupled to a bath, such that the polarization decays with a nonzero rate $\gamma = T_2^{-1}$, i.e., the so-called *homogeneous* ensemble. It has the same polarization and optical linear spectrum as that given by (9.23) and (9.24), respectively.

Imagine still another ensemble, which is composed of two-level absorbers having individual transition energies distributed around $\hbar\omega_0$ according to the Lorentzian distribution function which equals that given in (9.23). For this *inhomogeneous* ensemble, the linear optical polarization as well as the linear spectrum are in the limit $T_2 \to \infty$ again given by (9.24) and (9.23), respectively,

These examples show that linear optical transients and spectra cannot distinguish between physically quite different ensembles. In particular, homogeneously and inhomogeneously broadened ensembles and an ensemble consisting of systems with a suitably chosen continuum of excited states all have the same linear optical polarizations and spectra. The analysis in the following chapters shows that, in contrast to linear optics, the nonlinear optical response can be used to distinguish clearly between the underlying physically quite distinct cases.

9.3 Linear Response for Fano Systems

A special and interesting case of ensembles consisting of discrete and continuous transitions is that of Fano systems which have been introduced in Sect. 3.4. There, we discussed the Hamiltonian of the system and that of the light–matter interaction. In order to obtain the linear optical spectrum, one can apply two alternative approaches. One either starts from the Hamiltonian written in the "old" basis, (3.17), and the appropriate light–matter interaction (3.18). Alternatively, one diagonalizes the excited-state part, (3.19), and calculates the optical spectrum in terms of this "new" basis with the light–matter interaction given by (3.20) and (3.21). While for simple model systems, it is possible to calculate the spectrum analytically using this second approach, more general cases can only be treated numerically using the "old" basis states. We first follow this alternative, since it nicely illustrates the method for the calculation of optical properties of systems that are described by nondiagonal system Hamiltonians. Afterwards, we present without derivation the analytical expressions for the linear optical polarization which follow from the diagonalization of the model Hamiltonian. These are used to gain additional insight into the numerical results.

The temporal evolution of the linear optical polarization and the linear optical spectrum can be obtained from (8.64), which has to be adapted for the present model. We do not consider the Coulomb interaction and, thus, the term proportional to V_{12} is absent. Furthermore, we have only a single ground state $|h\rangle = |g\rangle$ to which we assign the energy $\epsilon_g = 0$. Therefore, the term containing T^h is absent as well. However, the Hamiltonian matrix for the excited states, T^e, has diagonal as well as nondiagonal elements. The diagonal elements are given by the energies $\epsilon^c_{\lambda'}$ of the M' discrete states $|\lambda'\rangle$ and the energies $\epsilon_{\nu'}$ of the states $|\nu'c\rangle$ forming the quasi-continuum. The nondiagonal elements are represented by the coupling matrix elements $V_{\nu'\lambda'}$.

The equations of motion for the linear optical polarization of this system read

$$-i\hbar\frac{d}{dt}p_{g\lambda'} = -\epsilon^c_{\lambda'}p_{g\lambda'} - \sum_{\nu'}V_{\lambda'\nu'}p_{g\nu'} + E(t)(\mu_{g\lambda'})^*$$

$$-i\hbar\frac{d}{dt}p_{g\nu'} = -\epsilon_{\nu'}p_{g\nu'} - \sum_{\lambda'=1}^{M'}V_{\nu'\lambda'}p_{g\lambda'} + E(t)(\mu_{g\nu'})^*. \qquad (9.25)$$

From the numerical solution of these differential equations, we calculate the observable linear optical polarization $P(t)$, i.e.,

$$\bar{P}(t) = \sum_{\lambda'=1}^{M'} \mu_{g\lambda'} p_{g\lambda'} + \sum_{\nu'} \mu_{g\nu'} p_{g\nu'}. \tag{9.26}$$

To specify the optical dipole matrix elements, we assume that all transitions into the continuum have equal weight, i.e., we have an unstructured continuum with

$$\mu_{g\nu'} = \mu_0. \tag{9.27}$$

The discrete transitions are given specific real values $\mu_{g\lambda'}$ in order to model different situations.

We further assume that the density of continuum states $g(\epsilon_{\nu'}) = g_0$ is constant and that the coupling of the discrete states $|\lambda'\rangle$ to the continuum is equal for all continuum states $|\nu'\rangle$, i.e., $V_{\nu'\lambda'} = V_{\lambda'}$.

The Fano model is a generalization of the coherent tunneling model. Instead of just two coupled states, we now have M' discrete states coupled to a quasi-continuum. We show below that, due to the coupling $V_{\lambda'}$, a population placed into one of the discrete states $|\lambda'\rangle$ decays into the continuum states $|\nu'c\rangle$. This decay is described by a finite-linewidth parameter defined by

$$\Gamma_{\lambda'} = 2\pi V_{\lambda'}^2 g_0. \tag{9.28}$$

We also define the parameter

$$q_{\lambda'} = \left(\frac{2}{\pi \Gamma_{\lambda'} g_0}\right)^{1/2} \frac{\mu_{g\lambda'}}{\mu_0}, \tag{9.29}$$

which quantifies the strength of the discrete optical transition relative to the continuum ones.

9.3.1 Numerical Results

We start the discussion with the classical Fano system, where a single discrete state $|1\rangle$ with energy ϵ_1^c is coupled (by V_1) to a continuum. The numerical result for the linear absorption spectrum is shown in Fig. 9.3, where

$$E = \frac{\hbar\omega - \epsilon_1^c}{\Gamma_1/2}. \tag{9.30}$$

The spectrum is a superposition of the constant spectrum of the continuum alone, the contribution due to the discrete state, and its coupling to the continuum. This latter contribution is a Lorentzian line with linewidth Γ_1 as long as there is no excitation into the continuum, i.e., if $\mu_0 \rightarrow 0$. If the excitation into the continuum is present, interference between the two excitation paths (due to discrete-state and continuum excitation) produces a Fano antiresonance with zero absorption at a particular energy which distorts the purely Lorentzian line. This is often called "Fano resonance". For this result to be

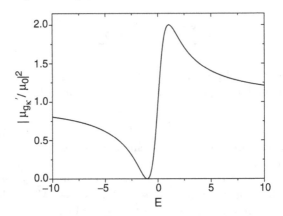

Fig. 9.3. Fano spectrum for the situation with a single state and $q_1 = 1$. The spectrum is normalized to that of the unstructured continuum

valid, it is important that we use an optical matrix element μ_{g1} that is real, which is often the case, in particular, if the states are nondegenerate. A complex matrix element would lead to an incomplete antiresonance.

Another widely observed Fano interference phenomenon exists in cases where more than a single discrete state and a continuum are present. This is called interference narrowing and, in other contexts, also excitation trapping.

Consider as a simple example two discrete states $|1\rangle$ and $|2\rangle$ with energies ϵ_1^c and ϵ_2^c, respectively, see Fig. 3.6. Their energy separation is

$$\Delta = \epsilon_2^c - \epsilon_1^c. \tag{9.31}$$

The two states can, e.g., represent the eigenstates of a double quantum well. We first assume that the broadening factors and the dipole matrix elements are equal, i.e., $\Gamma_1 = \Gamma_2 = \Gamma$ and $q_1 = q_2 = q$, respectively. For reasons that will soon become clear, we call the situations $\Delta > \Gamma$, $\Delta < \Gamma$, and $\Delta = \Gamma$ the underdamped, the overdamped, and the critical regimes, respectively.

The numerically calculated temporal evolution of the linear optical polarization $P(t)$ after optical $\delta(t)$ excitation is shown in Figs. 9.4, 9.5, and 9.6 for the three cases defined above. The temporal behavior of the polarization is dramatically different for the three regimes. While for the underdamped case oscillations can be seen, no such oscillations are present in the overdamped traces. Note that in the overdamped case there is a fast initial decay, followed by a π-phase shift seen as a sharp dip and a slowly decaying long-time regime. This behavior is called "excitation trapping". Also shown in the figures are traces for the case where the two discrete states are coupled to two different continua. In this situation, interference is absent and there are always oscillations.

It is remarkable that the form of the decaying traces of the linear optical polarization does not depend on q. This is a consequence of the $\delta(t)$ excitation

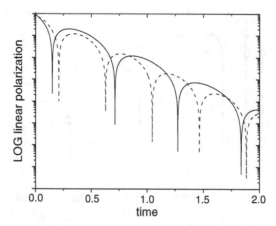

Fig. 9.4. Temporal dynamics of the linear optical polarization for a Fano system with two discrete states for the underdamped situation using $\Delta = 15$ and $\Gamma = 10$. The *solid line* shows $|P(t)|$ on a logarithmic scale. The *dashed lines* correspond to a decay into two different continua. In this case, the excitations into the discrete states do no longer interact. Beats are seen in any case. [From T. Meier et al., Phys. Rev. **B51**, 13977 (1995)]

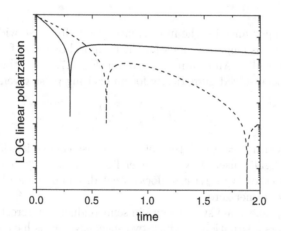

Fig. 9.5. Temporal dynamics of the linear optical polarization for a Fano system with two discrete states for the overdamped situation using $\Delta = 5$ and $\Gamma = 10$. The *solid line* shows $|P(t)|$ on a logarithmic scale. Note, the extremely slow decay for larger times if compared to that for short times. The *dashed lines* correspond to a decay into two different continua. In this case, the excitations into the discrete states no longer interact and, therefore, beats are seen in this case only. Additionally, there is also no slow decay component present for this noninteracting case. [From T. Meier et al., Phys. Rev. **B51**, 13977 (1995)]

implemented in the numerical solution of the equations of motion. Fano interference occurs between transitions $|g\rangle \rightarrow |1\rangle$ and $|g\rangle \rightarrow |2\rangle$ and is mediated by the common continuum. Whether or not the continuum is directly excited as

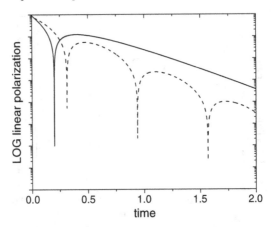

Fig. 9.6. Temporal dynamics of the linear optical polarization for a Fano system with two discrete states for the critical situation using $\Delta = 10$ and $\Gamma = 10$. The *solid line* shows $|P(t)|$ on a logarithmic scale. The *dashed lines* correspond to a decay into two different continua. In this case, the excitations into the discrete states no longer interact and beats are seen in this case only. [From T. Meier et al., Phys. Rev. **B51**, 13977 (1995)]

well is of no importance for the interference. For finite pulse widths, this feature is preserved as long as we look for the temporal behavior for times longer than the pulse width. An example is given in Fig. 9.7. Here, the equations of motion have been solved numerically for an exciting pulse given by

$$E_l(t) \propto \frac{1}{\delta} e^{-t^2/\delta^2}, \tag{9.32}$$

with a pulse width $\delta = 0.1$. A relic of the π-phase change is still clearly seen for this short light pulse. Only for longer light pulses ($\delta \geq 0.5$) and weaker transitions into the discrete states, does a weak dip remind one of the temporal position of the former zero.

In atomic (molecular) systems and in semiconductor heterostructures, one often encounters a situation in which two successive levels have opposite parity. Starting from a common ground state one transition is then allowed (say, q_1 is finite) while the other transition is forbidden ($q_2 = 0$). For such a situation, we show in Fig. 9.8 the trace of the linear optical polarization computed for the overdamped case. There are still two regimes and the polarization consists of a fast initial and a slow long-time component. However, these two regimes are no longer separated by a dip or zero, even for the short pulse width $\delta = 0.03$ considered in Fig. 9.8.

The Fourier transforms of the temporal traces yield the linear optical spectra which are displayed in Figs. 9.9, 9.10, and 9.11 for the three cases. While the temporal traces show an extremely different behavior for the underdamped, the overdamped and the critical regimes, the corresponding spectra

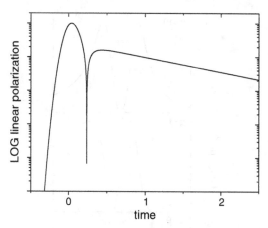

Fig. 9.7. Temporal dynamics of the linear optical polarization for a Fano system with two discrete states excited by a pulse of finite width of $\delta = 0.1$. The further parameters are $\Delta = 5$, $q_1 = q_2 = q = 10$, and $\Gamma_1 = \Gamma_2 = \Gamma = 10$, i.e., this situation corresponds to the strongly overdamped case. [From T. Meier et al., Phys. Rev. **B51**, 13977 (1995)]

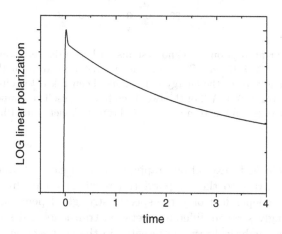

Fig. 9.8. Temporal dynamics of the linear optical polarization for a Fano system with two discrete states, excited by a pulse of finite width of $\delta = 0.03$. The further parameters are $\Delta = 0.25$, $q_1 = 0$, $q_2 = 10$, and $\Gamma_1 = \Gamma_2 = \Gamma = 0.5$, i.e., this situation corresponds to the strongly overdamped case. Note that the dip between the fast and the slow portion of the trace is absent in this case. [From T. Meier et al., Phys. Rev. **B51**, 13977 (1995)]

look quite similar. In particular, from the spectra it cannot be seen whether or not the temporal traces show oscillations.

The slow long-time decay observed in the trace for the overdamped case manifests itself as a sharp spectral feature, see Fig. 9.10. Therefore, the

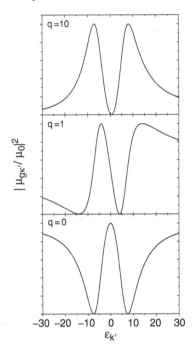

Fig. 9.9. Linear absorption for Fano systems with two discrete states in the underdamped case ($\Delta = 15$ and $\Gamma = 10$). The spectra are shown for three values of $q = q_1 = q_2$. The zero of the energy scale ω has been taken to coincide with the average $(\epsilon_1^c + \epsilon_2^c)/2 = \epsilon_1^c + \Delta/2$ of the two discrete energies. The temporal evolution of the linear polarization is shown in Fig. 9.4. [From T. Meier et al., Phys. Rev. **B51**, 13977 (1995)]

physical effect called "excitation trapping" is also called "interference narrowing". In contrast to the temporal traces which (apart from a complex prefactor) are independent of q, the spectra strongly depend on the value of q. This seemingly strange difference between transients and spectra will be resolved by the analytical results presented in the next section.

9.3.2 Analytical Discussion

All these observation can be understood if we follow Fano and diagonalize the upper-state part of the system Hamiltonian which leads to (3.19), (3.20), and (3.21). These equations described the system by a M-level absorber with a particular form of the matrix elements. From the discussion of M-level systems, we learned that the linear optical spectrum is given by the matrix elements squared times the (constant) density of states, see (9.22),

$$\chi''(\omega) = \pi g_0 |\mu_{g\kappa'}|^2. \tag{9.33}$$

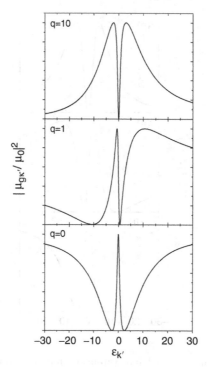

Fig. 9.10. Linear absorption for Fano systems with two discrete states in the over-damped case ($\Delta = 5$ and $\Gamma = 10$). The spectra are shown for three values of $q = q_1 = q_2$. Note the *sharp feature* at $\omega = 0$! The temporal evolution of the linear polarization is shown in Fig. 9.5. [From T. Meier et al., Phys. Rev. **B51**, 13977 (1995)]

Here, we do not present Fano's derivation. This is by no means trivial but lies outside the scope of this book. Instead, we quote and discuss the results. Fano found that the optical spectrum for M' discrete transitions degenerate with transitions into a continuum is given by

$$|\mu_{g\kappa'}|^2 = \frac{\left(1 + \sum_{\lambda'} q_{\lambda'} \frac{\Gamma_{\lambda'}/2}{\epsilon_{\kappa'} - \epsilon_{\lambda'}}\right)^2}{1 + \left(\sum_{\lambda'} \frac{\Gamma_{\lambda'}/2}{\epsilon_{\kappa'} - \epsilon_{\lambda'}}\right)^2} |\mu_0|^2. \tag{9.34}$$

For a single discrete state, this can be written as

$$|\mu_{g\kappa'}|^2 = \frac{(q_1 + E)^2}{1 + E^2} |\mu_0|^2, \tag{9.35}$$

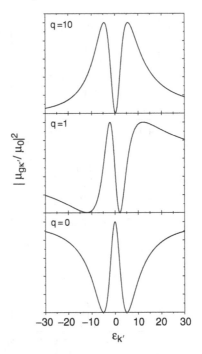

Fig. 9.11. Linear absorption for Fano systems with two discrete states in the critical case ($\Delta = 10$ and $\Gamma = 10$). The spectra are shown for three values of $q = q_1 = q_2$. The temporal evolution of the linear polarization is shown in Fig. 9.6. [From T. Meier et al., Phys. Rev. **B51**, 13977 (1995)]

with

$$E = \frac{\epsilon_{\kappa'} - \epsilon_1}{\Gamma_1/2}, \tag{9.36}$$

where $\epsilon_{\kappa'}$ has to be identified with the photon frequency $\hbar\omega$.

From these results, we can calculate the temporal traces $P(t)$ by Fourier-transforming (9.34) and (9.35) into the time domain, see (9.21). The result can be written, for a constant density of states function $g(E) = g_0$, as

$$\bar{P}(t) = -i\frac{\eta}{\hbar}\theta(t)g_0\mu_0^2[2\pi\hbar\delta(t) + L(t)], \tag{9.37}$$

where the absorption into the unstructured continuum is represented by the δ-function contribution to the above expression, while the second term $L(t)$ contains the transitions into the discrete states and the contribution due to interference.

For the case with a single discrete transition, we have

$$L(t) = \pi\frac{\Gamma_1}{2}(q_1 + i)^2 e^{i\epsilon_1 t/\hbar - \frac{\Gamma_1}{2}t/\hbar}. \tag{9.38}$$

For $\mu_0 \rightarrow 0$, the optical transitions into the unstructured continuum are switched off and the δ contribution vanishes. In this limit, we obtain

$$g_0 \mu_0^2 (q_1 + i)^2 \frac{\Gamma_1}{2} \pi \rightarrow \mu_{g1}^2, \tag{9.39}$$

i.e., we obtain the response to a δ-pulse excitation of an exponentially damped single absorber corresponding to a Lorentzian linear optical spectrum. This explains why Γ has been called the linewidth parameter.

For $\Gamma_1 \rightarrow 0$, we arrive at another trivial limit. Here, (9.39) is again valid, and the linear polarization is a superposition of the δ contribution due to the continuum and a nondecaying signal due to the discrete state.

For finite μ_0, the optical excitation into the continuum interferes with that of the state $|1\rangle$. This interference is described by the term $(q_1 + i)^2$. It is instructive to compare the functions $L(t)$ for the interfering Fano situation and for a noninterfering system ($V_1 = 0$) described by a Lorentzian line superimposed to a continuum. In the latter case, $L(t)$ has the same form as above, except that the term $+i$ is missing. The interference effects are most important for small q_1 values. For large q_1, we get $(q_1 + i)^2 \approx q_1^2$ and the Fano line as well as the corresponding $\bar{P}(t)$ traces recover the form characteristic for the noninterfering superposition.

The state $|1\rangle$ influences the optical properties and the dynamics of the system even if the transition $|g\rangle \rightarrow |1\rangle$ is optically forbidden ($q_1 = 0$). In this case, the linear spectrum shows a Lorentzian hole with zero absorption at ϵ_1 due to the admixture of the forbidden transition into the continuum absorption. In the time-dependent signal, there appears a π-phase shift between the first and the second term in (9.37). For a $\delta(t)$ excitation, this phase shift cannot be seen. However, for an excitation with a finite pulse width, we have the chance to obtain a cross-over from negative to positive values of $\bar{P}(t)$ at a finite time. This is demonstrated below when four-wave-mixing traces for Fano systems are discussed.

We now turn to the case of more than a single discrete transition and take $\Gamma_1 = \Gamma_2 = \Gamma$ and $q_1 = q_2 = q$. We start by looking for the zeros of the denominator of (9.34). We find two zeros which correspond to two poles in the relevant upper complex frequency plane (and two complex conjugate poles in the lower one). Note that in (9.21) t is larger than zero and the integral can be closed in the upper complex energy plane. The poles are given by

$$E_{1,2} = \epsilon_1^c + \frac{\Delta}{2} \pm \frac{1}{2} \sqrt{\Delta^2 - \Gamma^2} + i \frac{\Gamma}{2}. \tag{9.40}$$

We can now discuss the three regimes we have introduced above. The first one is the underdamped regime where $\Delta > \Gamma$. In this case, we find two different real parts of the poles and a single imaginary part. The real parts define two different energies, $\epsilon_1^c + \frac{\Delta}{2} \pm \frac{1}{2} \sqrt{\Delta^2 - \Gamma^2}$, and the imaginary part determines the damping constant $\Gamma/2$.

The second regime is the overdamped regime with $\Delta < \Gamma$. Here, we find only a single energy $\epsilon_1^c + \Delta/2$, but two different damping constants, $\Gamma/2 \pm \sqrt{\Gamma^2 - \Delta^2}/2$. In this case there is a contribution which has a damping constant $\Gamma/2 - \sqrt{\Gamma^2 - \Delta^2}/2$ describing the decay into the continuum. This damping decreases with decreasing Δ, or, in other words, if $\Delta \to 0$ the lifetime of this contribution goes to infinity and the corresponding linewidth to zero.

Defining real quantities

$$W_1 = \sqrt{\Delta^2 - \Gamma^2} \tag{9.41}$$

for the underdamped case and

$$W_2 = \sqrt{\Gamma^2 - \Delta^2} \tag{9.42}$$

for the overdamped case, we obtain in the time domain

$$L(t) = \frac{\Gamma}{2}\pi(q+i)^2 \left(1 + \frac{i\Gamma}{W_1}\right) e^{i(\epsilon_1^c + \Delta/2 + W_1/2)t/\hbar - (\Gamma/2)t/\hbar}$$
$$+ (W_1 \to -W_1),$$
$$L(t) = \frac{\Gamma}{2}\pi(q+i)^2 \left(1 + \frac{\Gamma}{W_2}\right) e^{i(\epsilon_1^c + \Delta/2)t/\hbar - (\Gamma/2 + W_2)t/\hbar}$$
$$+ (W_2 \to -W_2), \tag{9.43}$$

for the underdamped and overdamped cases, respectively.

It is interesting to compare this result to the case of two discrete transitions that decay into two independent continua, such that all interference effects are absent. For this case, we have

$$L_0(t) = \frac{\Gamma}{2}\pi(q+i)^2 \left(e^{i(\epsilon_1 + \Delta)t - (\Gamma/2)t} + e^{i\epsilon_1 t - (\Gamma/2)t}\right). \tag{9.44}$$

The analytical results agree nearly quantitatively with the numerical ones presented above and give additional insight into the interference phenomena inherent in Fano situations. However, the numerical calculations based on the equation of motion approach are more flexible and can easily be applied to model cases that cannot be treated analytically. One might, e.g., be interested in cases where the coupling into the continuum depends on the state of the continuum. Remember, here, we have always assumed that Γ depends solely on the discrete state. In realistic situations, the density of states $g(E)$ of the continuum alone is not independent of energy as assumed here and the same is true for the optical dipole matrix elements $\mu_0(\omega)$. In the analytical calculation of the time dependence of the optical polarization, we have assumed that the optical excitation is given by a $\delta(t)$ pulse. This is an idealization that is not applicable in many experimental situations. Furthermore, the analytical analysis, due to Fano, neglects energy shifts that are related to the broadening of the spectral lines. All these effects are fully included in the numerical approach.

9.4 Linear Response of the Semiconductor

The linear optical spectrum of the semiconductor model follows from the equation of motion of the interband coherence $p_{ij}^{he}(t)$ if a $\delta(t)$ pulse is used as excitation. Omitting all nonlinear terms, we have the Wannier equation (8.64) for the exciton. In the following, we discuss the optical spectrum for the ordered and the disordered cases. For this discussion, we ignore the two-fold degeneracy of both the valence and the conduction band and also the vector character of the light field and the optical dipole matrix element μ. In this limit, the equation of motion for the interband coherence has the simpler form

$$-i\hbar \frac{d}{dt} p_{12} = -\sum_j T_{2j}^e p_{1j} - \sum_i T_{i1}^h p_{i2} + V_{12} p_{12}$$
$$+E(t) \cdot (\mu_{12})^*, \tag{9.45}$$

with
$$\mu_{12} = \mu_0 \delta_{12}, \tag{9.46}$$

i.e., the optical dipole matrix element is diagonal in the site indices.

9.4.1 Ordered Semiconductor

The previous discussion of the optical response of level systems is meant to serve as a tutorial basis. Therefore, the results presented in these sections have been performed for sets of parameters that show the spectral and/or temporal signatures of the response most clearly. Physical units have been suppressed. The results may be quantified by scaling the parameters according to any desired set of physical units.

In the following, we would like to refer to the real world whenever a model for a one-dimensional solid is considered. This requires us to introduce reasonable units for the model parameters. As discussed in Chap. 6, the parameters should be chosen such that effective masses and gaps have values comparable to those in real semiconductors or semiconductor heterostructures. In addition, since we now consider the many-particle interaction, the exciton binding energy should be adjusted, such that it resembles typical values of realistic structures. Unfortunately, however, in our numerical evaluations, the sample size given by the number of sites N cannot be too large. Therefore, we have to check for finite size effects and perform studies for different system sizes in order to make sure that the results are converged.

To be specific, depending on the situation considered, some or all of the following physical parameters have to be given reasonable values:

- the effective mass of the electrons,
- the effective mass of the holes,

- the excitonic binding energy, which is determined by U_0 and the regularization parameter a_0 (taken to be $a_0 = 0.5a$ throughout, where a is the lattice constant),
- the disorder parameter in the conduction band W^e,
- the disorder parameter in the valence band W^h,
- the length scale of the disorder potential L,
- the phenomenological dephasing rate $\gamma = T_2^{-1}$,
- the radius of the ring R, with $2\pi R = Na$.

Once N and R are specified, J^e and J^h are determined by the effective electron and hole masses, respectively, see (6.19).

The gap energy given by ϵ_0 does not need to be specified. As we are applying the rotating-wave approximation, in most cases, only the detuning is relevant and the gap can be accounted for by choosing the central laser frequency accordingly. In the following, $\hbar\omega = 0$ often refers to the fundamental gap in the noninteracting case or to the excitonic resonance in the interacting case.

Once this set of parameters has been chosen, the number of sites N has to be taken large enough to achieve convergence. The results for the polarization are given in arbitrary units, as we do not specify the amplitude of the exciting laser field and the spectra are also given in arbitrary units since their scale is determined by the optical dipole matrix elements μ_0, which are input parameters into the model.

Let us first neglect the Coulomb interaction, i.e., set $U_0 = 0$. In this case, the optical spectrum is easily calculated using the representation in k space. The eigenvalues are given by (6.13) and the optical spectrum by (6.26). This spectrum was already presented schematically in Fig. 6.4. Here, the result of a calculation is given in Fig. 9.12 for the indicated parameters. The van

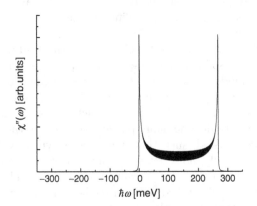

Fig. 9.12. Calculated linear optical spectrum for $N = 500$, $J^e = 50\,\text{meV}$, $J^h = 16.7\,\text{meV}$, $\gamma = 0.5\,\text{meV}$, and $U_0 = 0$. The spectrum consists of $N = 500$ broadened peaks, which give rise to very closely spaced wiggles in the center of the spectrum, which are not resolved. For $N \to \infty$ the spectrum would show a line in the middle of the black region

Hove singularities at the band edges, which are typical for one-dimensional ordered semiconductor systems, are somewhat broadened due to the finite homogeneous broadening used in the calculation.

Due to translational (actually rotational) symmetry, there is a quantum number k, which is conserved in optical transitions (since we neglect the small momentum of the photon). The optical spectrum can therefore be viewed as an inhomogeneous line composed of independent two-level absorbers characterized by their respective k values.

The Coulomb interaction enters the linear response equation of motion only via the direct electron–hole attraction and leads to new optical resonances, the exciton lines. Figure 9.13 shows the linear spectrum, which has a discrete excitonic resonance below $\hbar\omega = 0$. This latter energy corresponds to the fundamental band edge in the interaction-free case. The interband continuum no longer shows the singularities which dominate the interaction-free spectrum. In addition, the high-energy part of the continuum is remarkably reduced in height (not shown here). Actually, because of the finite number of sites N, we have a quasi-continuum. Therefore, a small damping term γ is added to the equation of motion which smears out the discrete resonances.

Figure 9.14 shows an enlarged plot of the spectral region around the band edge for a calculation where we have chosen a larger $\gamma = 0.5\,\mathrm{meV}$. We notice a second discrete peak just below the band edge of the noninteracting model, which is the second optically active excitonic resonance. Actually, there are more discrete resonances which, however, tend to merge into the onset of the continuum. For a similar situation, however, with a larger excitonic binding energy, see Fig. 14.1, which also shows a distinct third resonance.

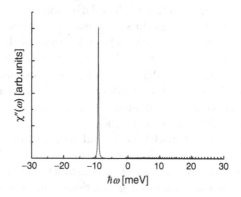

Fig. 9.13. Linear optical spectrum for an ordered semiconductor model showing the excitonic resonance. Here, a small phenomenological dephasing rate γ has been introduced, which leads to a small broadening of the discrete excitonic resonances. Clearly seen is the lowest resonance close to $-9\,\mathrm{meV}$ and the second-lowest resonance close to zero, which tends to merge with the onset of the (quasi-)continuum. Due to the rather long dephasing time, i.e., small γ of $0.1\,\mathrm{meV}$, the (quasi-)continuum is resolved as a succession of peaks. The parameters are $N = 500$, $J^e = 50\,\mathrm{meV}$, $J^h = 16.7\,\mathrm{meV}$, and $U_0 = 16\,\mathrm{meV}$

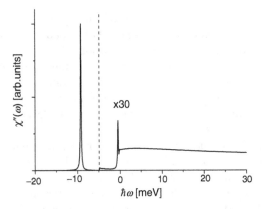

Fig. 9.14. Linear optical spectrum for an ordered semiconductor model showing the two lowest excitonic resonance more clearly. Here, a larger phenomenological dephasing rate $\gamma = 0.5\,\text{meV}$ has been introduced, which leads to a broadening of the discrete excitonic resonances. Note that the continuum absorption, which is amplified by a factor of 30, looks quite different as compared to that of the noninteracting case which is characterized by the van Hove singularities. The parameters are $N = 500$, $J^e = 50\,\text{meV}$, $J^h = 16.7\,\text{meV}$, and $U_0 = 16\,\text{meV}$

Whether this spectrum has to be classified as inhomogeneous or homogeneous cannot be decided on the basis of the linear analysis. This question has to be addressed when we discuss nonlinear optical properties.

When the excitonic resonance in Fig. 9.14 is excited by a sufficiently long laser pulse tuned to this resonance, the resulting linear optical polarization p_{ij} is determined by the wave function of the relative motion of the electron–hole pair in its lowest bound state. In three-dimensional systems this is called the $1s$ state in analogy to the classification of bound states of the hydrogen atom. Figure 9.15 shows the modulus $|p_{ij}^{he}|$ as a function of the relative coordinate $i - j$, which is discrete in our model. $|p_{ij}^{he}|$ looks like an exponential function $\propto \exp\left(-(i-j)/a_B\right)$ characterized by the Bohr radius a_B, known from the three-dimensional case. However, in our model it is given by a different function discussed in detail in Haug and Koch (2004).

For a three-dimensional model, analytical solutions of the exciton problem are available. These provide relations between the model parameters (effective masses and background dielectric constant) and the energies of the excitonic bound states and their wave functions. One finds that the states describing the relative motion of the bound and continuum regimes are essentially (up to different masses and background dielectric constant) identical to that of the hydrogen atom. In the excitonic case only those states can be excited optically, i.e., have allowed dipole transitions, for which the relative wave function has a finite value at the origin, i.e., the probability density for the electron and the hole at the same position is finite. In three-dimensional systems this is only possible for s states and all other states are optically

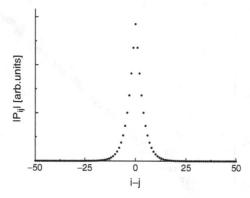

Fig. 9.15. The relative-coordinate wave function proportional to $|p_{ij}|$ for the bound electron–hole pair state, i.e., the "exciton" in the ordered semiconductor model. The parameters are $N = 200$, $J^e = 50\,\mathrm{meV}$, $J^h = 16.7\,\mathrm{meV}$, $\gamma = 0.5\,\mathrm{meV}$, and $U_0 = 16\,\mathrm{meV}$

inactive. In our present one-dimensional model, optically active excitons require a finite value for p_{ii} as well.

Besides the discrete resonances due to the bound states, the dramatic modification of the continuum, as evident by comparing Figs. 9.12 and 9.13, is also discussed in textbooks. In three-dimensional direct semiconductors, like GaAs (i.e., those having a direct gap at $k = 0$), the linear absorption edge is proportional to $\sqrt{\hbar\omega - E_{\mathrm{gap}}}$ if the Coulomb interaction is ignored. In the interacting case, this onset is modified by the so-called Sommerfeld factor, which removes the square-root behavior and leads to an enhanced absorption for photon energies just above the fundamental band edge. This enhanced absorption is due to electron–hole states that resemble a resonance, i.e., they are not strictly bound, but still are highly correlated due to their mutual interaction.

In our one-dimensional case, which in the interaction-free situation shows a singularity at the fundamental band edge, the electron–hole attraction reduces the absorption. This reduction is again the signature of the correlation between unbound electrons and holes.

Figure 9.16 shows the wave function for the relative motion in a different presentation. It has large amplitude only along the main diagonal, i.e., for electron and hole coordinates in close vicinity of each other. The width of this bar is determined by the Bohr radius of the exciton.

In addition to the excitonic resonances, which correspond to bound optically active electron–hole pairs, other bound electron–hole states also exist in our one-dimensional model, which cannot be excited optically because of symmetry reasons. Only if the system is perturbed by a symmetry breaking potential, e.g., due to disorder, do these resonances acquire a finite optical weight. For an example, see Chap. 16.

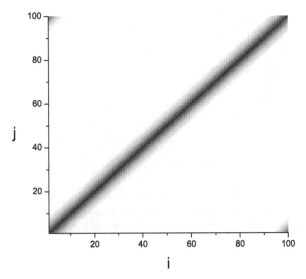

Fig. 9.16. Excitonic wave function $|p_{ij}|$ for the ordered semiconductor model as a contour plot. The parameters are $N = 100$, $U_0 = 16.7\,\text{meV}$, and $J^e = 3J^h = 50\,\text{meV}$

In Figs. 9.12–9.16, we have used $N = 500$ and $N = 100$. The radius of the ring is, therefore, much larger than the length scale determined by the interaction (i.e., the Bohr radius), which, as can be seen in Figs. 9.15 and 9.16, is less than 10 lattice constants. In this situation, we use the conventional Coulomb matrix element proportional to $1/(|i - j| + a_0/a)$ as given in (5.17). We later also consider semiconductor nanorings with radii comparable with the Bohr radius. In this case, (5.18) has to be applied for the Coulomb matrix element.

9.4.2 Dimerized Lattice

The Hamiltonian with the dimerization parameter δ from (6.40) is given by (5.8). The optical spectrum for the special case of equal couplings $J^e = J^h = J$ and dimerization parameters $\delta_e = \delta_h = \delta$ is shown by the solid line in Fig. 9.17, where we have taken $\delta = 3|J|$, a system of 200 sites, and a Coulomb interaction parameter of $U_0 = 80\,\text{meV}$. For comparison, we also plot the result for vanishing Coulomb interaction, i.e., $U_0 = 0$ (dotted line).

In Fig. 9.17, we see two groups of transitions, as expected for a dimerized structure. They show all the features known for the linear optical properties of a one-dimensional system: the dramatic Coulombic suppression of the free-particle continua with their singularities and very pronounced excitons. Most remarkably, we find, in addition to the two excitons $X^<$ and $X^>$, also an exciton in between $X^<$ and $X^>$, which has been denoted by $\bar{X}$ in Fig. 9.17.

In order to analyze these numerical results and to understand the origin of the third exciton, we now examine the band structure and the corresponding optical dipole matrix elements of this model. The lattice without dimerization has a lattice constant a. It changes to twice that value if the

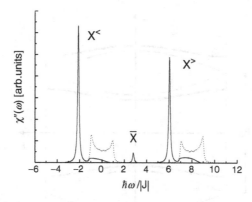

Fig. 9.17. Linear absorption spectrum of a dimerized semiconductor model. The coupling constants J as well as the distortion parameters $\delta = 3|J|$ are the same for both bands. A phenomenological dephasing rate has been introduced. The *dotted line* shows the spectrum calculated neglecting the Coulomb interaction, i.e., setting $U_0 = 0$. $X^<$ and $X^>$ are the regular excitons and $\bar{X}$ is the exciton related to the "forbidden" transitions. The parameters are $|J^e| = |J^h| = 80\,\text{meV}$, $T_2^{-1} = 3\,\text{meV}$, $a_0 = 1.7\,\text{nm}$, $a = 3.4\,\text{nm}$, $U_0 = 80\,\text{meV}$, and $N = 200$. [After K. Bott et al., Phys. Rev. **B56**, 12784 (1997)]

dimerization is introduced, since, in this case, every unit cell contains two sites, see Fig. 6.6. Correspondingly, the new Brillouin zone now contains k values between $[-\frac{\pi}{2a}, \frac{\pi}{2a}]$ and is thus called the reduced zone.

Of course, formally, one can also double the real-space unit cell even in the case without dimerization. The reduced zone then contains two valence bands and two conduction bands instead of just a single valence band and a single conduction band. They are degenerate at the boundaries of the reduced zone, see the thin lines in Fig. 9.18.

In this situation, single-particle optical transitions are only allowed for equal hole-k and electron-k. This implies that in the reduced zone, without dimerization only transitions between antiparallel bands are allowed if the electron and hole couplings are chosen to model a direct semiconductor, i.e., have the same sign (note that we use here $J^e = J^h$).

We expect and find that the dimerization lifts the degeneracy at the boundaries of the reduced zone, see thick lines in Fig. 9.18, such that we have a gap in the optical spectrum, in addition to the fundamental semiconductor gap. This additional gap is clearly visible in Fig. 9.17 for the single-particle transitions (dotted line).

The question arises, whether or not in the dimerized case there are transitions between parallel bands that contribute to spectral weight in the just mentioned gap where the third exciton $\bar{X}$ appears. To answer this question and to gain further insight into the single-particle transitions, we diagonalize the dimerized Hamiltonian analytically by applying a canonical transformation. Afterwards, the optical dipole matrix elements are analyzed in the eigenbasis.

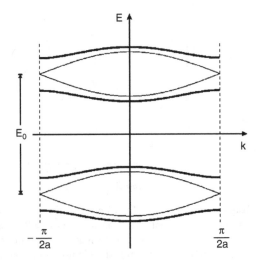

Fig. 9.18. Band structure of the dimerized semiconductor model. The *thin line* shows the cosine band structure of an isoenergetic distribution (no dimerization, $\delta = 0$) after doubling the unit cell and reducing the Brillouin zone. The finite distortion parameter $\delta = 3|J|$ leads to a splitting of both the conduction and the valence band as is shown by the *thick lines*. [After K. Bott et al., Phys. Rev. **B56**, 12784 (1997)]

The Hamiltonian, (5.8), is the sum of an electron Hamiltonian $\hat{H}_e$ and a hole Hamiltonian $\hat{H}_h$. We present the diagonalization only for the electron part,

$$\hat{H}_e = \sum_{l=1}^{N} \left(\epsilon_l^e c_l^\dagger c_l + J^e (c_l^\dagger c_{l+1} + c_{l+1}^\dagger c_l) \right). \tag{9.47}$$

Remember, we take J^e to be a negative quantity in order to have the band minimum in the center of the Brillouin zone. The dimerization for electrons is given by

$$\epsilon_l^e = \frac{1}{2}(\epsilon_0 + (-1)^l \delta_e). \tag{9.48}$$

Due to the electron–hole picture the same Hamiltonian applies for holes if we just interchange e with h and the operators c_l with d_l.

The total system is periodic; we therefore apply a Fourier transformation into k space. The new operators c_k are defined by

$$c_k = \frac{1}{\sqrt{N}} \sum_{l=1}^{N} e^{-ikla} c_l \tag{9.49}$$

and a corresponding equation holds for the adjoint operator $c_k^\dagger$. The inverse transform reads

$$c_l = \frac{1}{\sqrt{N}} \sum_{|k| \le \frac{\pi}{a}} e^{ikla} c_k, \tag{9.50}$$

where $\sum_{|k|\leq\frac{\pi}{a}}$ means summation over all k values in the nonreduced Brillouin zone, see (6.3). In order to obtain the inverse transform, we have made use of the orthogonality relation (6.5).

The reduced Brillouin zone for the dimerized model system only extends between the limits $-\frac{\pi}{2a}$ and $\frac{\pi}{2a}$, with the value $-\frac{\pi}{2a}$ excluded. Therefore, it is helpful to rewrite the Hamiltonian in k space with sums running only over the reduced zone. This yields the three terms

$$\frac{\epsilon_0}{2}\sum_{l=1}^{N}c_l^\dagger c_l = \frac{\epsilon_0}{2}\sum_{|k|\leq\frac{\pi}{2a}}(c_k^\dagger c_k + c_{k+\frac{\pi}{a}}^\dagger c_{k+\frac{\pi}{a}}),\qquad(9.51)$$

$$\frac{\delta_e}{2}\sum_{l=1}^{N}(-1)^l c_l^\dagger c_l = \frac{\delta_e}{2}\sum_{|k|\leq\frac{\pi}{2a}}(c_{k+\frac{\pi}{a}}^\dagger c_k + c_k^\dagger c_{k+\frac{\pi}{a}}),\qquad(9.52)$$

and

$$J^e\sum_{l=1}^{N}(c_l^\dagger c_{l+1} + c_{l+1}^\dagger c_l) = 2J^e\sum_{|k|\leq\frac{\pi}{2a}}\cos(ka)(c_k^\dagger c_k - c_{k+\frac{\pi}{a}}^\dagger c_{k+\frac{\pi}{a}}).\qquad(9.53)$$

For k values in the reduced zone, we define new fermionic operators $a_{k,+}$ and $a_{k,-}$ by the canonical transformation

$$a_{k,+} = \alpha_k^e c_k + \beta_k^e c_{k+\frac{\pi}{a}},$$
$$a_{k,-} = -\beta_k^e c_k + \alpha_k^e c_{k+\frac{\pi}{a}},\qquad(9.54)$$

with the inverse transformation

$$c_k = \alpha_k^e a_{k,+} - \beta_k^e a_{k,-},$$
$$c_{k+\frac{\pi}{a}} = \beta_k^e a_{k,+} + \alpha_k^e a_{k,-}.\qquad(9.55)$$

The new operators are required to obey the same fermionic anticommutation relations

$$\{a_{k,+}^\dagger, a_{k',+}\} = \delta_{kk'},\qquad(9.56)$$

etc., as the old ones

$$\{c_k^\dagger, c_{k'}\} = \delta_{kk'}.\qquad(9.57)$$

This imposes the relation

$$\alpha_k^{e2} + \beta_k^{e2} = 1\qquad(9.58)$$

for the coefficients which are otherwise still undetermined.

The operator products $c_k^\dagger c_{k'}$ appearing in the various terms of the Hamiltonian (9.51), (9.52), and (9.53) are given in terms of the new operators $a_{k,\pm}^\dagger a_{k,\pm}$ by

$$c_k^\dagger c_k = (\alpha_k^e a_{k,+}^\dagger - \beta_k^e a_{k,-}^\dagger)(\alpha_k^e a_{k,+} - \beta_k^e a_{k,-})$$
$$= \alpha_k^{e2} a_{k,+}^\dagger a_{k,+} + \beta_k^{e2} a_{k,-}^\dagger a_{k,-} - \alpha_k^e \beta_k^e (a_{k,+}^\dagger a_{k,-} + a_{k,-}^\dagger a_{k,+}),$$

$$c_{k+\frac{\pi}{a}}^\dagger c_{k+\frac{\pi}{a}} = (\beta_k^e a_{k,+}^\dagger + \alpha_k^e a_{k,-}^\dagger)(\beta_k^e a_{k,+} + \alpha_k^e a_{k,-})$$
$$= \beta_k^{e2} a_{k,+}^\dagger a_{k,+} + \alpha_k^{e2} a_{k,-}^\dagger a_{k,-} + \alpha_k^e \beta_k^e (a_{k,+}^\dagger a_{k,-} + a_{k,-}^\dagger a_{k,+}),$$

$$c_{k+\frac{\pi}{a}}^\dagger c_k = (\beta_k^e a_{k,+}^\dagger + \alpha_k^e a_{k,-}^\dagger)(\alpha_k^e a_{k,+} - \beta_k^e a_{k,-})$$
$$= \alpha_k^{e2} a_{k,-}^\dagger a_{k,+} - \beta_k^{e2} a_{k,+}^\dagger a_{k,-} + \alpha_k^e \beta_k^e (a_{k,+}^\dagger a_{k,+} - a_{k,-}^\dagger a_{k,-}),$$

$$c_k^\dagger c_{k+\frac{\pi}{a}} = (\alpha_k^e a_{k,+}^\dagger - \beta_k^e a_{k,-}^\dagger)(\beta_k^e a_{k,+} + \alpha_k^e a_{k,-})$$
$$= \alpha_k^{e2} a_{k,+}^\dagger a_{k,-} - \beta_k^{e2} a_{k,-}^\dagger a_{k,+} + \alpha_k^e \beta_k^e (a_{k,+}^\dagger a_{k,+} - a_{k,-}^\dagger a_{k,-}).$$

$$(9.59)$$

Now, we are ready to express the Hamiltonian in the reduced-zone k representation as

$$\hat{H}_e = \sum_{|k|\leq\frac{\pi}{2a}} \frac{\epsilon_0}{2}(a_{k,+}^\dagger a_{k,+} + a_{k,-}^\dagger a_{k,-})$$
$$+ \sum_{|k|\leq\frac{\pi}{2a}} \left[\frac{\delta_e}{2}2\alpha_k^e\beta_k^e + 2J^e\cos(ka)(\alpha_k^{e2} - \beta_k^{e2})\right]$$
$$\times (a_{k,+}^\dagger a_{k,+} - a_{k,-}^\dagger a_{k,-})$$
$$+ \sum_{|k|\leq\frac{\pi}{2a}} \left[\frac{\delta_e}{2}(\alpha_k^{e2} - \beta_k^{e2}) - 2J^e\cos(ka)2\alpha_k^e\beta_k^e\right]$$
$$\times (a_{k,+}^\dagger a_{k,-} + a_{k,-}^\dagger a_{k,+}).$$

$$(9.60)$$

Note that the third term couples the $+$ band to the $-$ band. If we choose

$$2\alpha_k^e\beta_k^e = \frac{\delta_e}{2u_e(k)} \tag{9.61}$$

and

$$\alpha_k^{e2} - \beta_k^{e2} = \frac{2J^e\cos(ka)}{u_e(k)} \tag{9.62}$$

with some function $u_e(k)$, this terms vanishes and the second term in (9.60) becomes

$$\sum_{|k|\leq\frac{\pi}{2a}} \frac{1}{u_e(k)}\left[\left(\frac{\delta_e}{2}\right)^2 + (2J^e\cos(ka))^2\right](a_{k,+}^\dagger a_{k,+} - a_{k,-}^\dagger a_{k,-}). \tag{9.63}$$

Equations (9.61), (9.62), and (9.58) can now be used to determine $u_e(k)$. We find

$$u_e(k) = \sqrt{\left(\frac{\delta_e}{2}\right)^2 + (2J^e\cos(ka))^2} \tag{9.64}$$

and, thus, the Hamiltonian $\hat{H}_e$ is diagonalized with the eigenenergies

$$E^e_\pm(k) = \frac{\epsilon_0}{2} \pm \sqrt{\left(\frac{\delta_e}{2}\right)^2 + (2J^e \cos{(ka)})^2}. \qquad (9.65)$$

For the holes, the calculation is completely analogous and, thus, we have two bands for the electrons and two bands for the holes. These are shown in Fig. 9.17 as thick lines.

Now, we can evaluate the optical dipole matrix elements. The light–matter interaction in the real-space representation reads as usual

$$\hat{H}_L = -E(t) \sum_{l=1}^{N} \mu_0 (d_l c_l + c_l^\dagger d_l^\dagger). \qquad (9.66)$$

The Fourier transformed k-space operators for the holes are in complete analogy to those for the electrons

$$d_k = \frac{1}{\sqrt{N}} \sum_{l=1}^{N} e^{ikla} d_l \qquad (9.67)$$

and a corresponding equation holds for the adjoint operator $c_k^\dagger$. The inverse transform reads

$$d_l = \frac{1}{\sqrt{N}} \sum_{|k| \leq \frac{\pi}{a}} e^{-ikla} d_k. \qquad (9.68)$$

Let us consider the term

$$\sum_{l=1}^{N} d_l c_l = \sum_{|k| \leq \frac{\pi}{2a}} (d_k c_k + d_{k+\frac{\pi}{a}} c_{k+\frac{\pi}{a}}). \qquad (9.69)$$

The new fermionic operators for the holes read

$$b_{k,+} = \alpha_k^h d_k + \beta_k^h d_{k+\frac{\pi}{a}},$$
$$b_{k,-} = -\beta_k^h c_k + \alpha_k^h d_{k+\frac{\pi}{a}}. \qquad (9.70)$$

In terms of these operators, the light–matter interaction reads

$$\hat{H}_L = -\mu_0 E(t) \sum_{|k| \leq \frac{\pi}{a}} [\mu(k)(b_{k,+} a_{k,+} + b_{k,-} a_{k,-})$$
$$+ \bar{\mu}(k)(b_{k,+} a_{k,-} - b_{k,-} a_{k,+}) + H.C.], \qquad (9.71)$$

with the abbreviations

$$\mu(k) = \alpha_k^e \alpha_k^h + \beta_k^e \beta_k^h,$$
$$\bar{\mu}(k) = \alpha_k^e \beta_k^h - \alpha_k^h \beta_k^e. \qquad (9.72)$$

The matrix elements $\mu_0\mu(k)$ describe single-particle transitions between the parallel subbands, while $\mu_0\bar{\mu}(k)$ describe those between antiparallel subbands. Two further abbreviations

$$\eta_k^e = \left[1 + \left(\frac{\delta_e}{4J^e\cos(ka)}\right)^2\right]^{-\frac{1}{2}},$$

$$\eta_k^h = \left[1 + \left(\frac{\delta_h}{4J^h\cos(ka)}\right)^2\right]^{-\frac{1}{2}}, \qquad (9.73)$$

allow us to write these expressions in a particularly compact form, namely

$$\mu(k) = \frac{1}{2}\left([(1-\eta_k^e)(1-\eta_k^h)]^{\frac{1}{2}} + [(1+\eta_k^e)(1+\eta_k^h)]^{\frac{1}{2}}\right),$$

$$\bar{\mu}(k) = \frac{1}{2}\left([(1-\eta_k^e)(1+\eta_k^h)]^{\frac{1}{2}} - [(1+\eta_k^e)(1-\eta_k^h)]^{\frac{1}{2}}\right). \qquad (9.74)$$

We see that the matrix elements depend on the ratios δ_e/J^e and δ_h/J^h only. For a model without dimerization, i.e., $\delta_e = \delta_h = 0$, we have $\bar{\mu}(k) = 0$. This means that transitions between antiparallel bands are forbidden, as expected.

In the case with dimerization, those transitions become allowed in general. However, if, in particular, $\delta_e/J^e = \delta_h/J^h$, they remain forbidden. This is the situation underlying the numerical calculation shown in Fig. 9.17. There are no single-particle transitions between states of parallel bands which would contribute to the spectral region in the gap between the lower-energy and the upper-energy portion of the spectrum.

On the other hand, even in this situation, we observe an excitonic resonance in the presence of the Coulomb interaction, i.e., the peak $\bar{X}$, close to photon energies which correspond to transitions between the parallel bands. In fact, this resonance is located at an energy that corresponds to the separation of two parallel bands minus the exciton binding energy. Hence, the additional exciton is related to transitions between bands that are forbidden if Coulomb interaction is neglected.

More insight into the nature of the exciton $\bar{X}$ can be gained by examining the wave function for the relative motion. Figures 9.19 and 9.20 display the modulus of p_{ij} in the $i-j$ space in a contour plot. For this result the model has been designed such that the on-site energy separations between the electron and hole levels are taken to be the larger ones at sites $i,j = 1,3,5,\ldots$, and the smaller ones at sites $i,j = 2,4,6,\ldots$.

It is clearly seen in Figs. 9.19 and 9.20 that the amplitudes are high for the coordinates $i = j = 2,4,6,8,\ldots$, and $i = j = 1,3,5,7,\ldots$, respectively. This shows that the lower exciton $X^<$ is predominantly built from states belonging to the two bands with minimal spacing, which in turn are formed predominantly from sites $i = 2,4,6,\ldots$ having the small energy separation as well. On the other hand, the exciton $X^>$ is related to the two bands with maximum spacing and thus related to the sites $i = 1,3,5,\ldots$ with the larger energy

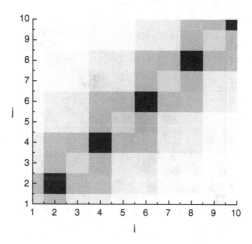

Fig. 9.19. Wave function $|p_{ij}|$ for the exciton $X^<$ for the dimerized model

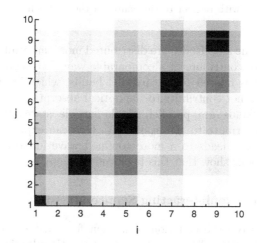

Fig. 9.20. Wave function $|p_{ij}|$ for the exciton $X^>$ for the dimerized model

separation. Between the maxima at the main diagonal, the wave function has low amplitude.

If we now turn to Fig. 9.21, which shows $|p_{ij}|$ for the additional exciton $\bar{X}$, we realize that, although the maxima appear at sites where $X^>$ has maxima, the amplitudes at other sites are comparable. In Fig. 9.21, we have also indicated the phases of p_{ij}. This shows that there are nodes between every two sites, i.e., in the center of the unit cell. Therefore, we conclude that this exciton has a strong contribution that is of odd symmetry with respect to the center of the unit cell.

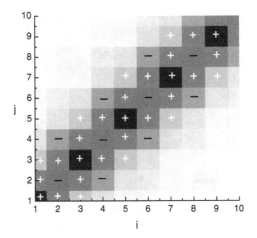

Fig. 9.21. Wave function $|p_{ij}|$ for the exciton $\bar{X}$ for the dimerized model. The phase of $-ip_{ij}$ is indicated by the $+$ and $-$ *signs* and shows the nearly odd-parity character of the wave function with respect to the center of the unit cell

The fact that the amplitudes are distributed more uniformly also indicates that to $\bar{X}$ all sites contribute in a comparable way. This again suggests that the $\bar{X}$ resonance is related to the parallel bands, which, without Coulomb interaction, would not contribute to the optical absorption.

Since these bands are parallel, single-particle states from the entire Brillouin zone contribute to the exciton $\bar{X}$. One can expect that this delocalization in k space leads to a more confined wave function in real space. Detailed calculations show that this is indeed the case.

9.4.3 Disordered Noninteracting Semiconductor

Again, we start from the equation of motion for p_{ij}, i.e., (9.45). However, we now assume that the diagonal elements of the Hamiltonian matrices, the single-site energies, are distributed according to a given distribution function. We still apply periodic boundary conditions, but there is no longer translational symmetry on length scales of the intersite separation. Consequently, the quantum number k is not conserved and an analytic calculation of the spectrum is no longer possible. However, using the equation of motion approach, we are able to obtain the spectrum numerically.

In the extreme case of vanishing intersite couplings J in both the conduction and the valence band, the optical spectrum is strictly inhomogeneous, since it is composed of independent two-level absorbers. The question, whether the spectrum of a disordered semiconductor with finite intersite couplings is a homogeneous or an inhomogeneous one and to what extent this classification is influenced by the Coulomb interaction, will be addressed below, after we have developed unambiguous criteria.

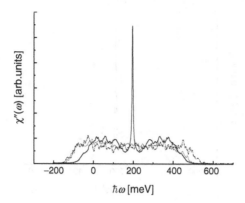

Fig. 9.22. Linear spectra for anticorrelated (*solid*), correlated (*dashed*), and uncorrelated (*dotted*) disorder, for equal electron and hole masses and disorder. The parameters are $J^e = 50\,\mathrm{meV}$, $J^h = 50\,\mathrm{meV}$, $W^e = 200\,\mathrm{meV}$, $W^h = 200\,\mathrm{meV}$, $U_0 = 0$, $\gamma = 3\,\mathrm{meV}$, and $N = 100$

Typically, the spectra for a disordered semiconductor with uncorrelated disorder and $U_0 = 0$ are broadened versions of those for the ordered case. The disorder leads to a smearing out of the van Hove singularities and to absorption tails extending into the band gap of the corresponding ordered model.

An interesting situation occurs if we consider anticorrelated disorder, equal electron and hole masses, and equal disorder amplitudes in the two bands. Figure 9.22 compares the spectra for equal masses and disorder. While the cases of uncorrelated and correlated disorder both yield similarly broadened spectra, the spectrum for anticorrelated disorder is symmetric around a certain energy, where we see a strong peak. This peak originates from transitions between states, which are situated in the band tails and are therefore localized more strongly as compared to states in the center of the bands. The localized transitions are similar to those in isolated two-level systems. In the anticorrelated disorder case, the two-level systems have equal energy separation at all sites. Therefore, the tail–tail transitions add up to the strong peak in the center of the spectrum. The symmetry around this peak is a consequence of the total electron–hole symmetry in the present case. Once the masses are different, even with a correspondingly scaled disorder, these features are absent.

Figure 9.23 compares the spectra for anticorrelated disorder considering equal and different masses. The disorder has been scaled accordingly. Only for equal masses are the peak and symmetry observed.

For correlated disorder, equal masses and equal disorder parameters, we have a strict selection rule, since the states in the conduction and valence bands are pairwise identical for equal electron and hole eigenenergies. The states we are referring to in our tight-binding model are actually described by envelope functions. Their scalar product determines the resulting dipole matrix element describing the transition between the eigenstates. Only for

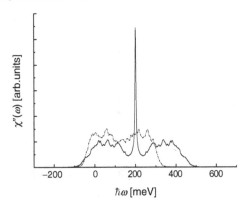

Fig. 9.23. Linear spectrum for anticorrelated and short-ranges disorder. *Solid line:* $J^e = 50\,\text{meV}$, $J^h = 50\,\text{meV}$, $W^e = 200\,\text{meV}$, $W^h = 200\,\text{meV}$, $U_0 = 0$, $\gamma = 3\,\text{meV}$, and $N = 100$. *Dashed line:* $J^e = 50\,\text{meV}$, $J^h = 16.7\,\text{meV}$, $W^e = 200\,\text{meV}$, $W^h = 200/3\,\text{meV}$, $U_0 = 0$, $\gamma = 3\,\text{meV}$, and $N = 100$

equal electron and hole energies are these scalar products equal to unity, and zero otherwise. Therefore, in this case the spectrum is determined up to a stretched energy scale by the single-particle spectrum.

9.4.4 Disordered Interacting Semiconductor

In the ordered case, the linear optical spectrum of the interacting model of the semiconductor is dominated by the excitonic resonance. It is thus an interesting question to what extent this resonance is disturbed by disorder. An overview is given in Fig. 9.24, which shows the spectra for the ordered, for a weakly, and for a strongly disordered case. In order to obtain a sufficiently smooth spectrum for $N = 100$ sites, a dephasing rate of $\gamma = 3\,\text{meV}$ has been used.

Since in typical semiconductor quantum wells made of GaAs, the effective masses of the (heavy) holes are about three times larger than those of the electrons, we have taken a ratio of three here. While weak disorder leads to a slightly broadened excitonic resonance and a continuum that remains nearly unchanged, for strong disorder, the excitonic resonance is no longer visible in the spectrum. Instead, in this case, many peaks originating from the particular local transitions appear.

Let us now concentrate on the excitonic resonance. In order to obtain smoother spectra configurational averages are calculated, i.e., from the given box-shaped distribution of width $W^{e,h}$ for the site energies a number of disorder realizations are randomly generated and the resulting spectra are computed. The total spectra are obtained by taking an average over these realizations.

Figure 9.25 shows averaged spectra for 20 realizations and a system size of $N = 100$. If long-range correlations are unimportant, the same spectra are expected to result for a single realization and $N = 2000$. To calculate

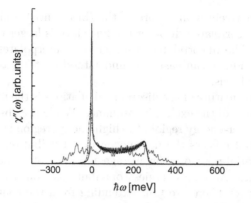

Fig. 9.24. Linear spectrum for the ordered (*solid*) semiconductor, $J^e = 50$ meV, $J^h = 16.7$ meV, $W^e = 0$, $W^h = 0$, $U_0 = 16$ meV, $\gamma = 3$ meV, $N = 100$, for the weakly disordered semiconductor (*dashed-dotted*), $J^e = 50$ meV, $J^h = 16.7$ meV, $W^e = 50$ meV, $W^h = 16.7$ meV, $U_0 = 16$ meV, $\gamma = 3$ meV, $N = 100$, and for the strongly disordered semiconductor (*dotted*), $J^e = 50$ meV, $J^h = 16.7$ meV, $W^e = 300$ meV, $W^h = 100$ meV, $U_0 = 16$ meV, $\gamma = 3$ meV, $N = 100$. Uncorrelated disorder has been used

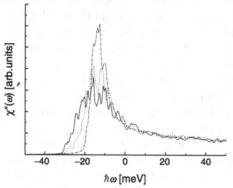

Fig. 9.25. The absorption at the excitonic resonance for short-range disorder and 20 realizations. *Solid line*: correlated disorder, $J^e = 50$ meV, $J^h = 16.7$ meV, $W^e = 50$ meV, $W^h = 16.7$ meV, $U_0 = 16$ meV, $\gamma = 0.5$ meV, $N = 100$. *Dashed-dotted line*: anticorrelated disorder, $J^e = 50$ meV, $J^h = 16.7$ meV, $W^e = 50$ meV, $W^h = 16.7$ meV, $U_0 = 16$ meV, $\gamma = 0.5$ meV, $N = 100$. *Dotted line*: uncorrelated disorder $J^e = 50$ meV, $J^h = 16.7$ meV, $W^e = 50$ meV, $W^h = 16.7$ meV, $U_0 = 16$ meV, $\gamma = 0.5$ meV, $N = 100$

spectra for such a model, however, requires numerically $(2000/100)^2 = 400$ times longer computer time. To obtain a criterion for the applicability of the averaging procedure, one has to analyze the relevant length scales of the problem. These are the excitonic Bohr radius a_B, the correlation length of the disorder potential L, and the length of the system Na. Averaging is allowed if Na is larger than both a_B and L. In Fig. 9.25, we have $L = a$, i.e., we have spatially uncorrelated disorder, which is called short-range disorder.

Due to the relatively weak disorder, the (broadened) excitonic resonance is still visible. For correlated disorder the linewidth is larger than for anticorrelated disorder. The uncorrelated disorder case interpolates between these extrema. These findings can easily be understood on the basis of the model of the site energy levels.

A characteristic feature of the disorder calculations is the unsymmetric line shape in the vicinity of the excitonic resonance. While the absorption shows a steep tail in the lower-energy region, the high-energy region is decaying slowly. The low-energy tail reflects the inhomogeneous distribution of site energies, i.e., in various spatial positions the average gap fluctuates due to fluctuations of the site energies. At high energies contributions from three sources have to be considered: (i) There are tails extending from the continuum into the excitonic region. (ii) There are higher bound states, which can contribute to the spectrum, because the disorder breaks the symmetry, which renders most of these bound states to be optically inactive in the ordered case. However, the most important contribution (iii) is due to the fact that in the disordered situation the center-of-mass momentum is no longer a good quantum number. In an ordered semiconductor, only excitons with center-of-mass momentum $\hbar K = 0$ can be optically excited by a spatially homogeneous field, because the photon momentum is (nearly) zero and the total momentum is conserved in the transition. This is the reason for the sharp excitonic resonance, which shows only the $K = 0$ states of the center-of-mass dispersion, which otherwise is parabolic at low energies. The $K \neq 0$ states exist, but are optically inactive because of the selection rule, which is a consequence of translational symmetry. In the disordered situation, there is no translational symmetry and the higher center-of-mass states become optically allowed. The extent to which this occurs depends on the range of center-of-mass wave vectors which contribute to a given excitonic state in the disordered case. This range is determined by the Fourier spectrum of the disorder potential. For a short-range disorder potential a larger range of wave vectors contributes, and thus the excitonic resonance is wider compared to the case of a long-range potential.

In the presence of a long-range disorder potential, the internal motion of the electron–hole pair is nearly unchanged and we can still classify the bound states in terms of the quantum numbers valid for the ordered case. In addition, however, this long-range potential defines potential wells which can localize the exciton as a whole. These potential wells can accommodate more than just a single localized center-of-mass state, which gives rise to more than a single excitonic line. The inhomogeneous excitonic spectrum can therefore be composed of sharp discrete lines due to the various spatial positions of the potential wells and in addition to the multiplicity of the localized states within a given well. While the former are statistically unrelated, the latter show energetic correlations due to the repulsion of levels in a given well.

For short-range disorder potentials, it makes no sense to distinguish between the relative and the center-of-mass motion. The disorder potential mixes up all discrete bound states, if it is strong enough (i.e., of the order of the

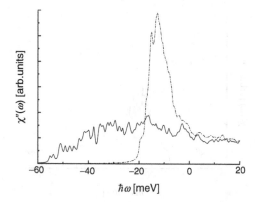

Fig. 9.26. The absorption at the excitonic resonance for long-range disorder, $L = 20a$. *Solid line*: correlated $J^e = 50\,\mathrm{meV}$, $J^h = 16.7\,\mathrm{meV}$, $W^e = 50\,\mathrm{meV}$, $W^h = 16.7\,\mathrm{meV}$, $U_0 = 16\,\mathrm{meV}$, $\gamma = 0.5\,\mathrm{meV}$, $N = 100$. *Dashed-dotted line*: anticorrelated $J^e = 50\,\mathrm{meV}$, $J^h = 16.7\,\mathrm{meV}$, $W^e = 50\,\mathrm{meV}$, $W^h = 16.7\,\mathrm{meV}$, $U_0 = 16\,\mathrm{meV}$, $\gamma = 0.5\,\mathrm{meV}$, $N = 100$

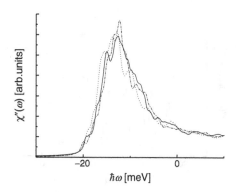

Fig. 9.27. The absorption at the excitonic resonance for anticorrelated disorder and various length scales. $J^e = 50\,\mathrm{meV}$, $J^h = 16.7\,\mathrm{meV}$, $W^e = 50\,\mathrm{meV}$, $W^h = 16.7\,\mathrm{meV}$, $U_0 = 16\,\mathrm{meV}$, $\gamma = 0.5\,\mathrm{meV}$, $N = 100$. *Solid line*: $L = 20a$; *dashed-dotted line*: $L = 5a$; and *dotted*: $L = a$

excitonic binding energy). On the other hand, for weak disorder the relative motion tends to average over the potential, leaving an effective potential characterized by a length scale of the order of the Bohr radius.

Figure 9.26 shows the effect of long-range disorder, $L = 20a$, on the excitonic linear spectrum. There is a large difference in the linewidth between energetically correlated and anticorrelated disorder, which again can be easily understood in terms of the site energy distribution, being parallel in the latter case and antiparallel in the former.

While the length scale has only a marginal effect on the excitonic linewidth for anticorrelated disorder, see Fig. 9.27, it has a profound influence for

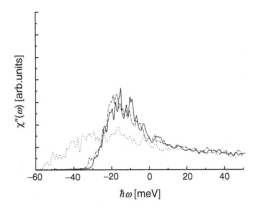

Fig. 9.28. The absorption at the excitonic resonance for correlated disorder and various length scales. $J^e = 50\,\text{meV}$, $J^h = 16.7$, $W^e = 50\,\text{meV}$, $W^h = 16.7\,\text{meV}$, $U_0 = 16\,\text{meV}$, $\gamma = 0.5\,\text{meV}$, $N = 100$. *Solid line:* $L = a$; *dashed-dotted line:* $L = 2a$; and *dotted:* $L = 20a$

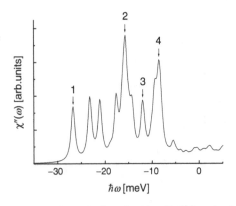

Fig. 9.29. Inhomogeneous excitonic line for the disordered semiconductor model with correlated disorder and length scale $L = 20a$. Only a single realization has been considered

correlated disorder, see Fig. 9.28. Clearly, in the latter case well-defined potential wells are formed for the center-of-mass motion for large L, which localize the center-of-mass motion of the excitonic resonances.

It is instructive to analyze the spatial localization of the excitons for a single realization of the correlated disorder. The spectrum is shown in Fig. 9.29. For the spectral positions of various peaks, the polarization $|p_{ij}|$ is plotted in the $i - j$ plane in Fig. 9.30. In case 1, 2, and 3, the polarization is strongly confined to the main diagonal, indicating bound electron–hole pairs. $|p_{ij}|$ has well-defined maxima at essentially just a single position, indicating the strong localization in local potential wells. In case 4 (the energetically highest strong peak in the spectrum), $|p_{ij}|$ has also a significant amplitude outside the main diagonal, indicating that electron and hole are no longer strongly bound to

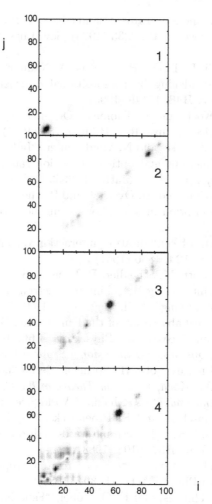

Fig. 9.30. Excitonic wave functions $|p_{ij}|$ for the disordered semiconductor model which show, depending on their spectral position, see Fig. 9.29, localization of the exciton at various places. Correlated disorder with a length scale $L = 20a$ is taken

each other. In this regime, there is a competition between the attractive Coulomb force and local minima in the single-particle site energies, which may lead to a dissociation of the exciton. In addition, for small i and j, continuum states can be seen as horizontally oriented clouds due to the local distortion of the gap.

9.5 Suggested Reading

1. S. Abe and Y. Toyozawa, "Interband absorption of disordered semiconductors in the coherent potential approximation", J. Phys. Soc. Jpn. **50**, 2185 (1981)

2. S.D. Baranovskii and A.L. Efros, "Band edge smearing in solid solutions", Fiz. Tekh. Poluprovodn. **12**, 2233 (1978) [Sov. Phys. Semicond. **12**, 1328 (1978)]

3. S.D. Baranovskii, U. Doerr, P. Thomas, A. Naumov, and W. Gebhardt, "Exciton line broadening by compositional disorder in alloy quantum wells", Phys. Rev. **B48**, 17149 (1993)

4. K. Bott, S.W. Koch, and P. Thomas, "Odd-parity excitons in semiconductor superlattices", Phys. Rev. B **56**, 12784 (1997)

5. H. Carruzo, K. Maschke, and N. Vandeventer, "Influence of chemical disorder on wavefunctions and optical transition rates in one-dimensional systems", J. Phys.: Condens. Matter **1**, 6633 (1989)

6. U. Dersch, M. Grünewald, H. Overhof, and P. Thomas, "Theoretical studies of optical absorption in amorphous semiconductors", J. Phys. C **20**, 121 (1987)

7. U. Fano, "Effects of configuration interaction on intensities and phase shifts", Phys. Rev. **124**, 1866 (1961)

8. F. Gebhard, K. Bott, M. Scheidler, P. Thomas, S.W. Koch, "Optical absorption of non-interacting tight-binding electrons in a Peierls-distorted chain at half band filling", Phil. Mag. B **75**, 1 (1997)

9. S. Glutsch, "Optical absorption of the Fano model: General case of many resonances and many continua", Phys. Rev. B **66**, 075310 (2002)

10. S. Glutsch, *Excitons in Low-Dimensional Semiconductors*, Springer Series in Solid-State Sciences, Vol. 141 (Springer, Berlin 2004)

11. H. Haug and S.W. Koch, *Quantum Theory of the Optical and Electronic Properties of Semiconductors*, 4th edn. (World Scientific, Singapore 2004)

12. J. Ingers, K. Maschke, and S. Proennecke, "Optical-transition-matrix elements between localized electronic states in disordered one-dimensional systems", Phys. Rev. **B37**, 6105 (1988)

13. K. Maschke and E. Mooser, "k-Space related properties in disordered solids", Helvetica Physica Acta **61**, 394 (1988)

14. K. Maschke, P. Thomas, and E. O. Göbel, "Fano interference in type-II semiconductor quantum-well structures", Phys. Rev. Lett. **67**, 2646 (1991)

15. T. Meier, A. Schulze, P. Thomas, H. Vaupel, and K. Maschke, "Signature of Fano-resonances in four-wave mixing experiments", Phys. Rev. **B51**, 13977 (1995)

16. H. Overhof and K. Maschke, "Exponential tails in optical absorption and electron-absorption of disordered systems – a one-dimensional model", J. Phys. Condens. Matter **1**, 431 (1989)

17. E. Runge and R. Zimmermann, "Spatially resolved spectra, effective mobility edge, and level repulsion in narrow quantum wells", Phys. Stat. Sol. B **206**, 167 (1998)

10

Coherent $\chi^{(3)}$ Processes for Level Systems

So far, we have computed only linear optical processes using the equation of motion for the interband coherence. In the following, we proceed to the nonlinear regime and analyze third-order coherent optical processes, which are studied in four-wave-mixing and pump–probe experiments, see Chap. 2. As already mentioned, their consistent theoretical description requires us to go beyond the Hartree–Fock approximation and to treat four-point correlations if the many-particle interaction is included. This results in a set of coupled equations of some complexity which has to be solved numerically. The resulting temporal traces of the observables show a number of features that have to be compared to experimental data in order to obtain a satisfactory interpretation of observed phenomena.

Before we proceed with the analysis of interacting many-body systems, it is instructive to investigate few-level systems in order to gain a pictorial understanding of the fundamental phenomena which occur in coherent nonlinear optical experiments. These models also allow us to derive the relevant equations in a quite transparent way and to introduce a number of notations, without cluttering the representation with the whole wealth of indices necessary to describe more realistic systems.

We consider the experimentally relevant situation where two or three short light pulses excite the system. These pulses propagate in the directions k_l with $l = 1, 2, 3$ and have relative delay times τ_1, τ_2. In cases where only two pulses are applied, one of them is counted twice, e.g., $k_1 = k_2$ and $\tau_1 = 0$. This is the typical pulse sequence of pump–probe experiments. In four-wave-mixing experiments performed in the self-diffraction configuration, the second and third pulses are identical.

For sufficiently weak incident fields, the solution of the equations of motion can be performed iteratively in powers of the field, starting with the equation for the interband coherence in linear response. The required final result is the interband coherence in third order in the light field. This is the source for the signal that is detected experimentally. In this book, we do not treat propagation effects. Therefore, we do not solve Maxwell's equations simultaneously with the material equations, but consider simply the third-order polarization as a representative for the experimental signal.

10.1 Coherent $\chi^{(3)}$ Processes for Two-Level Systems

We start with the analysis of coherent optical processes in ensembles of two-level systems. In order to derive analytical results we again assume that the optical pulses can be approximated as $\delta(t)$ functions which contain all frequencies with equal weight. In some cases, however, we are interested in an excitation of only part of the optical spectrum of the material system. In such situations, the application of analytical methods becomes cumbersome and, therefore, we solve the equations of motion numerically.

10.1.1 Pump–Probe Experiments

In typical pump–probe experiments a first pump pulse propagates in direction $\boldsymbol{k}_1$ and excites the system at time $t = 0$. At a later time $t = \tau > 0$, the system is probed by a weak pulse propagating in the direction $\boldsymbol{k}_2$, which is also the direction where the detector is placed. Since the intensity of the pump pulse is the relevant quantity, in a third-order analysis this first pulse enters twice, while the second pulse enters once in the iterative derivation of the equations of motion. Here, in contrast to our previous discussion of linear optical properties, the spatial phase factors of the different terms are important.

Since the various response functions now depend on the order and direction of the respective excitation pulses, we introduce the following notation: the superscript $(m|n)$ indicates spatial phase factors $\exp{(i(-m\boldsymbol{k}_1 - n\boldsymbol{k}_2) \cdot \boldsymbol{r})}$. A polarization having, e.g., a phase factor $\exp{(-i\boldsymbol{k}_1 \cdot \boldsymbol{r})}$ radiates into direction $\boldsymbol{k}_1$. As we are interested in a signal emanating from the sample into a given direction, we omit all terms that radiate into other than the direction where the detector is placed.

We start from the set of optical Bloch equations, (7.29), which we amend by phenomenological damping terms given by the population relaxation time T_1 and the dephasing time T_2. First, we solve these equations in first order in the pump pulse No. 1 with direction $\boldsymbol{k}_1$ at time $t = 0$. Since there is not yet a second pulse, we still have $\rho_{vv} = 1$ and $\rho_{cc} = 0$. Thus, assuming a $\delta(t)$ pulse,

$$E_1(t) = \eta_1 \delta(t), \tag{10.1}$$

we obtain from

$$\hbar \frac{d\bar{\rho}_{vc}^{(1|0)}}{dt} - i\hbar(\omega_{cv} - \omega_L)\bar{\rho}_{vc}^{(1|0)} + \frac{\hbar}{T_2}\bar{\rho}_{vc}^{(1|0)} = -iE_1^*(t)e^{-i\boldsymbol{k}_1 \cdot \boldsymbol{r}}\mu_{vc} \tag{10.2}$$

the first-order result

$$\bar{\rho}_{vc}^{(1|0)} = -\frac{i}{\hbar}\eta_1^*\mu_{vc}\theta(t)e^{-i\boldsymbol{k}_1\cdot\boldsymbol{r}}e^{i(\omega_{cv}-\omega_L)t-t/T_2}. \tag{10.3}$$

Hence, the polarization $\bar{P}_1$ created by the first pulse is given by

$$\bar{P}_1^{(1|0)}(t) = -\frac{i}{\hbar}\eta_1^*\theta(t)e^{-i\boldsymbol{k}_1\cdot\boldsymbol{r}-i\omega_L t}|\mu_{vc}|^2 e^{i\omega_{cv}t-t/T_2}. \tag{10.4}$$

According to the previous chapter, the linear optical spectrum is obtained (note that we have applied a $\delta(t)$ pulse) by transforming the polarization back into the stationary frame

$$P_1^{(1|0)}(t) = \bar{P}_1^{(1|0)}(t)e^{i\omega_L t}, \tag{10.5}$$

taking the complex conjugate, and Fourier transforming into the frequency domain, see (9.7), as

$$\chi(\omega) = \int_{-\infty}^{\infty} e^{i\omega t}(P_1^{(1|0)}(t))^* dt. \tag{10.6}$$

The linear absorption spectrum $\alpha(\omega)$ is proportional to the imaginary part of the linear susceptibility $\chi(\omega)$, which is a Lorentzian line, see (9.14), i.e., we find

$$\alpha(\omega) \propto \frac{(\pi T_2)^{-1}}{(\omega - \omega_{cv})^2 + T_2^{-2}}. \tag{10.7}$$

In second order in the first pulse No. 1, we have to consider

$$\hbar\frac{d\rho_{cc}}{dt} + \frac{\hbar}{T_1}\rho_{cc} = iE_1(t)(\mu_{vc})^*\bar{\rho}_{vc}^{(1|0)}e^{i\boldsymbol{k}_1\cdot\boldsymbol{r}}$$
$$-iE_1^*(t)\mu_{vc}(\bar{\rho}_{vc}^{(1|0)})^*e^{-i\boldsymbol{k}_1\cdot\boldsymbol{r}}, \tag{10.8}$$

which is solved by (note that $\int \delta(t)\theta(t)dt = \theta(0) = 1/2$)

$$\rho_{cc}^{(0|0)} = |\eta_1|^2|\mu_{vc}|^2\theta(t)e^{-t/T_1}/\hbar^2 = \left(\frac{\Theta_1}{2}\right)^2\theta(t)e^{-t/T_1}. \tag{10.9}$$

For the present pump excitation condition, (10.1), $|\eta_1||\mu_{vc}|/\hbar$ is equal to half of the Rabi angle Θ_1, and (10.9) is the low excitation limit of (7.65).

Now, we apply the second (probe) pulse

$$E_2(t) = \eta_2\delta(t-\tau), \tag{10.10}$$

with $\tau > 0$. The equation of motion for the polarization in the probe direction, i.e., $\boldsymbol{k}_2$, up to third order reads

$$\hbar\frac{d\bar{\rho}_{vc}^{(0|1)}}{dt} - i\hbar(\omega_{cv} - \omega_L)\bar{\rho}_{vc}^{(0|1)} + \frac{\hbar}{T_2}\bar{\rho}_{vc}^{(0|1)} = -iE_2^*(t)e^{-i\boldsymbol{k}_2\cdot\boldsymbol{r}}\mu_{vc}(1 - 2\rho_{cc}^{(0|0)}). \tag{10.11}$$

Its solution is

$$\bar{\rho}_{vc}^{(0|1)} = -\frac{i}{\hbar}\eta_2^*\theta(\tau)\theta(t-\tau)\mu_{vc}e^{-i\boldsymbol{k}_2\cdot\boldsymbol{r}}\left(1-\frac{\Theta_1^2}{2}e^{-\tau/T_1}\right)e^{i(\omega_{cv}-\omega_L)(t-\tau)}e^{-(t-\tau)/T_2}.$$
(10.12)

Finally, we obtain, for the polarization $\bar{P}^{(0|1)}$ radiating into direction $\boldsymbol{k}_2$,

$$\bar{P}^{(0|1)}(t,\tau) = \left(1-\frac{\Theta_1^2}{2}e^{-\tau/T_1}\right)\theta(\tau)\bar{P}_2^{(0|1)}(t-\tau),$$
(10.13)

where $\bar{P}_2^{(0|1)}(t-\tau)$ is the linear response to pulse No. 2 alone, i.e.,

$$\bar{P}_2^{(0|1)}(t-\tau) = -\frac{i}{\hbar}\eta_2^*\theta(t-\tau)|\mu_{vc}|^2 e^{-i\boldsymbol{k}_2\cdot\boldsymbol{r}}e^{i(\omega_{cv}-\omega_L)(t-\tau)}e^{-(t-\tau)/T_2}.$$
(10.14)

The pump–probe signal is therefore given by the linear response to pulse No. 2, $\bar{P}_2^{(0|1)}$, and an additional term, $\propto -\frac{\Theta_1^2}{2}\bar{P}_2^{(0|1)}$, which is the product of a reduction factor $-\frac{\Theta_1^2}{2}$, which is second order in the light field times the linear response to pulse No. 2. This reduction factor describes the beginning inversion of the two-level system as a response to the pump pulse No. 1. It is a nonlinear effect due to phase-space filling which is a consequence of the Pauli principle. The differential absorption $\delta\alpha(\omega,\tau>0)$ is solely represented by this second term.

In order to obtain the differential absorption, we only have to calculate the linear optical spectrum as a response to pulse No. 2 alone and to multiply this result by the reduction factor. Note that the linear spectrum in question is given by the linear response $P_2^{(0|1)}(t-\tau)$ to the $E_2(t) = \eta_2\delta(t-\tau)$-pulse No. 2 arriving at time $t=\tau$ at the system, i.e.,

$$P_2^{(0|1)}(t-\tau) = \eta_2\chi(t-\tau),$$
(10.15)

with (omitting now the spatial phase factor)

$$\chi(t) = \frac{i}{\hbar}\theta(t)|\mu_{vc}|^2 e^{-i\omega_{cv}t-t/T_2}.$$
(10.16)

Following again the steps introduced in the previous chapter, we find for the differential absorption

$$\delta\alpha(\omega,\tau>0) \propto -\frac{\Theta_1^2}{2\hbar}e^{-\tau/T_1}\frac{(\pi T_2)^{-1}}{(\omega-\omega_{cv})^2+T_2^{-2}}.$$
(10.17)

Hence, the differential absorption of a homogeneously broadened ensemble of identical two-level absorbers is given by a negative Lorentzian. This kind of signal is characteristic for so-called *bleaching*. The strength of the bleaching

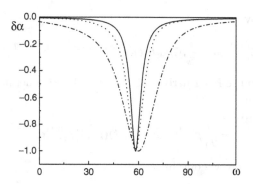

Fig. 10.1. *Solid line*: differential absorption of a homogeneously broadened ensemble of two-level absorbers. The resonance frequency is $\omega_{cv} = 60$ and $T_2^{-1} = 7$. The pump and probe fields have Gaussian envelopes with widths $\delta_1 = 1$ and $\delta_2 = 0.0012$, respectively. *Dotted line*: bleaching of an ensemble of inhomogeneously broadened two-level absorbers, see Fig. 10.5, with otherwise identical model parameters. Note the larger width for the inhomogeneous situation. In both cases, the time delay between pump and probe pulses is $\tau = 5$, i.e., there is no temporal pulse overlap. The *dash-dotted line* corresponds to the same system used for the *dotted line*, however, now the pulses have temporal overlap, $\tau = 1$. The additional broadening is due to the optical Stark effect which contributes to the differential absorption in this situation

increases with increasing Rabi angle Θ_1 of the pump pulse, i.e., it is proportional to the intensity of the pump pulse. With increasing delay τ, the bleaching decreases due to the finite lifetime T_1 of the population. The width of the bleaching spectrum is determined by the dephasing time T_2 for a purely homogeneously broadened ensemble. A numerically calculated example is shown in Fig. 10.1.

10.1.2 Coherent Spectral Oscillations

So far, we have considered the situation where the pump pulse arrives at $t = 0$ before the probe pulse, which arrives at $t = \tau > 0$. However, also the time-reversed situation yields a signal in the direction $\boldsymbol{k}_2$ of the probe pulse.

This time, we assume that the probe pulse arrives at time $t = 0$ from direction $\boldsymbol{k}_2$ and excites a linear polarization. The pump pulse now comes after the probe, i.e., at time $t = \tau > 0$ from the direction $\boldsymbol{k}_1$. Together with the polarization left from the probe, the delayed pump generates a third-order polarization contribution that also radiates into direction $\boldsymbol{k}_2$.

The linear polarization excited by the probe pulse, which we still call No. 2 even though it arrives first, is determined by

$$\hbar \frac{d\bar{\rho}_{vc}^{(0|1)}}{dt} - i\hbar(\omega_{cv} - \omega_L)\bar{\rho}_{vc}^{(0|1)} + \frac{\hbar}{T_2}\bar{\rho}_{vc}^{(0|1)} = -iE_2^*(t)e^{-i\boldsymbol{k}_2\cdot\boldsymbol{r}}\mu_{vc}, \qquad (10.18)$$

which is solved by

$$\bar{\rho}_{vc}^{(0|1)} = -\frac{i}{\hbar}\eta_2^*\mu_{vc}\theta(t)e^{-i\boldsymbol{k}_2\cdot\boldsymbol{r}}e^{i(\omega_{cv}-\omega_L)t-t/T_2}. \tag{10.19}$$

Now, the pump pulse No. 1 arrives at $t = \tau > 0$ and generates a population, which follows from

$$\hbar\frac{d\rho_{cc}^{(-1|1)}}{dt} + \frac{\hbar}{T_1}\rho_{cc}^{(-1|1)} = iE_1(t)(\mu_{vc})^*\bar{\rho}_{vc}^{(0|1)}e^{i\boldsymbol{k}_1\cdot\boldsymbol{r}}. \tag{10.20}$$

Solving this equation, we obtain

$$\rho_{cc}^{(-1|1)} = \frac{1}{\hbar^2}\eta_1\eta_2^*|\mu_{vc}|^2\theta(t-\tau)\theta(\tau)e^{i(\boldsymbol{k}_1-\boldsymbol{k}_2)\cdot\boldsymbol{r}}e^{-t/T_1}e^{i(\omega_{cv}-\omega_L)\tau-\tau/T_2}. \tag{10.21}$$

Finally, the equation of motion for the differential interband coherence (second order in the pump and first order in the probe field) is determined by

$$\hbar\frac{d\delta\bar{\rho}_{vc}^{(0|1)}}{dt} - i\hbar(\omega_{cv} - \omega_L)\bar{\rho}_{vc}^{(0|1)} + \frac{\hbar}{T_2}\bar{\rho}_{vc}^{(0|1)} = -iE_1^*(t)e^{-i\boldsymbol{k}_1\cdot\boldsymbol{r}}\mu_{vc}(1 - 2\rho_{cc}^{(-1|1)}), \tag{10.22}$$

which yields

$$\delta\bar{\rho}_{vc}^{(0|1)} = -\frac{i}{\hbar^3}|\eta_1|^2\eta_2^*\theta(\tau)\theta(t-\tau)\mu_{vc}|\mu_{vc}|^2e^{-i\boldsymbol{k}_2\cdot\boldsymbol{r}}e^{-\tau/T_1}e^{i(\omega_{cv}-\omega_L)t}e^{-t/T_2}. \tag{10.23}$$

Transforming back into the fixed frame and to the frequency domain, we obtain for the differential susceptibility

$$\delta\chi(\omega) \propto i\int_{-\infty}^{\infty}e^{i\omega t}\theta(t-\tau)e^{-i\omega_{cv}t-t/T_2}dt$$
$$= i\frac{e^{i(\omega-\omega_{cv})\tau-\tau/T_2}}{i(\omega-\omega_{cv}) - \frac{1}{T_2}}. \tag{10.24}$$

Note that the lower limit of the integration is now τ rather than zero as for the "normal" temporal order of the pulses. This leads to an additional τ-dependent phase factor in the differential susceptibility. The imaginary part, i.e., the differential absorption $\delta\alpha$, is

$$\delta\alpha(\omega,\tau > 0) \propto \Im\left[i\frac{e^{i(\omega-\omega_{cv})\tau-\tau/T_2}}{i(\omega-\omega_{cv}) - \frac{1}{T_2}}\right]$$
$$= e^{-\frac{\tau}{T_2}}\frac{-\frac{1}{T_2}\cos\left((\omega_{cv}-\omega)\tau\right) + (\omega_{cv}-\omega)\sin\left((\omega_{cv}-\omega)\tau\right)}{(\omega_{cv}-\omega)^2 + (\frac{1}{T_2})^2}. \tag{10.25}$$

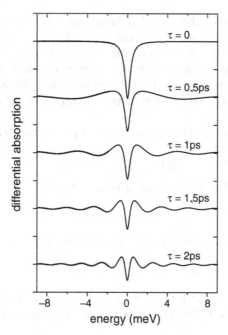

Fig. 10.2. Differential absorption spectra are shown for resonant pumping of a system of identical two-level absorbers for various time delays τ between the pump and the probe pulse. The spectra have been obtained by a numerical solution of the optical Bloch equations. When the probe pulse precedes the pump pulse the differential absorption exhibits spectral oscillations with a spectral period which is inversely proportional to the time delay τ. When τ approaches zero, these spectral oscillations vanish and the differential absorption develops into a purely negative bleaching. [After T. Meier and S.W. Koch, "Foundations of coherent transients in semiconductors", in *Encyclopedia of Modern Optics*, eds. B. Guenther, L. Bayvel, and D.G. Steel (Elsevier, Amsterdam, 2004) pp. 163–173]

We note the following features of this result, which are depicted in Fig. 10.2. The signal is symmetric around $\omega = \omega_{cv}$. For $\tau = 0$, a Lorentzian bleaching $\propto -(1/(\omega_{cv} - \omega)^2 + (1/T_2)^2)$ develops. It oscillates as a function of ω with a spectral period $\propto (1/\tau)$. These oscillations are called *coherent spectral oscillations*. They originate from the interaction of the pump field with the coherent probe polarization. On the other hand, the bleaching we have found for the "normal" temporal order of the pump and probe pulses is due to the interaction of the probe field with the pump-induced population.

10.1.3 Optical Stark Effect

So far we have discussed pump–probe situations where the pump is resonant with the two-level absorbers. There is another variant of pump–probe experiments, where the pump has a frequency that is detuned from the excitation energy of the two-level absorber. For large detuning, i.e., $(\omega_{cv} - \omega_L) \gg T_2^{-1}$ and $(\omega_{cv} - \omega_L) \gg \delta\omega_1$, where $\delta\omega_1$ is the spectral width of the pump pulse, the excitation of an isolated two-level absorber is only present during the time the pump pulse is acting on the system. This regime is called *adiabatic following*. If a probe pulse arrives during this time regime, the polarization in the direction of the probe pulse reflects the driven excitation of the system due to the pump pulse.

To illustrate this situation, let us assume that the pump light is stationary, i.e., given by the field

$$E(t) = E_1 e^{i\omega_L t} + c.c. \tag{10.26}$$

with constant E_1. The probe is given by a $\delta(t)$ pulse

$$E_2(t) = \eta_2 \delta(t), \tag{10.27}$$

with real η_2. We define the detuning of the pump field by

$$\Delta_\omega = \omega_{cv} - \omega_L \tag{10.28}$$

and assume the coherent limit $(2T_1 = T_2)$ and weak excitation intensities. This allows us to use (7.56) for the nonlinear polarization $\bar{\rho}_{vc}(t)$

$$\hbar\frac{d\bar{\rho}_{cv}}{dt} - i\hbar\Delta_\omega\bar{\rho}_{cv} + \frac{\hbar}{T_2}\bar{\rho}_{vc} = -iE_l^*(t)e^{-i\mathbf{k}_l\cdot\mathbf{r}}\mu_{vc}(1 - 2|\bar{\rho}_{vc}|^2). \tag{10.29}$$

The first term in brackets on the right-hand side of (10.29), i.e., $\propto 1$, is the linear polarization without the pump field. The remaining term, $\propto -2|\bar{\rho}_{vc}|^2$, determines the third-order differential polarization δP which induces the differential absorption given by $\delta\bar{\rho}_{cv}$. Thus, we have

$$\hbar\frac{d\delta\bar{\rho}_{cv}}{dt} - i\hbar\Delta_\omega\delta\bar{\rho}_{cv} + \frac{\hbar}{T_2}\delta\bar{\rho}_{vc} = 2iE_l^*(t)e^{-i\mathbf{k}_l\cdot\mathbf{r}}\mu_{vc}|\bar{\rho}_{vc}|^2. \tag{10.30}$$

There are two driving terms for $\delta\bar{\rho}_{vc}(t)$ radiating into direction $\mathbf{k}_2$:

(i) The stationary pump field E_1 creates a stationary linear polarization $\bar{\rho}_{vc}^{(1|0)}$, which enters as $|\bar{\rho}_{vc}^{(1|0)}|^2$ in the right-hand side of (10.30). In this case, we have to use $E_l^*(t) = E_2(t) = \eta_2\delta(t)$ with $\mathbf{k}_l = \mathbf{k}_2$.

(ii) The stationary pump field E_1 creates a stationary linear polarization $\bar{\rho}_{vc}^{(1|0)}$. Its complex conjugate is multiplied by $\bar{\rho}_{vc}^{(0|1)}$, which is due to the probe pulse. In this case, we have to use $E_l^*(t) = E_1$ with $\mathbf{k}_l = \mathbf{k}_1$.

In both cases, the spatial phase factors on the right-hand side of (10.30) combine to give $\exp(-i\mathbf{k}_2 \cdot \mathbf{r})$. Inserting the terms discussed in (i) and (ii), we have

$$\hbar\frac{d\delta\bar{\rho}_{cv}}{dt} - i\hbar\Delta_\omega\delta\bar{\rho}_{cv} + \frac{\hbar}{T_2}\delta\bar{\rho}_{vc} = 2i\mu_{vc}(E_1^* e^{-i\mathbf{k}_1\cdot\mathbf{r}}\bar{\rho}_{vc}^{(0|1)}(\bar{\rho}_{vc}^{(1|0)})^*$$
$$+ E_2^*(t)e^{-i\mathbf{k}_2\cdot\mathbf{r}}|\bar{\rho}_{vc}^{(1|0)}|^2). \tag{10.31}$$

From (10.3), we obtain the linear polarization as the response to the probe pulse as

$$\bar{\rho}_{vc}^{(0|1)} = -\frac{i}{\hbar}\eta_2\mu_{vc}\theta(t)e^{-i\boldsymbol{k}_2\cdot\boldsymbol{r}}e^{i\Delta_\omega t-t/T_2} \tag{10.32}$$

and the stationary linear response to the pump field is given by (note that $d\bar{\rho}_{vc}^{(1|0)}/dt = 0$)

$$-i\hbar\Delta_\omega\bar{\rho}_{vc}^{(1|0)} + \frac{\hbar}{T_2}\bar{\rho}_{vc}^{(1|0)} = -iE_1^*e^{-i\boldsymbol{k}_1\cdot\boldsymbol{r}}\mu_{vc}, \tag{10.33}$$

which leads to

$$\bar{\rho}_{vc}^{(1|0)} = \frac{1}{\hbar}E_1^*\mu_{vc}\frac{e^{-i\boldsymbol{k}_1\cdot\boldsymbol{r}}}{\Delta_\omega + \frac{i}{T_2}}. \tag{10.34}$$

With the Rabi frequency, see (7.42), $\Omega_1 = 2|E_1\mu_{vc}|/\hbar$, we have

$$|\bar{\rho}_{vc}^{(1|0)}|^2 = \frac{\Omega_1^2}{4}\frac{1}{\Delta_\omega^2 + \frac{1}{T_2^2}}. \tag{10.35}$$

Inserting these results into (10.31), we obtain

$$\frac{d\delta\bar{\rho}_{cv}}{dt} - i\Delta_\omega\delta\bar{\rho}_{cv} + \frac{1}{T_2}\delta\bar{\rho}_{vc} = \frac{\Omega_1^2\Theta_2}{4}e^{-i\boldsymbol{k}_2\cdot\boldsymbol{r}}\left(\theta(t)\frac{e^{i\Delta_\omega t-t/T_2}}{\Delta_\omega - \frac{i}{T_2}} + i\delta(t)\frac{1}{\Delta_\omega^2 + \frac{1}{T_2^2}}\right), \tag{10.36}$$

which is solved by

$$\delta\bar{\rho}_{vc} = \theta(t)\frac{\Omega_1^2\Theta_2}{4}e^{i\Delta_\omega t-t/T_2}e^{-i\boldsymbol{k}_2\cdot\boldsymbol{r}}$$

$$\times \int_{-\infty}^{t}\left(\theta(t')\frac{1}{\Delta_\omega - \frac{i}{T_2}} + i\delta(t')\frac{1}{\Delta_\omega^2 + \frac{1}{T_2^2}}\right)dt', \tag{10.37}$$

where we have introduced the Rabi angle $\Theta_2 = 2\eta_2\mu_{vc}/\hbar$. Evaluating the integral yields

$$\delta\bar{\rho}_{vc} = \theta(t)\frac{\Omega_1^2\Theta_2}{4}e^{i\Delta_\omega t-t/T_2}e^{-i\boldsymbol{k}_2\cdot\boldsymbol{r}}$$

$$\times \left(\frac{t}{\Delta_\omega - \frac{i}{T_2}} + \frac{i}{\Delta_\omega^2 + \frac{1}{T_2^2}}\right). \tag{10.38}$$

The spectrum is obtained by multiplying with $\mu_{vc}\exp(i\omega_L t)$, taking the complex conjugate, and Fourier transforming. Omitting the spatial phase factor $\exp(-i\boldsymbol{k}_2\cdot\boldsymbol{r})$, we obtain

$$\delta\alpha(\omega) \propto \Im \frac{\mu_{vc}\Omega_1^2\Theta_2}{4} \left(\frac{1}{\Delta_\omega + \frac{i}{T_2}} \int_0^\infty t e^{-i(\omega_{cv}-\omega)t-t/T_2} dt \right.$$

$$\left. - \frac{i}{\Delta_\omega^2 + \frac{1}{T_2^2}} \int_0^\infty e^{-i(\omega_{cv}-\omega)t-t/T_2} dt \right)$$

$$= \Im \frac{\mu_{vc}\Omega_1^2\Theta_2}{4} \left(\frac{-1}{(\Delta_\omega + \frac{i}{T_2})(\omega - \omega_{cv} + \frac{i}{T_2})^2} \right.$$

$$\left. - \frac{1}{(\Delta_\omega^2 + \frac{1}{T_2^2})(\omega - \omega_{cv} + \frac{i}{T_2})} \right). \tag{10.39}$$

The first term is proportional to

$$\Im \frac{-1}{(\Delta_\omega + \frac{i}{T_2})(\omega - \omega_{cv} + \frac{i}{T_2})^2} = \frac{((\omega - \omega_{cv})^2 - \frac{1}{T_2^2}) + 2(\omega - \omega_{cv})\Delta_\omega}{T_2(\Delta_\omega^2 + \frac{1}{T_2^2})\{(\omega - \omega_{cv})^2 + \frac{1}{T_2^2}\}^2} \tag{10.40}$$

and the second term is proportional to

$$\Im \frac{1}{(\Delta_\omega^2 + \frac{1}{T_2^2})(\omega - \omega_{cv} + \frac{i}{T_2})} = -\frac{\frac{1}{T_2}}{(\Delta_\omega^2 + \frac{1}{T_2^2})\{(\omega - \omega_{cv})^2 + \frac{1}{T_2^2}\}}. \tag{10.41}$$

We assume that the detuning of the pump field Δ_ω is large compared to the homogeneous width $1/T_2$, i.e., $\Delta_\omega T_2 \gg 1$. Apart from a bleaching term, (10.41), and a contribution symmetrical with respect to $\omega = \omega_{cv}$, (10.40), the dominant contribution to the differential absorption is a dispersive signal

$$\frac{2}{\Delta_\omega T_2} \frac{\omega - \omega_{cv}}{\{(\omega - \omega_{cv})^2 + \frac{1}{T_2^2}\}^2}. \tag{10.42}$$

Equation (10.42) corresponds to a differential blue shift since $\delta\alpha(\omega) > 0$ for $\omega > \omega_{cv}$ if $\Delta_\omega = \omega_{cv} - \omega_L > 0$, i.e., if the frequency of the pump field is lower than the resonance frequency of the two-level absorber. If one pumps at a frequency above that of the two-level absorber $\Delta_\omega = \omega_{cv} - \omega_L < 0$, we find a red shift. This result reminds one of the repulsion of quantum-mechanical levels due to a mutual interaction. In fact, there is a strong correspondence between quantum-mechanical level repulsion and the optical Stark effect, which becomes directly obvious if one uses a description with a quantized light field.

The coherent optical Stark effect is illustrated in Fig. 10.3. A homogeneously broadened ensemble of two-level absorbers is pumped by a spectrally narrow pulse energetically below the material resonance. The short probe pulse covers the entire spectral range of interest. The differential absorption has a dispersive character indicating a blue shift of the absorption line.

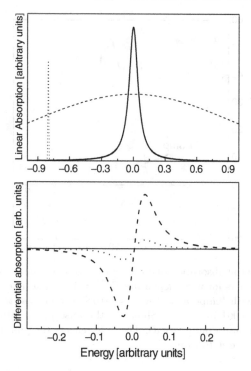

Fig. 10.3. Optical Stark effect for an ensemble of homogeneously broadened two-level absorbers. *Upper figure*: spectra of the pump (*dotted*) and probe (*dashed*) light fields, respectively, and the homogeneously broadened linear spectrum (*solid*). *Lower figure*: differential absorption showing the blue shift for two different excitation intensities

In Fig. 10.1, the differential absorption of an inhomogeneous ensemble of two-level absorbers, see Fig. 10.5, is shown for the case where the pump and probe pulses have temporal overlap. Compared to the situation with temporally separated pulses, an additional broadening is seen. This is due to the optical Stark effect, which is active if the pulses overlap. It leads to a blue (red) shift for systems that have resonance frequencies higher (lower) than the pump field, which results in an effective broadening of the line.

10.1.4 Hole Burning

In an ensemble of two-level systems which have transition energies distributed according to some distribution function, i.e., in an ensemble with an inhomogeneous spectrum, the excitation by $\delta(t)$ pulses leads to the same reduction factor for each system. On the other hand, if pulse No. 1 with a finite duration δ_1 is applied, it has a finite spectral width, such that only those systems are (partially) saturated which have resonances that overlap with

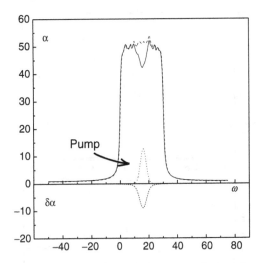

Fig. 10.4. Differential absorption of an inhomogeneously broadened system of $N = 500$ two-level absorbers for weak dephasing $T_2^{-1} = 0.7$ and a time delay of $\tau = 0.5$. The pump pulse with temporal width $\delta_1 = 0.5$ is centered at $\omega = 16$, the probe pulse with $\delta_2 = 0.001$ at $\omega = 0$. Shown is the absorption without pump pulse (*dashed*) and with pump pulse (*solid*). The *hole* reflects the pump pulse spectrum, which is shown by the *dotted line*

the spectral width of the pump pulse. The spectrum of the response to the second short pulse then shows a "hole" for those frequencies, which are contained in the Fourier transform of pulse No. 1. This phenomenon is called hole burning and is a typical signature of inhomogeneous ensembles of two-level absorbers. The origin of this spectral hole burning is the Pauli blocking effect, since those transitions that are (partially) saturated by the first pulse are no longer available for transitions tested by the probe. An example is shown in Fig. 10.4.

In Fig. 10.5, a hole-burning scenario is presented for an ensemble of inhomogeneously and homogeneously broadened two-level absorbers, where the same dephasing time $T_2 = 1/7$ is used as in Fig. 10.1. The laser spectrum of the pump has been taken to be narrower than the homogeneous width, see the dotted line. Again, we see a hole whose spectral width is in this case determined by the homogeneous broadening and not by the pump laser spectrum. However, the differential absorption has a larger width than that of the purely homogeneous ensemble underlying Fig. 10.1, although the same T_2 has been taken. This is due to the fact that not only the systems with resonance frequencies exactly at the laser frequency become influenced by the pump pulse, also those which, due to their finite homogeneous width, have some spectral overlap with the pump field, experience some degree of bleaching.

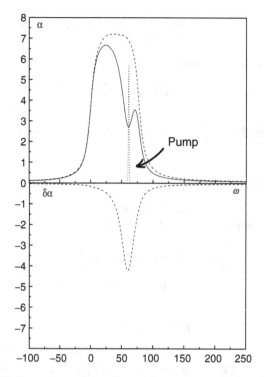

Fig. 10.5. *Solid line*: absorption of an inhomogeneously broadened system of $N = 500$ two-level absorbers with dephasing $T_2^{-1} = 7$ under the influence of the pump field. A hole is burned at the position of the pump laser spectrum $\omega_1 = 60$. The narrow pump laser spectrum corresponding to the large temporal width $\delta_1 = 1$ is indicated by the *dotted line*. The *dashed line* shows the absorption without pump field. In the *lower figure* the corresponding differential absorption $\delta\alpha$ is shown. Compare this with the bleaching of the purely homogeneous ensemble with otherwise identical parameters in Fig. 10.1

10.1.5 Four-Wave-Mixing Experiments

In the self-diffraction geometry of a four-wave-mixing experiment, the first pulse arrives at the sample from direction $\mathbf{k}_1$ at time $t = 0$. Later, at time $t = \tau$, a second pulse from direction $\mathbf{k}_2$ interacts with the already excited system. Experimentally, one measures the nonlinear signal which propagates into the direction $2\mathbf{k}_2 - \mathbf{k}_1$. Due to this direction selection of the measured signal, the first pulse enters once, while the second pulse enters the equations of motion twice.

We start from the set of (7.29) and solve it in first order in the first light pulse which arrives at time $t = 0$ from the direction $\mathbf{k}_1$. In this order, $\rho_{vv} = 1$ and $\rho_{cc} = 0$. Thus, from

$$\hbar \frac{d\bar{\rho}_{vc}^{(1|0)}}{dt} - i\hbar(\omega_{cv} - \omega_L)\bar{\rho}_{vc}^{(1|0)} + \frac{\hbar}{T_2}\bar{\rho}_{vc}^{(1|0)} = -iE_1^*(t)e^{-i\boldsymbol{k}_1\cdot\boldsymbol{r}}\mu_{vc} \qquad (10.43)$$

we find, using again a $\delta(t)$ pulse

$$E_1(t) = \eta_1\delta(t) \qquad (10.44)$$

that

$$\bar{\rho}_{vc}^{(1|0)} = -\frac{i}{\hbar}\eta_1^*\mu_{vc}\theta(t)e^{-i\boldsymbol{k}_1\cdot\boldsymbol{r}}e^{i(\omega_{cv}-\omega_L)t}e^{-t/T_2}, \qquad (10.45)$$

i.e.,

$$\bar{P}_1^{(1|0)}(t) = -\frac{i}{\hbar}\eta_1^*\theta(t)e^{-i\boldsymbol{k}_1\cdot\boldsymbol{r}-i\omega_L t}|\mu_{vc}|^2 e^{i\omega_{cv}t}e^{-t/T_2}. \qquad (10.46)$$

In second order (first pulse No. 1 with $\boldsymbol{k}_1$ at time $t = 0$, then second pulse No. 2 with $\boldsymbol{k}_2$ at time $t = \tau$), we have to solve

$$\hbar\frac{d\rho_{cc}}{dt} = iE_2(t)e^{i\boldsymbol{k}_2\cdot\boldsymbol{r}}(\mu_{vc})^*\bar{\rho}_{vc}^{(1|0)}$$
$$-iE_2^*(t)e^{-i\boldsymbol{k}_2\cdot\boldsymbol{r}}\mu_{vc}(\bar{\rho}_{vc}^{(1|0)})^*, \qquad (10.47)$$

with

$$E_2(t) = \eta_2\delta(t - \tau). \qquad (10.48)$$

Note that the first term on the right-hand side leads to a spatial phase factor $\exp\{i(\boldsymbol{k}_2 - \boldsymbol{k}_1)\cdot\boldsymbol{r}\}$, i.e., to a contribution $\rho_{cc}^{(1|-1)}$, while the second term yields $\rho_{cc}^{(-1|1)}$. It is only this second term which contributes to the signal which radiates into direction $2\boldsymbol{k}_2 - \boldsymbol{k}_1$, because the driving term in (7.29) has a phase factor $\exp(-i\boldsymbol{k}_2\cdot\boldsymbol{r})$ only for excitation by the second pulse. The corresponding solution of (10.47) reads

$$\rho_{cc}^{(-1|1)} = -\frac{i}{\hbar}\eta_2^*\theta(t - \tau)(\bar{\rho}_{vc}^{(1|0)}(\tau))^*\mu_{vc}e^{-i\boldsymbol{k}_2\cdot\boldsymbol{r}}. \qquad (10.49)$$

Finally, the equation of motion for the interband coherence to third order in the light field is

$$\hbar\frac{d\bar{\rho}_{vc}^{(-1|2)}}{dt} - i\hbar(\omega_{cv} - \omega_L)\bar{\rho}_{vc}^{(-1|2)} = -2iE_2^*(t)e^{-i\boldsymbol{k}_2\cdot\boldsymbol{r}}\mu_{vc}\rho_{cc}^{(-1|1)}, \qquad (10.50)$$

with the solution

$$\bar{\rho}_{vc}^{(-1|2)} = \frac{2}{\hbar^2}|\eta_2|^2\theta(t - \tau)e^{i(\omega_{cv}-\omega_L)(t-\tau)}\mu_{vc}^2(\bar{\rho}_{vc}^{(1|0)}(\tau))^*e^{-2i\boldsymbol{k}_2\cdot\boldsymbol{r}}$$
$$= 2\frac{i}{\hbar}\eta_2^*\bar{\rho}_{vc}^{(0|1)}(t - \tau)\mu_{vc}(\bar{\rho}_{vc}^{(1|0)}(\tau))^*e^{-i\boldsymbol{k}_2\cdot\boldsymbol{r}}. \qquad (10.51)$$

Note that the first term of (10.47) would have produced a contribution $\propto \exp(-i\boldsymbol{k}_1\cdot\boldsymbol{r})$ instead of the required one $\propto \exp(-i(2\boldsymbol{k}_2 - \boldsymbol{k}_1)\cdot\boldsymbol{r})$. Therefore, we obtain

$$\bar{P}^{(-1|2)}(t,\tau) = 2\frac{i}{\hbar}\eta_2^*\bar{P}_2^{(0|1)}(t-\tau)(\bar{P}_1^{(1|0)}(\tau))^*e^{-ik_2\cdot r}, \qquad (10.52)$$

which leads to a signal radiating into the direction $2k_2 - k_1$.

The time-integrated trace is obtained by integrating the detected intensity over time

$$TI \propto \int_{-\infty}^{\infty} |\bar{P}^{(-1|2)}(t,\tau)|^2 dt, \qquad (10.53)$$

which is now solely a function of τ, while the time-resolved trace is simply the intensity

$$TR \propto |\bar{P}^{(-1|2)}(t,\tau)|^2. \qquad (10.54)$$

This is a function which depends on time t and parametrically on the delay τ.

Assuming $\delta(t)$ pulses, the third-order polarization is a product of two first-order polarizations, see (10.52). The first one, $\bar{P}_1^{(1|0)}$, results from the first pulse and is taken at the time of the arrival of the second pulse, $t = \tau$. Furthermore, for radiation into direction $2k_2 - k_1$, the complex conjugate of this function enters. This corresponds to a phase conjugated contribution of the response to the first pulse to the final third-order polarization. This term is multiplied by $\bar{P}_2^{(0|1)}(t - \tau)$, i.e., by the linear response to the second pulse. The resulting contribution further develops in time and, thus, gives rise to the time-resolved trace. The τ-dependent factor $\bar{P}_1^{(1|0)}(\tau)$, on the other hand, determines the time-integrated trace, since $\bar{P}_2^{(0|1)}(t - \tau)$ is integrated from $t = \tau$ to $t \to \infty$ and is, thus, independent of τ.

We note that phase conjugation lies at the heart of four-wave mixing. It should not be confused with time reversal. This statement is going to become clear in Sect. 10.2.3, where we discuss quantum beats.

A slight modification of the four-wave-mixing experiment leads to a realization of a phase conjugating mirror. Instead of counting the second pulse in direction k_2 twice, as done here, one can apply a simultaneous pair of counterpropagating pulses, i.e., one in direction k_2 and another one in direction $-k_2$. The solution for $\bar{P}(t)$ then contains a term where these two spatial phase factors cancel: $k_2 + (-k_2)$ instead of $k_2 + k_2$. The resulting signal is emitted in the "phase conjugated" direction $-k_1$, i.e., it counterpropagates to the incoming first pulse.

Equation (10.52) is very instructive since it nicely shows that at least for a two-level system both the time-resolved and the time-integrated traces result from the linear polarization, i.e., they are identical up to factors. This feature is a consequence of the $\delta(t)$-pulse excitation assumed in the present analysis. For finite width of the light pulses, the proportionality between time-integrated and time-resolved signals is no longer valid and one has to solve the equations of motion numerically using a suitable code for evaluating the coupled differential equations. However, the fundamental features of the signals can be understood already on the basis of this simplified model.

10.1.6 Polarization Interference

Imagine an ensemble composed of independent two-level absorbers, which consists of two subensembles, one with resonance energies $\hbar\omega_1$ and the other one with resonance energies $\hbar\omega_2$. We consider a four-wave-mixing experiment with $\delta(t)$ pulses in the self-diffraction mode and study the signal in direction $2\mathbf{k}_2 - \mathbf{k}_1$.

Since the individual systems are completely independent of each other, we can solve the Bloch equation for each system separately and obtain a superposition of the two signals, i.e., from the subensemble with resonance frequency ω_1 and from the second one with resonance frequency ω_2. The result is the source of the radiation of all the systems in the ensemble, which is collected by the detector, i.e.,

$$\bar{P}^{(-1|2)}(t,\tau) \propto \bar{P}^{(0|1)}_{\omega_1}(t-\tau)(\bar{P}^{(1|0)}_{\omega_1}(\tau))^* + \bar{P}^{(0|1)}_{\omega_2}(t-\tau)(\bar{P}^{(1|0)}_{\omega_2}(\tau))^*, \quad (10.55)$$

where the subscripts ω_1 and ω_2 refer to the linear polarizations of the two subensembles. For the present system, the signal in the detector is given by

$$\bar{P}^{(-1|2)}(t,\tau) \propto 2e^{i\omega_0(t-2\tau)} \cos\frac{\Delta}{2}(t-2\tau)e^{-t/T_2}, \quad (10.56)$$

where we have used

$$\omega_0 = \frac{\omega_1 + \omega_2}{2},$$

$$\Delta = \omega_2 - \omega_1. \quad (10.57)$$

The signal seen by the detector is, therefore, modulated both as a function of t and of τ. The intensity proportional to $\cos^2\frac{\Delta}{2}(t-2\tau)$ is illustrated schematically as a function of t and τ in Fig. 10.6. It has a maximum at $t = 2\tau$ and along parallel lines, i.e., whenever

$$\frac{\Delta}{2}(t-2\tau) = n\pi, \quad (10.58)$$

where $n = 0, \pm1, \pm2, \ldots$. This modulation of the signal due to interference in the detector is called "polarization interference".

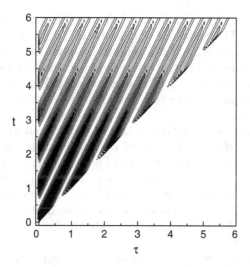

Fig. 10.6. t-τ plot for an ensemble consisting of two subensembles of two-level absorbers characterized by resonance frequencies ω_1 and ω_2, respectively. There are only signals above the diagonal $t = \tau$. The *parallel lines* indicate the maxima of $\cos^2 \frac{\Delta}{2}(t - 2\tau)$. Note that only for sufficiently small T_2, does the time-integrated signal show temporal modulations

10.1.7 Photon Echoes

Photon echoes of inhomogeneous ensembles of two-level systems are a direct analog of spin echoes of spin-1/2 systems. We again consider a four-wave-mixing experiment with $\delta(t)$ pulses performed on an ensemble of independent two-level systems with resonance frequencies ω_i distributed according to some distribution function $G(\omega_i)$, e.g., a Gaussian distribution. Again, we solve the Bloch equations for each subsystem separately and add up the resulting signal at the end. The result is similar to (10.55), except that now we have a sum over all frequencies which appear in the inhomogeneous ensemble, i.e.,

$$\bar{P}^{(-1|2)}(t,\tau) \propto \sum_i \bar{P}^{(0|1)}_{\omega_i}(t - \tau)(\bar{P}^{(1|0)}_{\omega_i}(\tau))^*, \qquad (10.59)$$

which in the continuum limit can be written as an integral

$$\bar{P}^{(-1|2)}(t,\tau) \propto \int G(\omega)\bar{P}^{(0|1)}(\omega, t - \tau)(\bar{P}^{(1|0)}(\omega,\tau))^* d\omega, \qquad (10.60)$$

where we have indicated the frequency in the argument of $\bar{P}$. Inserting the phase factors gives

$$\bar{P}^{(-1|2)}(t,\tau) \propto \int G(\omega)e^{i\omega(t-2\tau)}d\omega e^{-t/T_2}$$
$$= G(t - 2\tau)e^{-t/T_2}, \qquad (10.61)$$

where $G(t)$ is the Fourier transform of the distribution function $G(\omega)$. It is peaked at $t = 0$. Therefore, the signal seen by the detector is peaked at a time $t = 2\tau$. It is an echo which appears spontaneously and delayed after pulse No. 2. Its temporal width is inversely proportional to the spectral width of the distribution function or of the excitation pulse, if pulses are chosen that have a spectrum narrower than the inhomogeneous line of the ensemble of two-level systems.

The appearance of the echo can also be understood as a generalization of polarization interference. Instead of the single cosine function in (10.56) leading to parallel lines in the t-τ plot in Fig. 10.6, we now have many cosine functions in the sum over the amplitudes, see (10.59) or (10.60). These cosines have a dense spectrum of frequencies which lead to destructive interference for all times except at $t = 2\tau$, because here the values of the various frequencies do not matter at all. In this situation, only the line $t = 2\tau$ survives, see Fig. 10.7, which is the photon echo.

Note that, in contrast to the presentation of many textbooks about spin echoes, it is not necessary to apply π and $\pi/2$ pulses. The echo exists for any pulse intensity. For example, in our present treatment, we assumed low excitation levels (third-order theory), such that the pulses lead to Rabi angles much smaller than π.

Since the echo appears at the time $t = 2\tau$, its maximum intensity decays at $t = 2\tau$ according to

$$I_{\text{echo}}(t = 2\tau) \propto \left(e^{-\frac{2\tau}{T_2}}\right)^2 = e^{-4\tau/T_2}. \tag{10.62}$$

This result is also valid for the time-integrated signal, whereas the decay of the intensity I_{homog} from a homogeneous ensemble, i.e., all resonance frequencies are identical, is given by

$$I_{\text{homog}}(t) \propto e^{-2t/T_2}. \tag{10.63}$$

This results in a time-integrated trace that decays with the same exponential law $\propto \exp\left(-2\tau/T_2\right)$.

The gradual weakening of the photon echo with increasing time delay τ can be used to determine the dephasing time T_2, i.e., the homogeneous linewidth of inhomogeneously distributed ensembles. For the ensemble of two-level systems considered here, four-wave mixing thus gives the same information as hole burning in pump–probe experiments. However, the echo experiment is performed in the time domain, while hole-burning experiments provide spectral information.

At this point, we define what we understand when we use the term dephasing: The signature of optical dephasing is the decay of photon echoes with increasing delay time. We encounter situations where the equations of motion

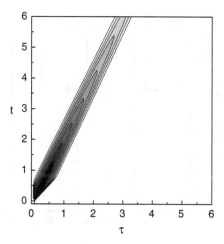

Fig. 10.7. t-τ plot for an ensemble consisting of subensembles of two-level absorbers characterized by resonance frequencies ω distributed according to a continuous distribution function $G(\omega)$. Only one line of maxima at $t = 2\tau$ survives due to destructive interference of all the other contributions for $n \neq 0$ in (10.58)

do not contain any phenomenological damping terms and still the echo intensity decays with increasing τ, however, often in a nonexponential manner. In this case, we also say that the ensemble in question shows optical dephasing, although the physically responsible dephasing mechanism may not be a trivial and obvious one.

10.2 Coherent $\chi^{(3)}$ Processes for Few-Level Systems

The coherent $\chi^{(3)}$ processes for few-level systems differ remarkably from that of ensembles of two-level systems. A prototype of such a system is the three-level system depicted in Fig. 10.8. Starting from a common ground state, optical excitation is possible into two excited states with resonance energies $\hbar\omega_1$ and $\hbar\omega_2$, and optical dipole matrix elements μ_{g1} and μ_{g2}, respectively. The linear optical spectrum of this system is exactly identical to that of two two-level systems, with the same resonance energies $\hbar\omega_1$ and $\hbar\omega_2$, and the same dipole matrix elements μ_{g1} and μ_{g2}, respectively. However, the pump–probe spectra and four-wave-mixing traces for the two systems are quite different. The same applies to systems with more than two excited states and, in particular, for systems with a continuous spectrum.

For ensembles of two-level systems, we have already studied the nonlinear optical response, both in the pump–probe and in the four-wave-mixing mode. To illustrate the corresponding features of few-level systems, we restrict ourselves to configurations where we have just one single ground state $|g\rangle$ but M' excited states. In this case, the equations of motion are, see (7.71) and (7.72),

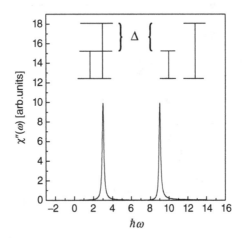

Fig. 10.8. A three-level absorber, a pair of two-level absorbers, and their identical linear spectra

$$\hbar\frac{d\bar{\rho}_{g\nu'}}{dt} - i\hbar(\omega_{\nu'g} - \omega_L)\bar{\rho}_{g\nu'} + \frac{\hbar}{T_2}\bar{\rho}_{g\nu'} = -iE_l^*(t)e^{-i\boldsymbol{k}_l\cdot\boldsymbol{r}}$$

$$\times(\mu_{g\nu'}\rho_{gg} - \sum_{\mu'}^{M'}\mu_{g\mu'}\rho_{\mu'\nu'}),$$

$$\hbar\frac{d\rho_{gg}}{dt} = iE_l^*(t)e^{-i\boldsymbol{k}_l\cdot\boldsymbol{r}}\sum_{\kappa'=1'}^{M'}\mu_{g\kappa'}(\bar{\rho}_{g\kappa'})^*$$

$$-iE_l(t)e^{i\boldsymbol{k}_l\cdot\boldsymbol{r}}\sum_{\kappa'=1'}^{M'}(\mu_{g\kappa'})^*\bar{\rho}_{g\kappa'}$$

$$\hbar\frac{d\rho_{\nu'\mu'}}{dt} - i(\epsilon_{\mu'} - \epsilon_{\nu'})\rho_{\nu'\mu'} = iE_l(t)e^{i\boldsymbol{k}_l\cdot\boldsymbol{r}}(\mu_{g\nu'})^*\bar{\rho}_{g\mu'},$$

$$-iE_l^*(t)e^{-i\boldsymbol{k}_l\cdot\boldsymbol{r}}\mu_{g\mu'}(\bar{\rho}_{g\nu'})^*.$$

$$(10.64)$$

10.2.1 Pump–Probe Experiments

As an illustration, let us consider an ensemble composed of identical three-level systems, see left-hand inset in Fig. 10.8. We excite with a short pulse No. 1 at time $t = 0$ from direction $\boldsymbol{k}_1$. At $t = \tau$, we record the linear response to pulse No. 2 which comes from the direction $\boldsymbol{k}_2$. In order to be able to perform analytical calculations, we use $\delta(t)$ pulses, see (10.1) and (10.10). In this case, however, we encounter the undesirable feature that both transitions with energies ω_1 and ω_2 are excited and consequently partly inverted to a degree which is determined by the magnitude of the dipole matrix elements μ_{g1} and μ_{g2}. To avoid this effect, we use the following trick: when pulse No. 1 arrives, we switch off the transition into state $|2\rangle$ by setting $\mu_{g2} = 0$. However, at the time of arrival of pulse No. 2, this transition is allowed. We choose $\mu_{g1} = \mu_{g2} = \mu$ for the following analysis.

The linear response result for the interband coherence $\bar{\rho}_{g1}^{(1|0)}$ is given by (10.3), while $\bar{\rho}_{g2}^{(1|0)} = 0$. For the intraband quantities, we obtain

$$\rho_{gg}^{(0|0)} = 1 - |\eta_1|^2|\mu|^2\theta(t)/\hbar^2,$$
$$\rho_{11}^{(0|0)} = |\eta_1|^2|\mu|^2\theta(t)/\hbar^2,$$
$$\rho_{12}^{(0|0)} = 0,$$
$$\rho_{21}^{(0|0)} = 0,$$
$$\rho_{22}^{(0|0)} = 0, \tag{10.65}$$

due to our choice $\mu_{g2} = 0$ at the time of the arrival of pulse No. 1.

The linear response to pulse No. 2 is now calculated with $\mu_{g2} = \mu$ as

$$\bar{\rho}_{g1}^{(0|1)} = \theta(t-\tau)\frac{i}{\hbar}\frac{\eta_2^*}{\hbar^2}\mu e^{-i\mathbf{k}_2\cdot\mathbf{r}}e^{i(\omega_1-\omega_L)(t-\tau)}(1-2|\eta_1|^2|\mu|^2),$$
$$\bar{\rho}_{g2}^{(0|1)} = \theta(t-\tau)\frac{i}{\hbar}\frac{\eta_2^*}{\hbar^2}\mu e^{-i\mathbf{k}_2\cdot\mathbf{r}}e^{i(\omega_2-\omega_L)(t-\tau)}(1-|\eta_1|^2|\mu|^2). \tag{10.66}$$

Again, as in the pump–probe results for two-level systems, we obtain the usual expression for the linear response result, however, with bleaching due to the reduction terms $(1 - 2|\eta_1|^2|\mu|^2)$ and $(1 - |\eta_1|^2|\mu|^2)$ for the interband coherence pertaining to transitions $|g\rangle \rightarrow |1\rangle$ and $|g\rangle \rightarrow |2\rangle$, respectively. Note that although the second transition has not been excited by the first pulse, it nevertheless is bleached, however, only by half of the amount of the first transition.

If we had calculated the pump–probe signals for an ensemble of two-level systems consisting of two subensembles having resonance energies $\hbar\omega_1$ and $\hbar\omega_2$ and dipole matrix elements $\mu_{g1} = \mu$ and $\mu_{g2} = 0$, and μ, respectively, by the same method, the signal with frequency ω_2 would show no reduction factor. We see that the common ground state in the three-level system leads to a coupling of the transitions, which becomes visible in nonlinear optical response.

Of course, the situation studied here is somewhat artificial. In order to demonstrate the effect of the internal coupling, we should rather apply a first pulse that has a sufficiently narrow spectrum, i.e., is temporally long enough,

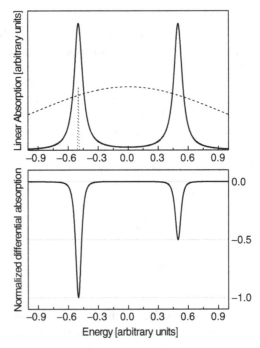

Fig. 10.9. Pump–probe differential absorption for an ensemble of identical three-level systems. The *upper figure* shows the linear absorption (*solid*), the spectrum of the pump (*dotted*) and that of the probe pulse (*dashed*), respectively. The *lower figure shows* the differential absorption. The high-energy transition which is not excited by the pump pulse is also bleached, however, only by half the amount of the low-energy transition

such that it only excites the transition $|g\rangle \rightarrow |1\rangle$ resonantly. The second pulse can be taken as a very short pulse covering both transitions simultaneously. This situation can, however, only be treated numerically. The result of such numerical calculations is shown in Fig. 10.9. We find a qualitatively similar result as in our analytical treatment, cf. (10.66).

Consider now a system that has many $M' \gg 1$ upper levels forming a quasi-continuum. Naively, one would expect that the first pulse burns a hole into the spectrum, corresponding to its own spectral width, see the discussion for ensembles of two-level systems. However, we demonstrate in Fig. 10.10 that *all* transitions are affected by a spectrally narrow pump, although most of them are not directly excited by the first pulse. This behavior is due to two reasons: (i) the internal coupling via the common ground state, cf. Fig. 10.9, and (ii) the coupling of the upper levels by the intraband coherences $\rho_{\nu'\mu'}$ for $\nu' \neq \mu'$.

The additional bleaching in the spectral range of the pump pulse is clearly seen for small M', e.g., in Fig. 10.9 for $M' = 2$ and in Figs. 10.10a and b. If

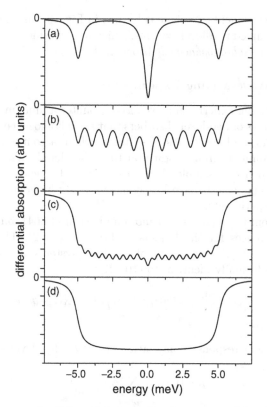

Fig. 10.10. Pump–probe differential absorption for an ensemble of systems having a single ground state and M' upper levels. (a) $M' = 3$, (b) $M' = 11$, (c) $M' = 21$, (d) $M' = 101$. For a small number of upper states, the bleaching, due to upper states populated by the pump pulse, is still clearly visible by the reduced absorption in the center of the spectrum, cf. also Fig 10.9 for $M' = 2$. This additional bleaching is reduced due to destructive interference mediated by the intraband coherences and, thus, vanishes for large M'. This results in a uniform bleaching of all transitions, even if most of them are not directly populated by the pump pulse. The parameters are: $t_1 = -6\,\mathrm{ps}$, $t_2 = 0\,\mathrm{ps}$, $\delta_1 = 2\,\mathrm{ps}$, $\delta_2 = 10\,\mathrm{fs}$, and $T_2 = 2\,\mathrm{ps}$

M' increases, destructive interference mediated by the intraband coherences reduces this feature. For large M', the uniform bleaching due to the depopulation of the ground state alone determines the differential absorption. This observation agrees with that of our simple model treated above. There the nondiagonal intraband coherences were all zero, see (10.65), since we used $\mu_{g2} = 0$.

In the limiting case of continua constructed by many $M' \gg 1$ densely spaced upper levels that are optically coupled to a common ground state, we obtain a behavior resembling *homogeneously* broadened optical lines in pump–probe experiments. No hole is burned in Fig. 10.10d, in contrast to ensembles of uncoupled two-level systems, which have the identical linear optical spectrum but are *inhomogeneously* broadened.

10.2.2 Four-Wave-Mixing Experiments

In four-wave-mixing experiments, we also encounter a fundamental difference between ensembles of two-level absorbers that are inhomogeneously broadened on the one hand, and ensembles of systems, where a continuum is optically coupled to a common ground state on the other hand. While the inhomogeneously broadened ensembles have been shown to yield echo signals, the continuum systems also behave like homogeneously broadened lines in four-wave-mixing experiments.

We start from the equations of motion given in (10.64), but calculate the interband coherences for the kinematic direction $(-1|2)$. The procedure is already known from the treatment of two-level systems, so we only present the result. It is formally identical to (10.52),

$$\bar{P}^{(-1|2)}(t,\tau) = \frac{2i}{\hbar}\eta_2^*\bar{P}_2^{(0|1)}(t-\tau)(\bar{P}_1^{(1|0)}(\tau))^*e^{-i\mathbf{k}_2\cdot\mathbf{r}}, \tag{10.67}$$

however, the linear response polarization of a given few-level system is now

$$\bar{P}^{(0|1)}(t) = \sum_{\nu'=1}^{M'} \mu_{g\nu'}^* \bar{\rho}_{g\nu'}^{(0|1)}. \tag{10.68}$$

This linear polarization enters (10.67) and determines the total four-wave-mixing polarization. This has to be contrasted with (10.59), where the sum over the nonlinear polarizations of the independent two-level absorbers has been taken.

10.2.3 Quantum Beats

It is instructive to study four-wave-mixing traces for an ensemble of three-level systems. We use (10.67) and (10.68) with $M' = 2$. The ensemble has a linear spectrum which is identical to that discussed in connection with polarization interference, see Sect. 10.1.6. However, for the four-wave-mixing polarization, we now obtain, instead of (10.56),

$$\bar{P}^{(-1|2)}(t,\tau) \propto 4e^{i\omega_0(t-2\tau)}\cos\frac{\Delta}{2}(t-\tau)\cos\frac{\Delta}{2}\tau e^{-t/T_2}, \tag{10.69}$$

where $\Delta = \omega_2 - \omega_1$ is the energy difference between the upper levels. This result is illustrated in Fig. 10.11 (compare with Fig. 10.6 for the polarization interference), where the maxima of the four-wave-mixing intensity are indicated

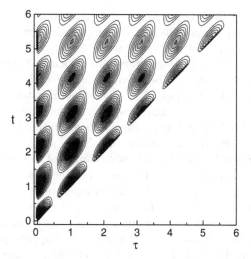

Fig. 10.11. t-τ plot for an ensemble consisting of three-level absorbers characterized by two resonance frequencies ω_1 and ω_2. There is only a signal above the diagonal $t = \tau$. Note that even for $T_2 = \infty$, the time-integrated signal shows temporal modulations

in the t-τ plane. As in the case of polarization interference, there are modulations which are determined by the splitting Δ. Since, for the present case, these modulations originate from a quantum-mechanically coupled system, they are called *quantum beats*.

Compared to polarization interference, the picture, however, looks quite different. There are certain values of delay τ

$$\tau = (2n+1)\pi/\Delta, \tag{10.70}$$

for $n = 0, 1, 2$, etc., at which there is no signal at any time t. The maximum signal is obtained for values of τ in between these zeros. The signal is modulated also as a function of t, as is the signal for polarization interference. In the latter case, the time-resolved signal has also modulations with the same frequency, however, its maximum amplitude is independent of τ and always of the same height (in the case $T_2 \to \infty$).

It is interesting to study the time-integrated signal as a function of τ if a phenomenological dephasing term $\propto 1/T_2$ is added to the equation of motion of the interband coherence. This produces an additional factor $\exp{(-t/T_2)}$ in the resulting polarization. The time-integrated trace for polarization interference is significantly modulated only for T_2 less than typically Δ^{-1}. The modulation depth decreases with increasing T_2. In contrast, for quantum beats, the time integrated trace always has maximal modulation depth. As a further difference, we note that time-resolved quantum beat signals always start at $t = \tau$ with a maximum, while polarization interferences start with any value between the maximum and zero, depending on τ.

A possible realization of quantum beat systems are, e.g., asymmetric semiconductor double quantum wells. If the energetically lowest levels in the two potential wells for the electrons are close in energy and the barrier between the wells is not too thick, the eigenstates are delocalized over both wells and we have the situation already discussed in Chap. 4 with respect to coherent tunneling. By designing the potential profile appropriately and applying an external bias, it is possible to excite only the left-hand electronic states by a short laser pulse No. 1 in such structures. This excitation is given by a superposition of the two electron eigenstates and also contains the hole state in the left-hand quantum well. The subsequent dynamics of the electron is the periodic motion discussed in Chap. 4. The period of the oscillation is determined by the energetic separation of the two upper eigenstates.

Pulse No. 2, having the same frequency as pulse No. 1 and arriving after a delay time τ, then leads to a nonlinear signal only if the amplitude of the electron state in the left-hand well is large. This happens periodically with the above mentioned period. The periodic modulation of the time-integrated four-wave-mixing trace is thus a signature of the periodic spatial motion of the electron amplitude reflecting coherent tunneling. On the other hand, one can view this scenario also as quantum beats of the three-level system composed of the hole state of the left-hand well and the two delocalized eigenstates of the electron.

We have seen that the occurrence of echoes in an inhomogeneous ensemble of two-level systems can be understood as destructive interference of the oscillating polarizations of the systems with differing resonance frequencies for all times except for $t \approx 2\tau$, where they interfere constructively. The seemingly analogous situation is an ensemble of three-level systems with Δ differing from system to system, see Fig. 10.12. We randomly take the splittings Δ of the upper levels from a uniform distribution in the interval $\Delta_0 - W < \Delta < \Delta_0 + W$. The four-wave-mixing trace for the subensemble of systems that have the same splitting 2Δ is the modulated signal given by (10.69). The total signal shows the characteristics of an echo. It is a superposition of all the amplitudes resulting in a signal that initially decays on a time scale of the reciprocal width of the distribution of the splittings Δ or of the reciprocal spectral width of the pulse, whatever is smaller. Its intensity has a constant long-time value of $1/2$. See Fig. 10.13 for an illustration of this situation.

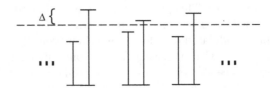

Fig. 10.12. An ensemble of three-level systems with varying energy splitting Δ

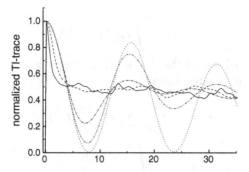

Fig. 10.13. Time-integrated signal for an ensemble of three-level absorbers with varying upper-level separation. The *lines* correspond to $W = 0$ (*dotted*), $W = 0.05$ (*dash-double-dotted*), $W = 0.1$ (*dash-dotted*), $W = 0.5$ (*dashed*), and $W = 1$ (*solid*). For vanishing variation $W = 0$, there are quantum beats which decay only because of the introduced dephasing rate $T_2^{-1} = 0.001$. For finite variation of the splitting of the upper levels $W \neq 0$, there is a decay towards a constant and nonzero value 0.5 (up to fluctuations due to the incomplete configurational average over $N = 800$ three-level systems)

10.2.4 Dephasing due to Continua

As a generalization, we consider more and more upper states with random energies. To be more specific, let us consider an ensemble of $M' + 1$-level systems. The ground state is optically coupled to M' upper states which have excitation energies $\omega_0 + \Delta_l^{(i)}$, where the $\Delta_l^{(i)}$ are drawn from a distribution function $G(\Delta)$. Here, i labels the individual system and $l = 1, \ldots, M'$ numbers the upper levels. The linear polarization of system i is (up to a certain prefactor A, which does not matter here)

$$\bar{P}_i^{(1|0)}(t) = A \sum_{l=1}^{M'} e^{i\Delta_l^{(i)}t}, \tag{10.71}$$

while the four-wave-mixing signal for this system is, see (10.67),

$$\bar{P}_i^{(-1|2)}(t,\tau) = A \sum_{l=1}^{M'} e^{i\Delta_l^{(i)}(t-\tau)} \sum_{k=1}^{M'} e^{-i\Delta_k^{(i)}\tau}. \tag{10.72}$$

At zero delay, $\tau = 0$, we have

$$\bar{P}_i^{(-1|2)}(t,0) = AM' \sum_l e^{i\Delta_l^{(i)}t}, \tag{10.73}$$

such that the total macroscopic polarization of the ensemble is

$$\sum_i \bar{P}_i^{(-1|2)}(t,0) = AM' \sum_l \langle e^{i\Delta_l^{(i)}t} \rangle = AM'^2 G(t), \tag{10.74}$$

where

$$G(t) = \int G(\Delta)e^{i\Delta t}d\Delta \tag{10.75}$$

is the Fourier transform of the distribution of upper level energies. Consequently, the time-integrated signal for $\tau = 0$ is

$$TI = A^2 \int_{\tau=0}^{\infty} |\sum_i \bar{P}_i^{(-1|2)}(t,0)|^2 dt \tag{10.76}$$

$$= A^2 M'^4 \int_0^{\infty} G^2(t)dt$$

$$= A^2 \frac{M'^4}{2} \int_{-\infty}^{\infty} G^2(t)dt.$$

On the other hand, for $\tau \neq 0$, we can write instead of (10.72)

$$\bar{P}_i^{(-1|2)}(t,\tau) = A \sum_l e^{i\Delta_l^{(i)}(t-2\tau)} + A \sum_{l \neq k} e^{i\Delta_l^{(i)}(t-\tau)-i\Delta_k^{(i)}\tau}. \tag{10.77}$$

From this expression, we obtain for the macroscopic polarization

$$\sum_i \bar{P}_i^{(-1|2)}(t,\tau) = AM'G(t-2\tau) + AM'^2 G(t-\tau)G(\tau). \tag{10.78}$$

Here, the second term vanishes for τ larger than the temporal width of $G(t)$, while the first term yields an echo at $t = 2\tau$ in the time-resolved trace. For $\tau \to \infty$, the time-integrated signal is given by

$$TI = A^2 \int_{\tau}^{\infty} |\sum_i \bar{P}_i^{(-1|2)}(t,\tau)|^2 dt \to A^2 M'^2 \int_{-\infty}^{\infty} G^2(t)dt. \tag{10.79}$$

The factor $1/2$ appearing in (10.77) is now missing since here the integration covers the whole temporal width of $G^2(t)$ due to the echo structure of the first term of (10.78). Therefore, we learn that the time-integrated signal approaches a saturation value for large τ that is by a factor $2/M'^2$ smaller than the initial value at $\tau = 0$. The more upper levels exist, the lower is the time-integrated signal for large τ. Figure 10.13 shows a special case for $M' = 2$.

In particular, for $M' \to \infty$, we may replace the sum over the M' upper levels in (10.71) by an integral over the continuum and obtain for the linear polarization

$$\bar{P}^{(1|0)}(t) = \int G(\omega)\mu^*(\omega)\bar{\rho}^{(1|0)}(\omega,t)d\omega, \tag{10.80}$$

using the same notation as in (10.59) and (10.60). Note that $G(\omega)|\mu^*(\omega)|^2$ is the linear optical spectrum of the ensemble. Equation (10.80) gives, up

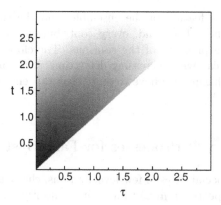

Fig. 10.14. t-τ plot for an ensemble consisting of many-level absorbers modeling an upper continuum. There is only a signal above the diagonal $t = \tau$. This signal decays both for $t \to \infty$ and $\tau \to \infty$, even without any phenomenological dephasing, i.e., for $T_2 \to \infty$. Thus, both the time-integrated and time-resolved traces are decaying due to intrinsic destructive interference among the optical excitations

to a factor, the Fourier transform of the linear spectrum $G(\omega)|\mu^*(\omega)|^2$ since $\bar{\rho}^{(0|1)}(\omega, t) \propto \exp(i\omega t)$. Assume that this spectrum has a Lorentzian form. Then, its Fourier transform describes an exponential decay $\propto \exp(-t/T_0)$, where T_0 is inversely proportional to the width of the spectrum. According to (10.67), the four-wave-mixing trace is then exponentially decaying for both increasing t and τ. There is only a negligible echo contribution. See also (10.78) in the limit $M' \to \infty$. For an illustration of this case see Fig. 10.14.

For other shapes of the linear spectrum, one can obtain decays which are, however, not simply exponential. The analysis presented here has, again, been performed assuming excitation $\delta(t)$ pulses, which do not have a finite spectral width. For realistic pulses with a finite spectral width, the decay of the four-wave-mixing signal is given precisely by the duration of the pulses, provided the linear spectrum of the system is broad and completely unstructured. This shows that four-wave-mixing traces obtained by exciting continua that are coupled optically to a common ground state only reflect the temporal shape of the excitation pulses.

We have now learned that the decay of optical signals can be due to continua, if a continuum of states is optically coupled to a common ground state. (Remember, a continuum given by an ensemble of independent two-level systems does not lead to decays, but to echoes.) The reason for this decay is destructive interference of internally coupled excitations with a dense spectrum of excitation energies. A particularly intriguing and not uncommon situation is the following: Consider an ensemble of few-level systems each having a single ground state and a continuous band of upper states with finite energetic width, i.e., a generalization of the ensemble shown in Fig. 10.12. The average excitation energy per system is assumed to be randomly distributed. This is an inhomogeneous feature of the ensemble, and the time-resolved traces show echoes. On the other hand, every contribution to the echo signal is decaying due to the presence of the continua. Therefore, we have, as a result, echoes that decay with increasing delay time τ. According to our definition of optical dephasing, we have to state that this ensemble shows optical dephasing.

10.3 Coherent $\chi^{(3)}$ Processes for Fano Systems

Fano systems are special cases of few-level systems, characterized by a particular linear optical spectrum $|\mu(E)|^2$ and, concomitantly, by a particular linear polarization $\bar{P}^{(1|0)}(t)$. For a pump–probe experiment, we therefore predict, on the basis of the previous discussion, that hole burning is not possible in an ensemble of identical Fano systems.

The time-integrated and time-resolved four-wave-mixing traces are both determined by the linear polarization, see (10.67), provided the excitation is performed by δ pulses, see Figs. 9.4, 9.5, and 9.6 for examples. For short excitation pulses of sufficiently large spectral width, additional features become visible, as, e.g., the π-phase shift in a Fano situation, where an optically forbidden discrete transition is coupled to an unstructured, optically allowed continuum, see the discussion in Sect. 9.3.1. This is demonstrated in Figs. 10.15 and 10.16, where both the time-resolved and the time-integrated traces are shown for excitation with Gaussian pulses having a width $\delta = 0.5$. For delay times τ that are sufficiently large to prevent significant pulse overlap, the calculated time-resolved trace consists of two distinct regimes. At short times, the continuum excitation produces a rapidly decaying structure reflecting the temporal pulse shape, while at later times, the signal is determined by the decay Γ appearing in the function $L(t)$, see (9.38). For a Fano parameter $q = 0$, these two regimes are separated by a zero, reflecting the π-phase shift due to the interference term $(q + i)^2$. For increasing q, the zero converts into a dip (see dashed line in Fig. 10.15), which disappears for large q. This zero or dip in the time-resolved trace is therefore the signature of the Fano interference for the single-state situation. For $\delta(t)$ pulses, this signature is absent, and the trace is given by a simple exponential, since in this situation the interference term $(q + i)^2$ only enters as a prefactor. For small τ, we see an additional zero at shorter times in Figs. 10.15, which is due to the pulse overlap. It is again produced by the π-phase shift occurring during the arrival of the second pulse.

The corresponding time-integrated trace is shown in Fig. 10.16. As pulse overlap effects are integrated over, the time-integrated trace approximately resembles the time-resolved trace for large τ. This emphasizes that the time-integrated trace more closely reflects the linear polarization than the time-resolved trace for finite pulse overlap.

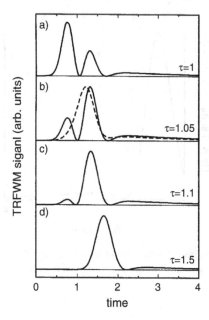

Fig. 10.15. Time-resolved traces for a Fano system with a single discrete state. The transition into this discrete state is taken to be forbidden, $q_1 = 0$. The coupling to the continuum is $\Gamma_1 = 0.5$ and the temporal pulse width is $\delta = 0.5$. Two zeros are seen and their temporal position depends on the delay between the pulses τ. The *dashed line* in (**b**) corresponds to $q_1 = 0.5$. [From T. Meier et al., Phys. Rev. **B51**, 13977 (1995)]

10.4 Homogeneous versus Inhomogeneous Ensembles

From our discussion of two- and few-level systems, we obtain a number of far-reaching insights into the nonlinear optical properties of such ensembles. These findings are summarized here.

- Ensembles
  - An "inhomogeneous ensemble" of two-level systems is realized if the optical resonance energies ω_i are randomly distributed according to a certain distribution function $G(\omega_i)$. Its spectrum is given by the distribution function $G(\omega_i)$ weighted by the respective optical dipole matrix elements squared.
  - We call an ensemble of identical two-level systems a "homogeneous ensemble" if the damping of its polarization is determined by a "dephasing time" T_2.
  - These two models can be combined, i.e., we can have an inhomogeneous ensemble of two-level systems all characterized by the same dephasing time. We call this ensemble a "mixed ensemble".

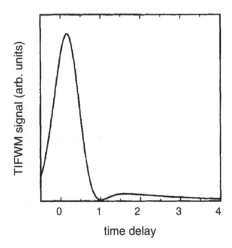

Fig. 10.16. Time-integrated trace for the situation of Fig. 10.15. The π-phase change is clearly seen as a minimum at $\tau = 1$. [From T. Meier et al., Phys. Rev. **B51**, 13977 (1995)]

- Pump–probe
  - In pump–probe experiments, it is possible to burn holes into the spectrum of inhomogeneous ensembles. The holes have a spectral width which is inversely proportional to the duration of the pump pulse, i.e., $\propto \delta^{-1}$. The hole burned into the spectrum of a mixed ensemble is either given by the dephasing time (if $T_2 < \delta$) or by the duration of the pulse (if $T_2 > \delta$).
  - No holes can be burned in spectra of homogeneous ensembles, provided the pump pulse is represented by a simple sinusoidal dependence on time multiplied by some Gaussian or similar positive definite envelope function. For other pump pulses (not discussed here), holes can also be burned in homogeneous lines.
  - Pump–probe experiments performed on ensembles of M'-level systems, where $M' \to \infty$, such that the upper states form a continuum, show exactly the same signatures as those performed on homogeneous ensembles. In particular, no holes can be burned into the spectrum.
- Four-wave mixing
  - Dephasing is defined by the decay of the time-integrated signal with increasing τ.
  - The time-resolved traces of inhomogeneous ensembles show echoes appearing at twice the delay time τ relative to the first pulse. Their temporal width is determined by the spectral width of the ensemble if the spectral pulse width δ^{-1} is larger than the spectral width of the ensemble. In the opposite case, the width is determined by δ^{-1}. Since the amplitude of the photon echoes is independent of the delay, also

the time-integrated traces are independent of the delay time τ (for τ larger than δ), i.e., there is no dephasing.

- Echoes can be understood to be due to polarization interference of many independent radiations from individual two-level systems. These interferences are destructive for all times t except those in the vicinity of twice the delay time $t \approx 2\tau$.

- For mixed ensembles, the amplitudes of the echoes decay with increasing delay time, i.e., the time-integrated trace decays. This ensemble shows dephasing given by the dephasing time T_2.

- Ensembles composed of identical three-level systems show quantum beats, which differ in a characteristic way from polarization interferences of ensembles of two-level systems having only two different resonance frequencies.

- The quantum beats always start with a maximum at $t = \tau$. Their amplitude is modulated as a function of τ, which is only partly true for polarization interference if there is a sufficiently small dephasing time.

- This observation clearly shows that the four-wave-mixing signals are not due to time-reversal (but due to phase conjugation).

- The time-integrated and time-resolved signals both show a decay for ensembles composed of $M' \to \infty$ systems that model a continuum of excited states and there are no echoes. According to our definition of dephasing, this ensemble shows dephasing and, therefore, looks like a homogeneous ensemble, although no phenomenological dephasing time is introduced.

- The dephasing can be understood to be a consequence of destructive interference of quantum beats for times larger than the reciprocal width of the excited spectrum.

- Ensembles composed of few-level systems with different upper-level energies show both echoes and a decay with increasing τ, at least initially (depending on the number of optically excited upper levels). Therefore, they behave like mixed ensembles.

It is evident that the identification of ensembles in terms of "homogeneous" or "inhomogeneous" ensembles from coherent nonlinear optical measurements is, even for these relatively simple examples, not unique. Quite different physical situations have similar signatures in their experimental data. On the other hand, ensembles that seem to be not so different from each other can lead to extremely different experimental signatures. Note that we are able to trace back the appearance both of echoes and of dephasing (in ensembles without phenomenological dephasing) to destructive interference, in the first case, of polarization interference and, in the second case, of quantum beats.

In the following, we extend the discussion to the semiconductor model and consider both disorder and many-particle interactions. In this case, the identification of the model in terms of a "homogeneous" or "inhomogeneous" ensemble is even more problematic. This is related to the fact that a clear distinction of polarization interference and quantum beats is generally not possible in a strict manner. These notations can only be applied as limiting cases in realistic semiconductor models.

10.5 Suggested Reading

1. I.D. Abella, N.A. Kurnit, and S.R. Hartmann, "Photon echoes", Phys. Rev. **141**, 391 (1966)
2. R. Binder, S.W. Koch, M. Lindberg, W. Schäfer, and F. Jahnke, "Transient many-body effects in the semiconductor optical Stark effect: A numerical study", Phys. Rev. **B43**, 6520 (1991)
3. K. Bott, O. Heller, D. Bennhardt, S.T. Cundiff, P. Thomas, E.J. Mayer, G.O. Smith, R. Eccleston, J. Kuhl, and K. Ploog, "Influence of exciton–exciton interactions on the coherent optical response in GaAs quantum wells", Phys. Rev. **B48**, 17418 (1993)
4. K. Bott, E.J. Mayer, G.O. Smith, V. Heuckeroth, M. Hübner, J. Kuhl, T. Meier, A. Schulze, M. Lindberg, S.W. Koch, P. Thomas, R. Hey, K. Ploog, "Dephasing of interacting heavy-hole and light-hole excitons in GaAs quantum wells", J. Opt. Soc. Am. B **13**, 1026 (1996)
5. S.W. Koch, N. Peyghambarian, and M. Lindberg, "Transient and steady state optical nonlinearities in semiconductors", J. Phys. C: Solid State Phys. **21**, 5229 (1988)
6. M. Koch, J. Feldmann, G. von Plessen, E.O. Göbel, and P. Thomas, "Quantum beats versus polarization interference: An experimental distinction", Phys. Rev. Lett. **69**, 3631 (1992)
7. J. Kuhl, E.J. Mayer, G. Smith, R. Eccleston, D. Bennhardt, P. Thomas, K. Bott, and O. Heller, "Polarization dependence of degenerate four-wave-mixing on 2D excitons in GaAs quantum wells", Semicond. Sci. Technol. **9**, 429 (1994)
8. N.A. Kurnit, I.D. Abella, and S.R. Hartmann, "Observation of a photon echo", Phys. Rev. Lett. **13**, 567 (1964)
9. K. Leo, J. Shah, T. C. Damen, A. Schulze, T. Meier, S. Schmitt-Rink, P. Thomas, E.O. Göbel, S.L.Chuang, M.S. C. Luo, W. Schäfer, K. Köhler, and P. Ganser, "Dissipative dynamics of an electronic wavepacket in a semiconductor double well potential", IEEE J. Quantum Electronics **28**, 2498 (1992)
10. M. Lindberg and S.W. Koch, "Coherent oscillations and dynamic Stark effect in semiconductors", Phys. Rev. **B38**, 7607 (1988)
11. M. Lindberg and S.W. Koch, "Theory of coherent transients in semiconductor pump–probe spectroscopy", J. Opt. Soc. Am. **B5**, 139 (1988)

12. M. Lindberg and S.W. Koch, "Theory of the optical Stark effect in semiconductors under ultrashort-pulse excitation", Phys. Stat. Sol. **B150**, 379 (1988)

13. V.G. Lyssenko, J. Erland, I. Balslev, K.-H. Pantke, B.S. Razbirin, and J. M. Hvam, "Nature of nonlinear four-wave-mixing beats in semiconductors", Phys. Rev. **B 48**, 5720-5723 (1993)

14. E.J. Mayer, G.O. Smith, V. Heuckeroth, J. Kuhl, K. Bott, A. Schulze, T. Meier, D. Bennhardt, S.W. Koch, P. Thomas, R. Hey, and K. Ploog, "Evidence of biexcitonic contributions to four-wave mixing in GaAs quantum wells", Phys. Rev. **B50**, 14730 (1994)

15. E.J. Mayer, G.O. Smith, V. Heuckeroth, J. Kuhl, K. Bott, A. Schulze, T. Meier, S.W. Koch, P. Thomas, R. Hey, and K. Ploog, "Polarization dependence of beating phenomena at the energetically lowest exciton transition in GaAs quantum wells", Phys. Rev. **B51**, 10909 (1995)

16. T. Meier, A. Schulze, P. Thomas, H. Vaupel, and K. Maschke, "Signature of Fano-resonances in four-wave mixing experiments", Phys. Rev. **B51**, 13977 (1995)

17. G. Noll, U. Siegner, S. Shevel, and E.O. Göbel, "Picosecond stimulated photon echo due to intrinsic excitations in semiconductor mixed crystals", Phys. Rev. Lett. **64**, 792 (1990)

18. M. Reichelt, C. Sieh, T. Meier, and S.W. Koch, "Comparison of the differential absorption obtained within a few-level model and the microscopic density-matrix theory", Phys. Stat. Sol. **B221**, 249 (2000)

19. S. Schmitt-Rink, D. Bennhardt, V. Heuckeroth, P. Thomas, P. Haring G. Maidorn, H. Bakker, K. Leo, D.S. Kim, J. Shah, and K. Köhler, "Polarization dependence of heavy- and light-hole quantum beats", Phys. Rev. **B46**, 10460 (1992)

20. G.O. Smith, E.J. Mayer, V. Heuckeroth, J. Kuhl, K. Bott, T. Meier, A. Schulze, D. Bennhardt, S.W. Koch, P. Thomas, R. Hey, and K. Ploog, "Polarization selection rules for quantum beating between light and heavy hole excitons in GaAs quantum wells", Solid State Commun. **94**, 373 (1995)

21. A. Wurger, *From Coherent Tunnelling to Relaxation: Dissipative Quantum Dynamics of Interacting Defects*, Springer Tracts in Modern Physics, Vol. 135 (Springer, Berlin 1997)

22. T. Yajima and Y. Taira, "Spatial optical parametric coupling of picosecond light pulses and transverse effect in resonant media", J. Phys. Soc. Jpn. **47**, 1620 (1979)

11

Coherent $\chi^{(3)}$ and $\chi^{(5)}$ Processes in Ordered Semiconductors

In the previous chapter, we analyzed coherent nonlinear optical processes for level systems. For these simple systems, the various dynamical features can be understood relatively easily. In nature, such systems are realized, e.g., in molecularly doped solids and, *cum grano salis*, in semiconductors with dominant excitonic resonances or those possessing strong disorder.

It is the purpose of the present chapter to learn to what extent the simple concepts and features discussed for level systems are relevant for ordered semiconductors. Semiconductor optics cannot be described properly without the Coulomb interaction. In the ordered case, the linear spectrum already shows excitonic resonances, as is evident from the Wannier equation, see (8.64) and (9.45) and also Fig. 9.13. When calculating coherent $\chi^{(3)}$ signals, we have to use the equations of motion (8.54) and (8.56), or alternatively (8.58) and (8.59), which have been derived in Sect. 8.2.3. These sets of equations extend the semiconductor Bloch equations beyond the Hartree–Fock approximation, which is given in (8.62). However, in certain cases, already the Hartree–Fock equations give reasonable results.

A remark is in order to clarify the notion of "excitonic resonance" or "exciton". What is meant here in most cases is the optical resonance related to the lowest lying bound electron–hole state in the ordered case. In three-dimensional semiconductors, this state is classified by the quantum number $1s$ and the higher ones by $2s$, $3s$, etc., see Sect. 9.4.1. In our one-dimensional model, the notion of s states is not as well defined as in three- or two-dimensional systems, and the relative-motion wave functions are given by Whittaker functions. To avoid confusion, we therefore refer to the energetically lowest resonance as the "lowest exciton".

The ordered tight-binding semiconductor model possesses translational symmetry, the single-particle eigenstates are Bloch functions, and the single-particle eigenvalues follow a cosine band structure, see Sect. 6.1. Although in this case, it is possible to use the k representation, we continue to apply the real-space representation, which is a more natural representation for disordered and spatially inhomogeneous semiconductors.

The model is defined by the four-band Hamiltonian, (5.8), with the many-particle interaction given by (5.13) and (5.17), and the optical dipole matrix elements of (5.23). For historical reasons, one typically uses the equations of motion for the dynamical variables $p_{ij}^{he}(t)$, etc., as complex conjugate to those describing the dynamics of $\rho_{he}(t)$; compare, e.g., (7.29) with (8.54). Obviously, this does not have any consequences for the description of the physics, which is given by the real part of the dynamic quantities.

Although our one-dimensional tight-binding model does not provide a quantitative description of a real semiconductor, it can be used to qualitatively model the optical excitations in the energetic vicinity of the fundamental gap in a reasonable way. To make contact with reality, we choose the parameters in such a way that known features like the exciton binding energy are reproduced. Also the realistic effective masses of electrons and holes can be represented by a suitable choice of the model parameters, see the discussion in Chap. 6. We also, except for some fundamental questions, treat the spin-dependent selection rules carefully by considering the 2×2-band model for heavy-hole electron transitions. Thus, in contrast to the previous chapter, our physical parameters now have real units. Therefore, the results can directly be compared with experimental findings, as has been done in many cases.

The evaluation of the equations of motion for the interacting semiconductor models can no longer be performed analytically as in the case of level systems. Therefore, most of the results presented here have been obtained by numerical calculations. The equations of motion have already been derived in Chap. 8. Therefore, in the remainder of this book, the density of equations in the presentation is considerably reduced in comparison to previous chapters. Instead, we focus on presenting figures which show the numerically obtained results and their physical interpretation.

11.1 Noninteracting Particles

For optical single-particle transitions in an ordered semiconductor, the k vector is conserved. These transitions are therefore independent of each other, but have different resonance frequencies (up to spin- and symmetry-induced degeneracies). This system represents an inhomogeneous ensemble. Therefore, pump–probe experiments can be used to burn holes and four-wave-mixing experiments yield echoes. It needs, however, to be critically addressed, whether such a description, i.e., without considering the many-body Coulomb interaction, is adequate.

11.2 Interacting Particles

It is useful to illustrate the various transitions involved in the set of equations (8.54) and (8.56), or alternatively (8.58) and (8.59) by the simplified scheme shown in Fig. 11.1. Considering an excitation with two optical

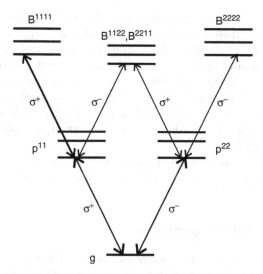

Fig. 11.1. Schematic drawing of the relevant dynamic variables p (single excitons) and B (biexcitons). Here, the *superscripts* label the involved electron and hole bands. The two types of possible circularly polarized optical transitions are indicated by σ^+ and σ^-. The ground state is denoted by g. [From C. Sieh et al., Eur. Phys. J. B **11**, 407 (1999)]

pulses, pump–probe and four-wave-mixing signals can be described, as discussed earlier in Chap. 10.

In first order in the light field, only p^{11} and p^{22} can be excited. This is due to the fact that the optical matrix elements for p^{12} and p^{21} vanish and that in the linear regime no coupling between the subspaces of different spin states exists within our model.

In second order in the light field, depending on the polarization of the incident laser pulses, different types of biexciton states can be excited. At this point a remark is in order. We will in the following repeatedly talk about "biexcitons". This notion is used to characterize two electrons and two holes that are correlated by the Coulomb interaction. However, it does not always mean that these objects have a bound state, in the sense of a bound hydrogen molecule. The notion of a "bound biexciton" is explicitly applied if the latter is the case.

If we consider two interactions with two co-circularly polarized pulses, B^{1111} or B^{2222} are relevant. $B^{h'e'he}_{abcd}$ is antisymmetric with respect to interchanging the band and real-space indices of the two electrons and the two holes, respectively, i.e., $B^{he'h'e}_{cbad} = B^{h'ehe'}_{adcb} = -B^{h'e'he}_{abcd}$. Since the contributions of the two electrons and the two holes which enter into B^{1111} and B^{2222} come from the same bands, the real-space part of the corresponding B's has to be antisymmetric. Such biexciton states typically do not include bound states. If we consider two interactions with oppositely circularly polarized pulses, B^{1122} and B^{2211} are created in second order. For these B's no definite symmetry for the real-space part of B exists and, therefore, the real-space part may be symmetric, which allows for bound biexciton states. Note the analogy with the description of hydrogen molecules in the Heitler–London approach, where also the symmetric real-space states (and antisymmetric spin-singlet states) are those which lead to the covalent bonding between the molecules. In the numerical treatment, it is sufficient to calculate only either B^{1122} or B^{2211} since the value of the other one can be determined using the antisymmetry $B^{1122}_{abcd} = B^{2211}_{cdab}$.

Due to the selection rules also in third order p^{12} and p^{21} do not contribute to the optical response. Therefore, within $\chi^{(3)}$, the response is fully determined by p^{11} and p^{22}.

11.2.1 Pump–Probe Experiments

We assume excitation with two pulses and use the rotating-wave approximation, i.e., the light field is given by the first term in (3.11) with

$$\boldsymbol{E}_l(t) = E_l(t)\boldsymbol{e}_l. \tag{11.1}$$

We consider two pulses, i.e., $l = 1, 2$. The temporal envelope of the pump field is

$$E_1(t) \propto e^{-((t+\tau)/\delta_1)^2} \tag{11.2}$$

and that of the probe field

$$E_2(t) \propto e^{-(t/\delta_2)^2}. \tag{11.3}$$

Their Gaussian widths are δ_1 and δ_2, respectively, and temporally the pulses are centered at $t = -\tau$ and $t = 0$. The polarization directions are denoted by $\boldsymbol{e}_1$ and $\boldsymbol{e}_2$, and ω_1 and ω_2 are their central frequencies, respectively. A positive time delay τ corresponds to the pump pulse arriving before the probe pulse. Equations (8.54) and (8.56) are solved numerically in the time domain up to third order in the optical field ($\chi^{(3)}$ limit). We obtain $\delta\boldsymbol{P}(t,\tau)$, which is the time-domain polarization in differential absorption geometry, by considering all contributions which (i) propagate in the direction of the probe pulse ($\boldsymbol{E}_2$), and (ii) include two interactions with the pump pulse ($\boldsymbol{E}_1$) and are linear in the probe pulse. Since the probe pulse is taken to be spectrally broad, the differential absorption can be determined via

$$\delta\alpha(\omega,\tau) \propto \Im\left[\int dt \ (\boldsymbol{e}_2)^* \cdot \delta\boldsymbol{P}(t,\tau)e^{i\omega t}\right], \tag{11.4}$$

where $\boldsymbol{e}_2$ denotes the polarization direction of the probe pulse.

We start by discussing the differential absorption for resonant pump excitation considering co-circular and opposite circular polarization of the pump and probe pulses. The model parameters are chosen such that the exciton binding energy roughly resembles that of experimentally investigated semiconductor quantum wells. In particular, $J^e = 15$ meV and $J^h = 1.5$ meV, the interaction parameters are $U_0 = 15$ meV and $a_0/a = 0.5$ with the lattice constant $a = 5$ nm. This yields an excitonic binding energy $E_b = 15.1$ meV for the heavy-hole exciton. In order to obtain smooth results for our small system of $N = 10$ sites, we introduce damping parameters $T_{2p} = 3$ ps and $T_{2B} = 1.5$ ps into the equations for p and B, respectively.

Computed differential absorption spectra for resonant excitation at the exciton resonance with co-circularly polarized pump and probe pulses, i.e., $e_1 = e_2 = \sigma^+$, using a pump pulse with $\delta_1 = 1$ ps, for different pump–probe time delays are shown in Fig. 11.2a, c, and e (note that the zero of the energy scale is chosen to coincide with the energy of the lowest exciton). The spectral width of the pump-pulse is much narrower than the exciton binding

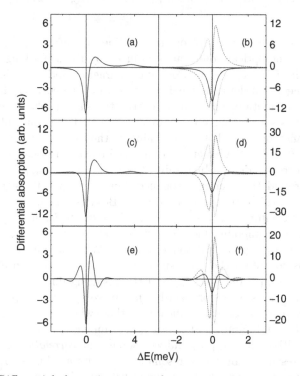

Fig. 11.2. Differential absorption spectra for excitation resonant with the exciton with co-circularly polarized pump (σ^+) and probe (σ^+) for different time delays. The total spectrum is displayed in (**a**) for $\tau = 2$ ps, (**c**) $\tau = 0$ ps, and (**e**) $\tau = -2$ ps. (**b**), (**d**), and (**f**) show the three contributions to the differential absorption $\delta\alpha$: $\delta\alpha_{pb}$ multiplied by 5 (*solid*), $\delta\alpha_{CI,1st}$ (*dashed*), and $\delta\alpha_{CI,corr}$ (*dotted*). Here, the lowest excitonic resonance is taken as the zero of the energy scale. [From C. Sieh et al., Eur. Phys. J. B **11**, 407 (1999)]

energy of 15.1 meV. For the probe pulse, $\delta_2 = 10\,\mathrm{fs}$ has been used, corresponding to a spectrally flat probe spectrum in the frequency region of interest. For excitation with $e_1 = e_2 = \sigma^+$, $\delta P(t, \tau)$ also has the same polarization as the pulses. Figure 11.2a shows $\delta\alpha(\omega)$ for a positive delay of $\tau = 2\,\mathrm{ps}$. The differential absorption is strictly negative in the vicinity of the exciton resonance corresponding to a pump-pulse-induced bleaching. The positive contributions to $\delta\alpha(\omega)$ are related to excited state absorption which is induced by transitions to biexciton states. For energies larger than the exciton energy, we see in Fig. 11.2a two positive peaks indicating some structure in the biexciton continuum.

The total differential absorption can be written as the sum of three contributions (see discussion following (8.54))

$$\delta\alpha(\omega, \tau) = \delta\alpha_{\mathrm{pb}}(\omega, \tau) + \delta\alpha_{\mathrm{CI,1st}}(\omega, \tau) + \delta\alpha_{\mathrm{CI,corr}}(\omega, \tau), \qquad (11.5)$$

indicating the three sources of optical nonlinearity in the $\chi^{(3)}$ limit. Here, "pb" denotes the optical nonlinearity induced by Pauli blocking. The terms denoted with "CI" are due to Coulomb interaction induced nonlinearities. "CI,1st" is the first-order (Hartree–Fock) term, and "CI,corr" the higher-order correlation contribution. These three contributions are displayed separately in Fig. 11.2b. It is shown that $\delta\alpha_{\mathrm{pb}}$ is weak (note that it is multiplied by 5 in Fig. 11.2b) and corresponds to a bleaching at the exciton resonance. $\delta\alpha_{\mathrm{CI,1st}}$ is very strong and antisymmetric around the exciton resonance. Its dispersive shape corresponds to a blue shift of the exciton. $\delta\alpha_{\mathrm{CI,corr}}$ is also mainly dispersive around the exciton resonance, but with opposite sign compared to $\delta\alpha_{\mathrm{CI,1st}}$, i.e., this term yields a red shift. Besides contributions with resonances at the exciton energy, $\delta\alpha_{\mathrm{CI,corr}}$ also includes terms having resonances at the energies of unbound biexciton states. Therefore, it is not completely antisymmetric around the exciton resonance. When adding up these three contributions to obtain the total signal via (11.5), strong cancellations occur between $\delta\alpha_{\mathrm{CI,1st}}$ and $\delta\alpha_{\mathrm{CI,corr}}$ and the resulting differential absorption, see Fig. 11.2a, shows a predominantly absorptive spectral shape around the exciton resonance. As already indicated earlier, when comparing (8.54) and (8.56) to (8.58) and (8.59), part of $\delta\alpha_{\mathrm{CI,corr}}$ compensates the first-order term $\delta\alpha_{\mathrm{CI,1st}}$. Additionally, $\delta\alpha_{\mathrm{CI,corr}}$ induces some real *correlation* contributions due to the presence of biexciton resonances and it contributes to the bleaching at the exciton. In fact, the bleaching at the exciton resonance in Fig. 11.2a is dominated by $\delta\alpha_{\mathrm{CI,1st}} + \delta\alpha_{\mathrm{CI,corr}}$ and only weakly enhanced by $\delta\alpha_{\mathrm{pb}}$.

For $\tau = 0$, see Fig. 11.2c and d, we find differential absorption spectra that are qualitatively similar to those at $\tau = 2\,\mathrm{ps}$. The main differences can be attributed to weak positive contributions for energies below the exciton energy. These are related to the coherent oscillations, see Sect. 10.1.2, which become dominant for negative pump–probe delays. For $\tau = -2\,\mathrm{ps}$, see

Fig. 11.2e, coherent oscillations dominate the differential absorption spectra. Figure 11.2f shows that they are present in all three contributions.

We now consider resonant excitation with oppositely circularly polarized pump and probe pulses, $e_1 = \sigma^+$ and $e_2 = \sigma^-$. For this polarization geometry, $\delta\alpha_{\mathrm{pb}}$ vanishes, and also $\delta\alpha_{\mathrm{CI,1st}}$ vanishes as long as the system is homogeneous. This is due to the fact that none of these contributions introduces any coupling between the subspaces of different spin states. Therefore, for this polarization geometry, the total signal is identical to the correlation contribution, $\delta P = \delta P_{\mathrm{CI,corr}}$. It can easily be seen that $\delta P_{\mathrm{CI,corr}}$ is polarized in the same way as the σ^- probe pulse.

The same model parameters as those leading to Fig. 11.2 have been assumed. We obtain the spectra displayed in Fig. 11.3a–c. For positive and zero delay, see Fig. 11.3a and b, we again find (negative) bleaching at the exciton and (positive) excited-state absorption due to transitions to biexciton states. Whereas for co-circularly polarized excitation only contributions from unbound biexciton states are present, now there is a clear signature of a bound biexciton in the differential absorption spectra, appearing about 2.7 meV below the excitonic resonance. For negative delay, see Fig. 11.3c, coherent oscillations also appear in this case.

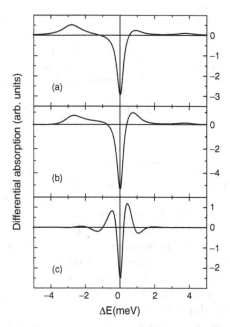

Fig. 11.3. Differential absorption spectra for resonant excitation at the exciton resonance with opposite circular polarized pump (σ^+) and probe (σ^-) pulses for different time delays. The total spectrum is displayed in (**a**) for $\tau = 2\,\mathrm{ps}$, (**b**) $\tau = 0\,\mathrm{ps}$, and (**c**) $\tau = -2\,\mathrm{ps}$. The lowest excitonic resonance is taken as the zero of the energy scale. [From C. Sieh et al., Eur. Phys. J. B **11**, 407 (1999)]

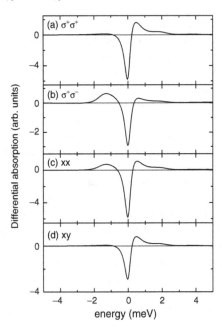

Fig. 11.4. Differential absorption spectra $\delta\alpha$ for resonant excitation at the exciton resonance and a time delay of $\tau = 0$ ps. (**a**) Co-circular ($\sigma^+\sigma^+$), (**b**) opposite-circular ($\sigma^+\sigma^-$), (**c**) linear parallel (xx), and (**d**) linear perpendicular (xy) perpendicular polarized pump and probe pulses, respectively. Note that in this figure we use $J^e = 8$ meV, $J^h = 0.8$ meV, and $U_0 = 8$ meV. These parameters reduce the exciton binding energy to about 8 meV and that of the biexciton to about 1.5 meV. [From T. Meier and S.W. Koch, in *Ultrafast Physical Processes in Semiconductors*, special issue of Series *Semiconductors and Semimetals*, Vol. 67, pp. 231–313 (Academic Press, New York, 2001)]

A comparison between the differential absorption spectra for four different excitation configurations is given in Fig. 11.4 where a time delay $\tau = 0$ ps between the pump and probe pulses has been used. Figure 11.4 shows that the line shape for co-circularly polarized excitation is close to that for linear perpendicular excitation, and that opposite-circularly polarized excitation is close to linear parallel excitation. Induced absorption caused by transitions to the biexciton shows up only for opposite-circularly and linear parallel polarized excitations. (Note that some small positive $\delta\alpha$ below the exciton resonance in Fig. 11.4a and d is induced only by the overlap of the pump and the probe pulses.) A simple understanding of the selection rules governing the appearance of the biexciton contributions in Fig. 11.4 is provided by the reduced level scheme, see Fig. 11.5, which shows that the circular selection rules can be transformed into linear selection rules. The equivalent level schemes displayed in Fig. 11.5 explain that for linearly polarized pump and probe pulses a bound biexciton is excited only if both pulses are not orthogonally polarized.

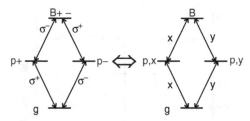

Fig. 11.5. Reduced level scheme including the two spin-degenerate excitons $p+$ and $p-$ as well as the bound biexciton $B + -$. The model with circularly polarized transition matrix elements is equivalent to the model with linear polarized selection rules. [From T. Meier and S.W. Koch, in *Ultrafast Physical Processes in Semiconductors*, special issue of Series *Semiconductors and Semimetals*, Vol. 67, pp. 231–313 (Academic Press, New York, 2001)]

Thus, in agreement with the numerical results of Fig. 11.4 the bound biexciton appears in the differential absorption spectra for $\sigma^+\sigma^-$ and xx excitation, but does not contribute for $\sigma^+\sigma^+$ and xy.

We now discuss the influence of detuning of the pump pulse below the exciton resonance for co-circularly polarized excitation again for the same parameters. Figure 11.6 shows the resulting differential absorption spectra for various detunings of -1.51 meV, -3.02 meV, -7.5 meV, and -22.6 meV of the pump-pulse energy below the exciton energy. The higher detunings (-7.5 meV and -22.6 meV) are much larger than both the spectral width of the pump pulse and the homogeneous width of the exciton resonance. The solid line gives the full $\delta\alpha$, whereas the dashed line gives the result within the time-dependent Hartree–Fock approximation where correlations are neglected, i.e., $\delta\alpha_{\mathrm{HF}} = \delta\alpha_{\mathrm{pb}} + \delta\alpha_{\mathrm{CI,1st}}$. With increasing detuning, the bleaching at the exciton resonance develops into a dispersive shape corresponding to a blue shift. Furthermore, it is shown in Fig. 11.6 that the importance of the carrier correlations diminishes rapidly with increasing detuning. We thus conclude that the time-dependent Hartree–Fock approximation gives a good description for this polarization configuration as long as off-resonant excitation is considered. Higher-order Coulomb correlation effects can be expected to be significant if we consider a detuned pump pulse and opposite circularly polarized pump and probe pulses. Let us use a pump pulse that is tuned 4.5 meV below the exciton resonance. Figures 11.7a–d display the results for $\sigma^+\sigma^+$ and $\sigma^+\sigma^-$ excitation. In the $\sigma^+\sigma^+$ configuration, see Fig. 11.7a and c (and also for xy and xx excitation, not shown), the differential absorption corresponds to a blue shift as seen in the calculation presented above. Most remarkably, however, for $\sigma^+\sigma^-$, see Figs. 11.7b and d, there is a red shift!

Note that for $\sigma^+\sigma^-$ excitation both the Pauli blocking and the first-order Coulomb-induced nonlinearity, i.e., the Hartree–Fock contribution, vanish identically, because neither one of these terms leads to a coupling among

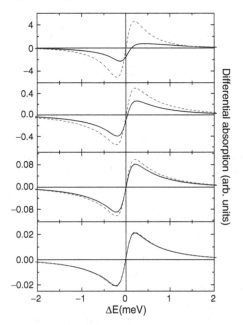

Fig. 11.6. Differential absorption spectra for various detunings of the pump pulse below the lowest exciton resonance with co-circularly polarized pump (σ^+) and probe pulses (σ^+) for zero time delay. The central frequency of the pump pulse is chosen -1.51 meV, -3.02 meV, -7.5 meV, and -22.6 meV below the energy of the lowest exciton resonance (from above). The *solid line* gives the full $\delta\alpha$, and the *dashed line* is the result of a Hartree–Fock calculation ($\delta\alpha_{\mathrm{pb}} + \delta\alpha_{\mathrm{CI,1st}}$) neglecting correlations ($\delta\alpha_{\mathrm{CI,corr}}$). [From C. Sieh et al., Eur. Phys. J. B **11**, 407 (1999)]

the subspaces of different spin states within $\chi^{(3)}$. Thus, in this case, the total signal is entirely due to the Coulomb correlations.

One might be tempted to relate the unexpected red shift, which has also been seen experimentally, to the presence of bound biexcitonic states. Let us, therefore, analyze the origin of the red shift by looking at the individual contributions to the signal. We find that for $\sigma^+\sigma^+$ polarization and strong detuning below the exciton the Pauli blocking and the first-order Coulomb term always induce a blue shift, whereas the Coulomb correlations always correspond to a red shift. The fact that also for $\sigma^+\sigma^+$ excitation, where no bound biexcitons are excited, the correlation term alone still corresponds to a red shift demonstrates that this shift is not directly related to the existence of a bound biexciton.

To support this conclusion, additional calculations of the differential absorption spectra in the $\sigma^+\sigma^-$ configuration are performed. To eliminate the bound biexciton contribution also for the $\sigma^+\sigma^-$ excitation, the six terms containing the attractive and repulsive Coulomb terms for two electrons and two holes have been artificially dropped from the homogeneous part of the

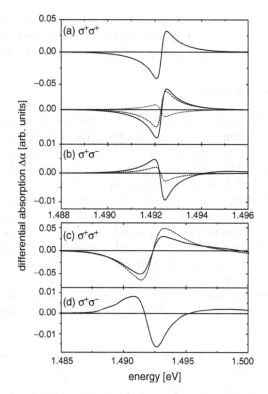

Fig. 11.7. Calculated differential absorption spectra for excitation 4.5 meV below the exciton and $\tau = 0$ ps. (a) One-dimensional model, co-circularly polarized pump and probe pulses ($\sigma^+\sigma^+$), the *lower panel* shows the three contributions to the signal (*solid*: Pauli-blocking; *dashed*: first-order Coulomb; and *dotted*: Coulomb correlation). (b) One-dimensional model, opposite-circularly polarized pump and probe pulses ($\sigma^+\sigma^-$). The *dashed line* represents the result of a calculation without biexcitons, as discussed in the text. (c) same as (a) and (d) same as (b), respectively, but for the full two-dimensional quantum-well model. The *dashed line* in (c) displays the result of a Hartree–Fock calculation neglecting Coulomb correlations. [From C. Sieh et al., Phys. Rev. Lett. **82**, 3112 (1999)]

equation of motion for the biexciton amplitude $\bar{B}$. In this case, also for $\sigma^+\sigma^-$ excitation, no bound biexcitons exist, but the result displayed in Fig. 11.7b shows that the signal amplitude is somewhat reduced. The red shift, however, clearly persists. The same calculations performed for $\sigma^+\sigma^+$ excitation (not shown in the figure) where no biexciton is excited yield a similar result. The signal amplitude of the correlation term is reduced, but its red shift persists. This demonstrates that for the occurrence of the red shift the exact structure of the biexciton states is unimportant.

Generally, one expects that Coulomb effects are more prominent in a one-dimensional model if compared to a two-dimensional model more appropriate for quantum-well systems. Therefore, calculations for a two-dimensional model have also been performed. As can be seen in Fig. 11.7, however, qualitatively the same features as in the one-dimensional case are seen.

11.2.2 Four-Wave Mixing

In this section, we analyze the influence of correlations on four-wave-mixing spectroscopy. As outlined above for the case of pump–probe experiments, also four-wave-mixing signals depend sensitively on the polarization directions of the incident pulses. Due to the propagation directions of the signals monitored in pump–probe and four-wave mixing, k_2 and $2k_2 - k_1$, respectively, the biexciton manifold representing carrier correlations is probed in different ways in both experiments. Note that in contrast to Chap. 10 the notation $(m|n)$ now means a spatial phase factor $\exp{(i(mk_1 + nk_2) \cdot r)}$ because for historical reasons we are here considering the complex conjugate polarization.

Again, we apply two incident pulses ($l = 1, 2$) and use the rotating-wave approximation. In first order in the optical field, one needs to consider the interband coherences which are generated by each one of the two pulses via the linear source term $E_i(t) \cdot \mu^*$. The linear equations of motion are

$$
\begin{aligned}
-i\hbar \frac{d}{dt} p_{12}^{he(1|0)} = &-\sum_j T_{2j}^e p_{1j}^{he(1|0)} - \sum_i T_{i1}^h p_{i2}^{he(1|0)} \\
&+ V_{12} p_{12}^{he(1|0)} + E_1(t) \cdot (\mu_{12}^{he})^*,
\end{aligned}
\tag{11.6}
$$

$$
\begin{aligned}
-i\hbar \frac{d}{dt} p_{12}^{he(0|1)} = &-\sum_j T_{2j}^e p_{1j}^{he(0|1)} - \sum_i T_{i1}^h p_{i2}^{he(0|1)} \\
&+ V_{12} p_{12}^{he(0|1)} + E_2(t) \cdot (\mu_{12}^{he})^*,
\end{aligned}
\tag{11.7}
$$

where the superscript $(n|m)$ denotes the kinematic direction $nk_1 + mk_2$.

In second order in the optical field, the biexciton amplitudes B can be excited. Keeping only those terms that lead to a contribution in a third-order analysis of four-wave mixing (i.e., terms with superscript $(-1|2)$ in third order), the relevant equation of motion in second order is

$$
\begin{aligned}
-i\hbar \frac{d}{dt} B_{ba12}^{h'e'he(0|2)} = &-\sum_i (T_{2i}^e B_{ba1i}^{h'e'he(0|2)} + T_{i1}^h B_{bai2}^{h'e'he(0|2)} \\
&+ T_{ai}^e B_{bi12}^{h'e'he(0|2)} + T_{ib}^h B_{ia12}^{h'e'he(0|2)})
\end{aligned}
$$

$$+(V_{ba} + V_{b2} + V_{1a} + V_{12} - V_{b1} - V_{a2})B_{ba12}^{h'e'he(0|2)}$$
$$-(V_{ba} + V_{12} - V_{b1} - V_{a2})p_{1a}^{he'(0|1)} p_{b2}^{h'e(0|1)}$$
$$+(V_{1a} + V_{b2} - V_{b1} - V_{a2})p_{ba}^{h'e'(0|1)} p_{12}^{he(0|1)}. \tag{11.8}$$

Furthermore, we have to include all lower-order contributions which generate a single-exciton amplitude associated with the direction $2k_2 - k_1$

$$-i\hbar\frac{d}{dt}p_{12}^{he(-1|2)} = -\sum_j T_{2j}^e p_{1j}^{he(-1|2)} - \sum_i T_{i1}^h p_{i2}^{he(-1|2)}$$
$$+V_{12}p_{12}^{he(-1|2)}$$
$$+ \sum_{abv'c'} (V_{a2} - V_{a1} - V_{b2} + V_{b1})[(p_{ba}^{h'e'(1|0)})^* p_{b2}^{h'e(0|1)} p_{1a}^{he(0|1)}$$
$$-(p_{ba}^{h'e'(1|0)})^* p_{ba}^{h'e'(0|1)} p_{12}^{he(0|1)} - (p_{ba}^{h'e'(1|0)})^* B_{ba12}^{h'e'he(0|2)}]$$
$$-E_2(t) \cdot \left[\sum_{abh'e'} ((\mu_{1b}^{he'})^*(p_{ab}^{h'e'(1|0)})^* p_{a2}^{h'e(0|1)} \right.$$
$$\left. +(\mu_{b2}^{h'e})^*(p_{ba}^{h'e'(1|0)})^* p_{1a}^{he'(0|1)}) \right]. \tag{11.9}$$

Again, we have explicitly included only terms that contribute to four-wave mixing in third order. Note that in (11.9), $(p^{(1|0)})^*$ corresponds to the kinematic direction $-k_1$.

The time-resolved four-wave-mixing *amplitude* depending on real time t and time delay τ is given by

$$P_{\mathrm{FWM}}(t,\tau) = \sum_{ijhe} \mu_{ij}^{he}(p_{ij}^{he})^{(-1|2)}(t,\tau). \tag{11.10}$$

In analogy to the situation of the pump–probe experiments, this amplitude is induced by optical nonlinearities corresponding to Pauli blocking and first and higher-order Coulomb terms. Therefore, it can be split into the sum over the individual amplitudes

$$P_{\mathrm{FWM}}(t,\tau) = \sum_n P_{\mathrm{FWM}}^n(t,\tau) \tag{11.11}$$

resulting from these different optical nonlinearities. In particular, we decompose

$$P_{\mathrm{FWM}}(t,\tau) = P_{\mathrm{FWM}}^{\mathrm{pb}}(t,\tau) + P_{\mathrm{FWM}}^{\mathrm{CI,first}}(t,\tau) + P_{\mathrm{FWM}}^{\mathrm{CI,corr}}(t,\tau), \qquad (11.12)$$

where the superscript "pb" denotes the optical nonlinearity due to Pauli blocking, "CI,first" is the first-order (Hartree–Fock) term and "CI,corr" is the higher-order correlation contribution.

Pump–probe and four-wave-mixing experiments involve different pulse sequences and signal directions, leading to different polarization dependencies for the excitation of biexcitons. Whereas pump–probe experiments performed with opposite circularly polarized pulses allow one to concentrate on effects induced by bound biexcitons, for this situation the lowest order four-wave-mixing signal vanishes. On the other hand, for both linear parallel and perpendicular polarized excitation, the biexciton does contribute to four-wave-mixing signals, but is seen in pump–probe experiments only for linear parallel but not for linear perpendicular polarized pulses. As is discussed in more detail below, in four-wave-mixing experiments performed with linear perpendicular polarized pulses, the spectral components at the exciton resonance are strongly reduced and the signal is dominated by transitions at frequencies between the exciton and biexciton states. In pump–probe experiments, such a selective reduction of the contributions due to exciton bleaching is not possible. Hence, analyzing the polarization dependencies of four-wave-mixing and pump–probe signals makes it possible to investigate the dynamics of different correlation-induced effects.

In the following, we analyze exciton–biexciton beats in time-resolved and time-integrated four-wave-mixing traces within the coherent limit. For this purpose, we compare the four-wave-mixing signals obtained for linear parallel and linear-perpendicular polarized incident pulses. Our results show that for a homogeneous system in the time-integrated four-wave-mixing traces biexciton-induced beats are absent for positive delays in the $\chi^{(3)}$ limit. They are, however, induced by either higher intensities ($\chi^{(5)}$) or by disorder.

In the following analysis, we consider all terms arising in the coherent $\chi^{(5)}$ limit, i.e., we include one-, two-, and three-exciton amplitudes represented by p, B, and W. For numerical convenience, we reduce the system size to $N = 6$, which introduces some well-known finite size effects, for example an enhanced biexciton binding energy (3.3 meV instead of 2.6 meV for a larger system). Since, however, the features of the four-wave-mixing response reported here are in agreement with calculations for larger systems and with experimental observations, we assume for numerical convenience that even this small system size is sufficient for a qualitative description.

The other parameters are identical to those used in the previous section: $J^e = 15$ meV, $J^h = 1.5$ meV, $U_0 = 15$ meV, and $a_0/a = 0.5$. This results in an excitonic binding energy of $E_b = 15$ meV and a biexcitonic binding energy of $E_{Bb} = 3.3$ meV.

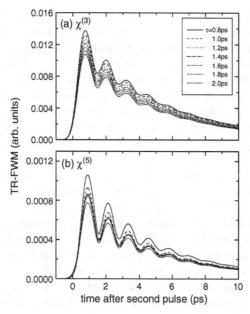

Fig. 11.8. Time-resolved four-wave mixing versus real time for delays from $\tau = 0.8$ ps to $\tau = 2.0$ ps as indicated in the graph for excitation with linear parallel polarized pulses: (**a**) pure $\chi^{(3)}$ and (**b**) pure $\chi^{(5)}$. The temporal width of the exciting pulses was chosen as $\delta_1 = \delta_2 = 500$ fs and both pulses are tuned 1 meV below the exciton resonance. The phenomenologically introduced dephasing rates are $\gamma_p = 1/5$ ps, $\gamma_B = 2\gamma_p = 1/2.5$ ps, and $\gamma_W = 3\gamma_p = 1/1.667$ ps. [From T. Meier and S.W. Koch, in *Ultrafast Physical Processes in Semiconductors*, special issue of Series *Semiconductors and Semimetals*, Vol. 67, pp. 231–313 (Academic Press, New York, 2001)]

Figure 11.8 displays numerical results on the time-resolved four-wave-mixing intensities for various time delays between $\tau = 0.8$ and $\tau = 2.0$ ps after excitation with two linear parallel polarized pulses (xx). In Fig. 11.8a the $\chi^{(3)}$ result is displayed. It shows temporal beats with a period of 1.2 ps. This beat period is just the inverse of the biexciton binding energy of about 3.3 meV which has also been obtained independently by calculating the differential absorption for excitation with opposite circular polarized pump and probe pulses.

The appearance of the exciton–biexciton beats in the time-resolved four-wave-mixing traces in Fig. 11.8a, can be understood by analyzing the schematic equations (8.60) and (8.61) that can be used to qualitatively describe the nonlinear response within the coherent $\chi^{(3)}$ limit. Considering only a short-lived source term, e.g., $\mu E p^* p$ in (8.60), the exciton amplitude evolves freely and oscillates with the exciton frequency ω_x, i.e., $p(t) \propto \exp(i\omega_x t)$. However, due to the presence of another source term that itself is decaying slowly, as is the case for the Coulomb source $V p^* B$ in (8.60), one obtains an oscillation with a frequency that is given by the difference between ω_{2x}

and ω_x, i.e., $p^*B \propto \exp(i(\omega_{2x} - \omega_x)t)$. Solving the equation of motion for the exciton coherence p driven by such a source term, one finds that, in this case, p includes oscillations with both ω_x as well as $\omega_{2x} - \omega_x$, i.e., $p(t) \propto \alpha \exp(i\omega_x t) + \beta \exp(i(\omega_{2x} - \omega_x)t) = \exp(i\omega_x t)[\alpha + \beta \exp(i(\omega_{2x} - 2\omega_x)t)]$, where α and β determine the weight of the two spectral components. Consequently, the time-resolved intensity $|p(t)|^2$ shows temporal beating. If the excitation is chosen such that mainly the biexciton is excited, as is the case for the results displayed in Fig. 11.8, one thus expects beats in the time-resolved four-wave-mixing traces with a temporal period inversely proportional to the biexciton binding energy $\omega_{2x} - 2\omega_x$.

Since in Fig. 11.8 the time axis is chosen such that the last pulse, which is pulse No. 2 for positive time delay, always arrives at $t = 0$, the phase of the beating is independent of the time delay between the two pulses. This is a consequence of the fact that in four-wave-mixing experiments a biexciton amplitude B is only excited by pulse No. 2, see (11.8), which contributes in second order to the $\chi^{(3)}$-four-wave-mixing signal monitored in direction $2k_2 - k_1$. The arrival of pulse No. 2 thus defines the onset of the biexciton-induced beats. With increasing time delay τ, the time-resolved four-wave-mixing intensity displayed in Fig. 11.8a simply shows a decrease determined by the decay rate T_{2p}^{-1}, since the polarization induced by pulse No. 1 decays during the time interval τ until pulse No. 2 arrives and the four-wave-mixing response is generated. The fact that, for positive time delays, the phase of the biexciton beats is independent of τ automatically implies the absence of biexciton beating in the time-integrated four-wave-mixing traces. The time-integrated four-wave-mixing signal is the integral over the time-resolved four-wave-mixing intensity, see (10.53), and, thus, up to $\chi^{(3)}$, it shows only a simple decay but no modulation with increasing τ.

Compared to the $\chi^{(3)}$ results, the τ dependence is different for the pure $\chi^{(5)}$ contribution (i.e., the third-order contribution is omitted) to the time-resolved four-wave-mixing traces, see Fig. 11.8b. Although this contribution alone cannot be observed directly in experiments, it is instructive to consider it separately here. In the pure $\chi^{(5)}$ contribution, exciton–biexciton beats are present in the time-resolved four-wave-mixing traces, however, unlike in $\chi^{(3)}$, we find a nontrivial dependence of the amplitude of the time-resolved four-wave-mixing intensity on the time delay τ. Whereas within $\chi^{(3)}$, the amplitude simply decays monotonically (exponentially) with increasing τ, in $\chi^{(5)}$ the amplitude decay is modulated. Thus, for excitation with higher intensities, one has an additional dependence of the time-resolved four-wave-mixing response on time delay τ. As discussed below, this τ-dependent modulation of the time-resolved four-wave-mixing amplitude in $\chi^{(5)}$ introduces biexcitonic beats for higher intensities in the time-integrated four-wave-mixing traces.

Figure 11.9 displays numerical results for time-resolved four-wave-mixing intensities at various time delays between $\tau = 0.8$ and $\tau = 2.0\,\text{ps}$ after excitation with two linear perpendicular polarized pulses (xy). As for xx excitation, see Fig. 11.8, also for xy excitation beats appear in time-resolved

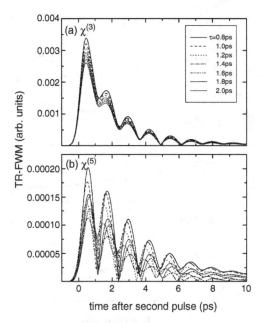

Fig. 11.9. Time-resolved four-wave-mixing versus real time for delays from $\tau = 0.8$ ps to $\tau = 2.0$ ps for excitation with linear perpendicular polarized pulses: **(a)** pure $\chi^{(3)}$ and **(b)** pure $\chi^{(5)}$. [From T. Meier and S.W. Koch, in *Ultrafast Physical Processes in Semiconductors*, special issue of Series *Semiconductors and Semimetals*, Vol. 67, pp. 231–313 (Academic Press, New York, 2001)]

four-wave-mixing traces. In the case of xy excitation, the signal intensity is reduced and the modulation depth is stronger than for xx. Both these effects are due to a strong suppression of spectral four-wave-mixing components at the exciton frequency.

The modulations seen in Fig. 11.9a correspond to the inverse of the difference between the bound biexciton and the lowest unbound biexciton state energies. The latter is energetically situated slightly above twice the energy of the lowest exciton resonance. Note that within the time-dependent Hartree–Fock approximation there is no difference between the four-wave-mixing signal for xx and xy excitation, thus, the polarization dependence of the $\chi^{(3)}$ response, see Figs. 11.8a and 11.9a, is solely induced by Coulomb correlations. As for xx excitation also for xy excitation within $\chi^{(3)}$, the phase of the beats is independent of the time delay and the amplitude decays monotonically with increasing time delay, see Fig. 11.9a.

If we analyze the pure $\chi^{(5)}$ contribution to the time-resolved four-wave-mixing trace for xy, see Fig. 11.9b, we find that both the amplitude of the exciton–biexciton beats and their phase change with time delay. This is different compared to xx excitation, see Fig. 11.8b, where only the amplitude of the time-resolved four-wave-mixing traces was modulated with τ.

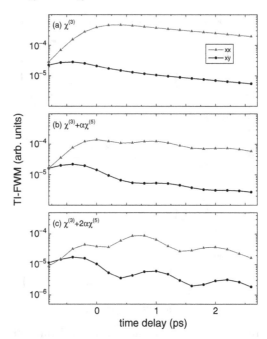

Fig. 11.10. Time-integrated four-wave-mixing versus time delay for linear parallel (*triangles up*) and linear perpendicular (*circles*) excitation: (a) pure $\chi^{(3)}$, (b) $\chi^{(3)} + \alpha\chi^{(5)}$, and (c) $\chi^{(3)} + 2\alpha\chi^{(5)}$ with a finite α. The *lines* are guides for the eye. [From T. Meier and S.W. Koch, in *Ultrafast Physical Processes in Semiconductors*, special issue of Series *Semiconductors and Semimetals*, Vol. 67, pp. 231–313 (Academic Press, New York, 2001)]

Due to the interesting dynamics of the $\chi^{(5)}$ contributions, we can expect some nontrivial behavior of the intensity-dependent time-integrated four-wave-mixing signal, which is analyzed below.

Figure 11.10 shows numerical results on the intensity dependence of time-integrated four-wave-mixing traces for xx and xy excitations. As already explained above, on the basis of the time-resolved four-wave-mixing signals, in the coherent $\chi^{(3)}$ limit no exciton–biexciton beats are present for positive time delays in time-integrated four-wave-mixing traces for either polarization configuration, see Fig. 11.10a. For higher excitation intensities, however, $\chi^{(5)}$ contributions become relevant and, consequently, for both xx and xy configurations beats appear, see Fig. 11.10b and c. These beats are a consequence of the amplitude modulations of the biexciton-induced beats appearing in the time-resolved four-wave-mixing traces in $\chi^{(5)}$ for the case of xx and also of the phase shifts in $\chi^{(5)}$ for xy excitation. The biexciton-induced beats in the time-integrated four-wave-mixing traces become stronger with increasing intensity. Furthermore, a weak polarization dependence of the beat phase is visible in Fig. 11.10c.

11.3 Suggested Reading

1. P. Brick, C. Ell, G. Khitrova, H.M. Gibbs, T. Meier, C. Sieh, and S.W. Koch, "The influence of light holes on the optical Stark effect of the heavy-hole exciton", Phys. Rev. B **64**, 075323 (2001)
2. J. Feldmann, T. Meier, G. von Plessen, M. Koch, E.O. Göbel, P. Thomas, G. Bacher, C. Hartmann, H. Schweizer, W. Schäfer, and H. Nickel, "Coherent dynamics of excitonic wave packets", Phys. Rev. Lett. **70**, 3027 (1993)
3. J. Feldmann, M. Koch, E.O. Göbel, F. Jahnke, T. Meier, W. Schäfer, P. Thomas, S.W. Koch, H. Nickel, S. Luttgen, and W. Stolz, "Coherent dynamics of exciton wave packets in semiconductor heterostructures", Semicond. Sci. Technol. **9**, 1965 (1994)
4. A. Knorr, R. Binder, E.M. Wright, and S.W. Koch, "Amplification, absorption and lossless propagation of femtosecond pulses in inverted semiconductors", Opt. Lett. **18**, 1538 (1993)
5. A. Knorr, T. Österreich, K. Schönhammer, R. Binder, and S.W. Koch, "Asymptotic analytic solution for Rabi-oscillations in a system of weakly excited excitons", Phys. Rev. **B49**, 14024 (1994)
6. S.W. Koch, A. Knorr, R. Binder, and M. Lindberg, "Microscopic theory of Rabi flopping, photon echo, and resonant pulse propagation in semiconductors", phys. stat. sol. **b173**, 177 (1992)
7. S.W. Koch, T. Meier, F. Jahnke, and P. Thomas, "Microscopic theory of optical dephasing in semiconductors", Appl. Phys. A **71**, 511 (2000)
8. S.W. Koch, M. Kira, and T. Meier, "Correlation effects in the optical properties of semiconductors", J. Optics B **3**, R29 (2001)
9. W. Langbein, T. Meier, S.W. Koch, and J.M. Hvam, "Spectral signatures of $\chi^{(5)}$ processes in four-wave mixing of homogeneously broadened excitons", J. Opt. Soc. Am. B **18**, 1318–1325 (2001)
10. M. Lindberg, R. Binder, and S.W. Koch, "Theory of the semiconductor photon echo", Phys. Rev. **A45**, 1865 (1992)
11. M. Lindberg, Y.Z. Hu, R. Binder, and S.W. Koch, "$\chi^{(3)}$ Formalism in optically excited semiconductors and its applications in four-wave-mixing spectroscopy", Phys. Rev. **B50**, 18060 (1994)
12. T. Meier, and S.W. Koch, "Excitons versus unbound electron–hole pairs and their influence on exciton bleaching: A model study", Phys. Rev. B **59**, 13202–13208 (1999)
13. T. Meier, S.W. Koch, P. Brick, C. Ell, G. Khitrova, and H.M. Gibbs, "Signatures of correlations in intensity dependent excitonic absorption changes", Phys. Rev. B **62**, 4218 (2000)
14. T. Meier, S.W. Koch, M. Phillips, and H. Wang, "Strong coupling of heavy- and light-hole excitons induced by many-body correlations", Phys. Rev. B **62**, 12605 (2000)
15. T. Meier and S.W. Koch, "Coulomb correlation signatures in the excitonic optical nonlinearities of semiconductors", in *Ultrafast Physical Processes*

in Semiconductors, special issue of Series *Semiconductors and Semimetals*, Vol. 67, p. 231 (Academic Press, New York 2001)

16. T. Meier, C. Sieh, E. Finger, W. Stolz, W.W. Rühle, P. Thomas, S.W. Koch and A.D. Wieck, "Signatures of biexcitons and triexcitons in coherent non-degenerate semiconductor optics" phys. stat. sol. b **238**, 537 (2003)

17. A. Schulze, A. Knorr, P. Thomas, and S.W. Koch, "Theoretical analysis of higher–order wave mixing in semiconductors", Solid State Commun. **94**, 911 (1995)

18. C. Sieh, T. Meier, F. Jahnke, A. Knorr, S.W. Koch, P. Brick, M. Hübner, C. Ell, J. Prineas, G. Khitrova, and H.M. Gibbs, "Coulomb memory signatures in the excitonic optical Stark effect", Phys. Rev. Lett. **82**, 3112 (1999)

19. C. Sieh, T. Meier, A. Knorr, F. Jahnke, P. Thomas, and S.W. Koch, "Influence of carrier correlations on the excitonic optical response including disorder and microcavity effects", Eur. Phys. J. B **11**, 407 (1999)

20. H.P. Wagner, H.-P. Tranitz, M. Reichelt, T. Meier, and S.W. Koch, "Coherent spectral oscillations in multi-wave mixing", Phys. Rev. B **64**, 233303 (2001)

Coherent $\chi^{(3)}$ and $\chi^{(5)}$ Processes in Disordered Semiconductors

In the previous chapter, we have seen that coherent nonlinear processes in ordered semiconductors show a number of features that need detailed and sometimes extensive calculations for their interpretation. The Coulomb many-particle interaction is an essential ingredient in the theory and is responsible for most of the features present in the nonlinear response.

We now turn to the discussion of the nonlinear optical response of disordered semiconductors. Here, in addition to the many-particle interaction, the disorder is expected to produce unique signatures in the nonlinear response. From the discussion of level systems, we may expect to find photon echoes, since disorder certainly contributes to the inhomogeneous character of the spectra. One might also suspect that disorder may be a cause of dephasing. In the perturbation-theory picture a Bloch state, which is the eigenstate of the ordered system, undergoes successive scattering, or even multiple scattering at the disorder potential. It is a common, although often incorrect belief, that scattering leads to dephasing. However, if one thinks in terms of the eigenstates of the disordered system, one cannot easily see why in this stationary situation dephasing should be present. Eigenstates are not scattered at the potential for which they are calculated. Therefore, a more detailed analysis is needed to answer the question whether static disorder leads to dephasing or not. We refer to this process as *disorder-induced dephasing*.

Notoriously, numerical calculations for nonlinear optical signals in models that contain both the many-particle interaction and the disorder on the level of $\chi^{(3)}$ processes are extremely hard to perform. Usually, one can only treat one-dimensional, and in addition rather small systems in a way to ensure convergence. In order to answer fundamental questions concerning the existence and the properties of disorder-induced dephasing, we first ignore all many-particle interactions. Afterwards, we include the Coulomb interaction together with the disorder.

12.1 Noninteracting Particles

12.1.1 Pump–Probe Experiments

Although calculations of pump–probe signals for the disordered, but interaction-free case, are not hard to perform, to our knowledge, nobody has looked at this case. In fact, this model is somewhat artificial. However, it yields the required answer to the question posed above, namely about the possibility and action of disorder-induced dephasing. Since this topic can be analyzed by calculating four-wave-mixing signals, pump–probe experiments are not considered here.

12.1.2 Four-Wave-Mixing Experiments

We consider a semiconductor model with energetically uncorrelated short-range disorder, characterized by equal electron and hole masses ($J^e = J^h = J$) and equal disorder parameter in the conduction and valence band ($W^e = W^h = W$). The dimensionless parameter characterizing the strength of disorder is $\eta = W/J$. We assume a short laser pulse centered energetically in the center of the spectrum and exciting transitions in the energetic range that is of the order of the optical band width. Furthermore, we use a simple two-band model in this discussion.

From the theory of disordered solids, we know that in one-dimensional systems with uncorrelated short-range disorder all electronic eigenstates are localized. The localization length depends on the strength of disorder W relative to the coupling J, i.e., on η. For $\eta = 0$, there is no disorder and the eigenstates are extended Bloch states, i.e., the localization length is infinite. For $\eta \to \infty$ (the completely disordered case), the two-level systems of which the semiconductor model is composed are decoupled ($J = 0$), and the eigenstates are the states at the isolated two-level systems, i.e., completely localized. In these two extreme cases, it is easy to infer the character of the nonlinear optical response.

- *Ordered systems, $\eta = 0$*
  Due to translational symmetry, there is a quantum number, the wave number k, which is conserved in the optical transition. This selection rule tells us that for any given frequency component of the laser pulse there are exactly two energetically degenerate transitions at k and $-k$ in one dimension. As the energy of these transitions depends on k, we have an inhomogeneous line which leads to photon echoes in the nonlinear time-resolved response. The echoes do not show any dephasing, i.e., their amplitude is independent of the delay time τ, once τ is larger than the temporal pulse width.
- *Completely disordered case, $\eta = \infty$*
  The system is an ensemble of decoupled two-level absorbers characterized by an inhomogeneous line. The nonlinear response shows echoes as well.

Their amplitude does not depend on τ, i.e., there is no dephasing in this model. Here, we have a local selection rule in real space, i.e., transitions are possible only within a given two-level absorber.

Thus, in the noninteracting model, we have no dephasing in the extreme cases $\eta = 0$ (order) and $\eta \to \infty$ (complete disorder) if we do not introduce mechanisms into the model Hamiltonian that are known to lead to dephasing and if we do not add a phenomenological dephasing rate T_2^{-1} to the equations of motion.

What happens for $0 < \eta < \infty$? To answer this question, we solve the optical Bloch equations which we obtain by neglecting all terms proportional to V and all terms B in (11.6)–(11.9). In this model, the nonlinearity is exclusively due to Pauli blocking. The solution of the equations of motion for the pertinent model is given in Fig. 12.1. Here, the time-integrated trace is shown in the left-hand side figure for various values of the disorder. A decay of the time-integrated trace with increasing delay time τ would be interpreted as dephasing. In accordance with our discussion of the extreme cases, there is no dephasing for $\eta = 0$ and $\eta \to \infty$. In between, however, we observe various degrees of decay. This decay is clearly not exponential and thus, a simple dephasing rate T_2^{-1} can therefore not be defined. In order to quantify the decay, we have taken the value of the time-integrated trace at $\tau = 4$ ps which is plotted on the right-hand side of Fig. 12.1 as a function of η. Starting from the ordered case, the signal first decreases strongly, which indicates rapid dephasing. This could roughly be explained by the scattering picture mentioned above. A more precise interpretation is given below.

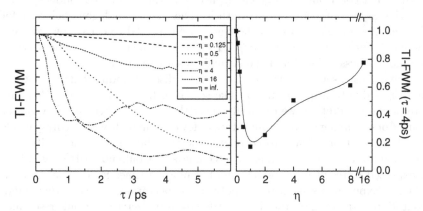

Fig. 12.1. *Left*: Time-integrated traces for the semiconductor model with uncorrelated disorder. The couplings $J^e = J^h$ and disorder parameters $W^e = W^h$ are identical for both bands. The Coulomb interaction is neglected. Excitation is in the center of the absorption band. For $\eta = W/J = 0$ and ∞, the photon echoes do not show any decay. The most pronounced (nonexponential) decay occurs at $\eta = 1$, as shown in the *right-hand side figure*

Starting from the completely disordered situation, the dephasing time also decreases if the coupling relative to the disorder increases. There is a minimum at η close to unity, i.e., when $W \approx J$, indicating the fastest dephasing.

The interpretation of these findings relies on the relaxation of selection rules. In the ordered case, we have a momentum selection rule, while in the completely disordered case, there is a local selection rule. Their violation (in the first case by increasing disorder, in the second case by increasing coupling) leads to dephasing.

Let us discuss the single-particle eigenstates. Starting from the ordered case, increasing disorder leads to an increasing mixing of various Bloch states. Starting from the completely disordered case and adding increasing coupling, the states of the individual levels become increasingly mixed due to this coupling. In both cases, a given hole state is, for $0 < \eta < \infty$ and uncorrelated disorder, no longer optically coupled to just one single electron state, and vice versa. This results in transitions that resemble N-level systems. The $\chi^{(3)}$ response of such systems shows quantum beats or a decay of the time-integrated trace, depending on the number of states involved in a given transition, see Sect. 10.4. Even if only a few eigenstates contribute to a given transition, the inhomogeneous ensemble of such N-level systems results in an initial decay, with a finite long-time value for the time-integrated trace, see the discussion in Sect. 10.1.5. Such behavior is seen for large disorder, i.e., for $\eta \gg 1$, in Fig. 12.1. In fact, for large disorder only few levels of neighboring states have the chance to be effectively coupled to form an eigenstate with small localization length. The inhomogeneous ensemble of such few-level systems leads to the initial decay. The long-time value is the lower, the larger the number of sites that contribute to a given eigenstate in the disordered model system, i.e., the larger the localization length is. There is no analogous behavior for the weakly disordered case, so the traces are expected to decay towards zero for small, but finite η. The value $\eta = 1$ seems to be a cross-over value between the more or less ordered and the strongly disordered regimes.

From this discussion, we learn that the disorder-induced dephasing for finite η is due to a quantum-beat-like scenario. If we consider the Coulomb interaction in addition, we can still expect that a similar scenario is valid for transitions coupling to continuum states. However, in reality the many-particle interaction not only leads to correlated electron–hole pairs, like excitons or continuum states, it also leads to screening and Coulomb scattering, which is expected to result in extremely short dephasing times for continuum excitations. Experiments seem to support this expectation. On the other hand, excitation in the range of the (in the ordered case) discrete excitonic resonance can be expected not to be influenced very strongly by effective Coulomb scattering. Therefore, for the interacting semiconductor model, four-wave-mixing traces have mainly been studied, both experimentally and theoretically, for excitation in the vicinity of the excitonic resonance.

Before we turn to this subject, some words about four-wave-mixing signals and their interpretation in higher-dimensional semiconductors are in order.

From the theory of Anderson localization, and, in particular, from scaling theory, it is known that in one-dimensional disordered systems (with uncorrelated short-range disorder and neglecting the many-particle interaction) all eigenstates are localized for arbitrary small strengths of disorder. In two-dimensional systems, all states are also localized, however, the nature of this localization (called weak localization) differs from that of the one-dimensional case if disorder is weak. In three-dimensional systems, it depends on η, whether all states are localized, or whether there is a range in the center of the band, where extended states exist. The localized states are found in the wings of the single-particle density of states function $\rho(E)$ (called tails) and they are separated by the "mobility edges" from the delocalized states. The Anderson transition is given by that value of η where the two mobility edges from the two extremities of the band coincide in the center, such that all states become localized. This critical value for η depends on the details of the model, but is close to $\eta \approx 15$. The microscopic theory of Anderson localization shows that this phenomenon is due to the wave nature of the particles and relies on coherence. Therefore, it is not surprising that close links exist between Anderson localization and disorder-induced optical dephasing of photon echoes, at least in the noninteracting situation. For interacting models, both the theory of Anderson localization and that of disorder-induced dephasing are not yet fully developed. However, a close link between these two phenomena can also be expected in the interacting situation.

12.2 Interacting Particles

12.2.1 Pump–Probe Experiments

We now discuss pump-probe results for our model system underlying the numerical calculations in Sect. 11.2 with the model parameters used in Fig. 11.2. Now, in addition to the many-particle Coulomb interaction, we also include disorder. We consider diagonal, i.e., energetic, disorder by assuming a Gaussian distribution of the electron site energy ϵ^e, while the hole energies ϵ^h remain unchanged, i.e., they are ordered. To achieve sufficient convergence, the numerical results are averaged over 120 realizations for a disorder of $W^e = 2.35\,\text{meV}$ (full width at half maximum of the Gaussian distribution) and over 180 realizations for a disorder of $4.70\,\text{meV}$, respectively.

Figure 12.2 shows the numerical results for excitation at the energetic position of the disorder-free exciton. First, the results for co-circularly polarized exciting pulses are discussed. Figure 12.2a displays $\delta\alpha$ for $\tau = -2\,\text{ps}$ (probe pulse preceding pump pulse) without disorder (solid), with a disorder of $2.35\,\text{meV}$ (dashed), and with a twice as strong disorder (dotted). It is shown that the amplitudes of the coherent oscillations which dominate the signal of the perfect sample, see Fig. 11.2e, are strongly reduced with increasing disorder. This can be easily understood by the fact that the disorder partially acts as an inhomogeneous broadening. If the coherent oscillations are

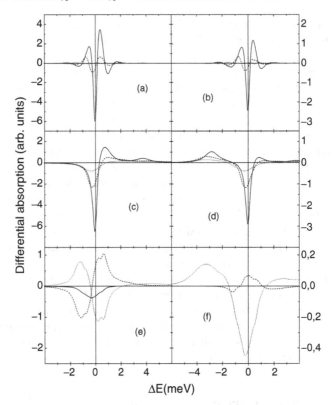

Fig. 12.2. Differential absorption spectra for excitation at the lowest exciton resonance. (**a**) Co-circularly polarized pump and probe pulses and $\tau = -2\,\mathrm{ps}$, *solid*: ordered system; *dashed*: 2.35 meV (FWHM) Gaussian disorder; *dotted*: 4.70 meV Gaussian disorder. (**b**) Same as (**a**), but with oppositely circularly polarized pump and probe pulses. (**c**) Same as (**a**) for $\tau = 2\,\mathrm{ps}$. (**d**) Same as (**b**) for $\tau = 2\,\mathrm{ps}$. (**e**) The three contributions to $\delta\alpha$ displayed in (**c**) for 4.70 meV disorder: $\delta\alpha_{\mathrm{pb}}$ (*solid*); $\delta\alpha_{\mathrm{CI,1st}}$ (*dashed*); and $\delta\alpha_{\mathrm{CI,corr}}$ (*dotted*). (**f**) The two contributions to $\delta\alpha$ displayed in (**d**) for 4.70 meV disorder: $\delta\alpha_{\mathrm{CI,1st}}$ (*dashed*) and $\delta\alpha_{\mathrm{CI,corr}}$ (*dotted*). [From C. Sieh et al., Eur. Phys. J. B **11**, 407 (1999)]

averaged over an inhomogeneous distribution with a width comparable to or larger than the energetic period of the coherent oscillations, the resulting $\delta\alpha$ becomes much smaller. Figure 12.2c displays $\delta\alpha$ for $\tau = 2\,\mathrm{ps}$. One can identify that the disorder induces a red shift of the bleaching maximum as well as some broadening of the exciton line. This red shift can be expected already from the linear optical spectrum, which develops tails into the gap region and a red shift for increasing disorder. Furthermore, the two-peak structure in the induced absorption present for the ordered case above the exciton resonance is averaged out with increasing disorder. The general line shape of $\delta\alpha$ for excitation with co-circularly polarized pulses, i.e., a bleaching at the exciton resonance and excited-state absorption for higher energies induced by unbound biexciton states, however, survives in the presence of disorder. The three contributions to $\delta\alpha$ for $\tau = 2\,\mathrm{ps}$ and strong disorder are displayed separately in Fig. 12.2e. As in the ordered case, both $\delta\alpha_{\mathrm{CI,1st}}$ and $\delta\alpha_{\mathrm{CI,corr}}$ are quite large and strongly compensate each other. Compared to the ordered case, the relative importance of $\delta\alpha_{\mathrm{pb}}$ becomes larger and, for the present parameters, is responsible for about half of the bleaching at the exciton resonance. This suggests that with increasing disorder the signatures of the Coulomb many-particle interaction are reduced compared to the Pauli-blocking signatures.

Next, we discuss the results obtained for oppositely circularly polarized pump–probe pulses. Figure 12.2b displays $\delta\alpha$ for $\tau = -2\,\mathrm{ps}$ without disorder (solid), with a disorder of $2.35\,\mathrm{meV}$ (dashed), and with a twice as strong disorder (dotted). Like Fig. 12.2a, Fig. 12.2b also shows that the coherent oscillations are strongly suppressed by the disorder. Figure 12.2d displays $\delta\alpha$ for $\tau = 2\,\mathrm{ps}$. As in Fig. 12.2c, one can identify that the disorder induces the expected red shift of the bleaching maximum as well as some broadening of the exciton line. Also, the positive low-energy biexciton peak, which in the ordered case appears $2.7\,\mathrm{meV}$ below the excitonic resonance, shifts towards lower energies and is broadened. The energetic shift of the biexciton line is approximately twice as large as the shift of the exciton. Even for the strongest disorder of $4.70\,\mathrm{meV}$ considered here, which is larger than the biexciton binding energy, a pronounced biexciton peak is still present in $\delta\alpha$. This is consistent with time-integrated four-wave-mixing experiments, which reported pronounced exciton–biexciton beats in extremely strongly disordered quantum-well samples, see Albrecht et al. (1996).

It can thus be concluded that for the model and parameters investigated here, the disorder, besides some red shifting of both the exciton as well as the biexciton contributions, mainly induces an inhomogeneous broadening of both the one- and two-exciton states. The two contributions to $\delta\alpha$ for $\tau = 2\,\mathrm{ps}$ for the present case of $4.70\,\mathrm{meV}$ disorder and excitation with oppositely circularly polarized pulses, where $\delta\alpha = \delta\alpha_{\mathrm{CI,1st}} + \delta\alpha_{\mathrm{CI,corr}}$ since $\delta\alpha_{\mathrm{pb}}$ vanishes, are displayed separately in Fig. 12.2f. The first-order Coulomb contribution which vanishes in the ordered case, becomes finite with disorder. Compared to the correlation contribution it is, however, still quite small, at least for the relatively weak disorder considered here ($W < J^e$).

12.2.2 Four-Wave Mixing in the $\chi^{(3)}$ Limit

Again we consider the model underlying Sect. 11.2. The calculations are performed for a system consisting of 10 sites using periodic boundary conditions. For the ordered case, it has been checked that all features present in the numerical results do not change for larger systems.

The envelopes of the incident pulses are again taken to be Gaussians, i.e., $E_l(t) = \exp(-((t - t_l)/\delta_l)^2)$, with a duration characterized by δ_l and the maximum at $t = t_l$. Here, we consider degenerate excitation, i.e., two pulses with the same central frequency ($\omega_1 = \omega_2$) and duration ($\delta_1 = \delta_2$). Furthermore, we set $t_2 = 0$ and $t_1 = -\tau$, where τ is the delay between the two pulses. Thus, $t = 0$ corresponds to the arrival of the second pulse. The duration of the pulses is taken as $\delta_1 = \delta_2 = 500\,\text{fs}$ and the central frequencies of the pulses are tuned $1\,\text{meV}$ below the energy of the $1s$ exciton in the ordered system in order to avoid excitation of the continuum.

To demonstrate the importance of correlations for resonant excitation of the lowest exciton, we start by discussing the time-resolved signal of the ordered model which serves as a reference for the calculations for the disordered model. In contrast to the more complete calculation in Sect. 11.2.2, we restrict the treatment of the Coulomb correlations up to the biexciton amplitudes B and neglect the three-exciton amplitudes W. Figure 12.3 displays the calculated time-resolved signal for (a) linear parallel (xx) and (b) linear perpendicular (xy) excitation. To get insight into the signatures of the different types of optical nonlinearities, we plot in Fig. 12.3 the total signal as well as the time-resolved signals induced purely by Pauli blocking (pb), first-order Coulomb (CI,1st), and higher order Coulomb correlations (CI,corr). Because the $500\,\text{fs}$ pulses excite predominantly the lowest exciton and no energetically higher exciton states, "pb" and "CI,1st" show no modulations. Furthermore, none of these two components depend on the polarization configuration, but give identical results for xx and xy. The "pb" term simply rises with the integral over the second pulse and then remains constant. For both excitation configurations, it yields only a small contribution to the total signal. The time-resolved intensity induced by the "CI,1st" term increases quadratically with time. As long as no dephasing is included in the calculation, this unphysical increase, which is induced by the local-field-like nonlinearity of the lowest exciton, continues infinitely. To obtain finite signals as well as a polarization dependence of the nonlinear optical response, it is essential to treat also the "CI,corr" term. As is shown in Fig. 12.3a and b, the time-resolved intensity of this term alone basically increases quadratically with time. On top of this increase, one can see polarization-dependent beats, which are induced by the excitation of bound and unbound biexciton states. When building the total signal from these three nonlinearities, the quadratic increase with time of "CI,1st" and "CI,corr" vanishes due to strong compensations among these terms, and the total signal remains finite, see the solid lines in Fig. 12.3.

For both the xx and xy configurations, the total time-resolved signal is modulated. The rapid modulations for xx correspond to exciton–biexciton beats and have a frequency determined by the inverse biexciton binding energy (about $2.7\,\text{meV}$ for the present parameters). The weaker slow modulations are due to excited unbound biexciton states. For xy polarized excitation, the modulation periods correspond to the inverse of the energy differences between biexciton states. Thus, the small modulation period is the inverse of the difference between the biexciton and the lowest unbound biexciton state, which in the present model is energetically situated slightly above twice the energy of the lowest exciton. Furthermore, we see, by comparing Figs. 12.3a

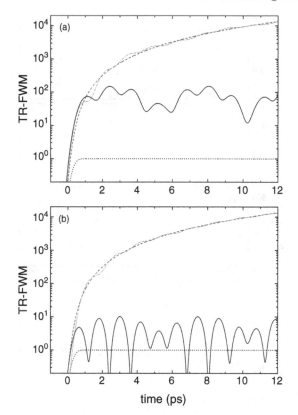

Fig. 12.3. Time-resolved four-wave-mixing traces for delay $\tau = 2\,\mathrm{ps}$ for the ordered model: (**a**) linear parallel xx and (**b**) linear perpendicular xy excitation. Besides the total signal (*solid*), the results for the three nonlinearities corresponding to Pauli blocking (*dotted*), first-order Coulomb (*dashed*), and higher order Coulomb correlations (*dashed-dotted*) are displayed separately. [From S. Weiser et al., Phys. Rev. **B61**, 13088 (2000)]

and b, that the signal for xy is weaker and more strongly modulated than for xx. This polarization dependence is solely induced by Coulomb correlations and has its origin in strong cancellations of the excitonic signal components for the xy configuration.

Next, results on time-resolved traces including disorder are presented. As an example, we consider here diagonal disorder in the conduction band $\delta\epsilon_i^c$ characterized by a Gaussian distribution $\exp(-(\delta\epsilon_i^e)^2/(2W^{e2}))$ with $W^e = 2.5\,\mathrm{meV}$, i.e., W^e is about 1/6th of the exciton binding energy. The disorder is assumed to be uncorrelated in real space, such that the energies for all sites are taken independently of the distribution. Figures 12.4a–e display time-resolved signals for three time delays. To obtain the four-wave-mixing signals with disorder, we have averaged over 100 disorder realizations. It has been checked

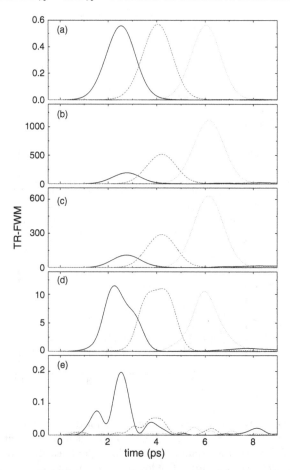

Fig. 12.4. Time-resolved four-wave-mixing traces for delays $\tau = 2$, 4, and 6 ps for conduction-band disorder of strength $W^e = 2.5$ meV: (**a**) only Pauli blocking ($xx = xy$); (**b**) Hartree–Fock calculation given by Pauli blocking plus first-order Coulomb terms for xx; (**c**) Hartree–Fock calculation given by Pauli blocking plus first-order Coulomb terms for xy; (**d**) total signal including correlations for xx excitation; and (**e**) total signal including correlations for xy excitation. [From S. Weiser et al., Phys. Rev. **B61**, 13088 (2000)]

for some representative calculations that, by adding more realizations, no qualitative change of the signals occurs. In Fig. 12.4a, only the "pb" term is considered which gives identical results for xx and xy excitation. It is shown that the disorder-induced inhomogeneous broadening of the energy of the lowest exciton leads to the emission of the signal as a photon echo. No obvious dependence of the time-resolved intensity on the time delay is present in Fig. 12.4a. Figures 12.4b and c show the time-resolved signals obtained at the time-dependent Hartree–Fock level, which is given by considering the "pb" and the "CI,1st" terms, for xx and xy excitation, respectively. Except for a different overall amplitude, the signals in Fig. 12.4b and c are very similar. As for the "pb" term alone, also within the time-dependent Hartree–Fock calculations, the disorder induces a photon echo emission of the time-resolved traces. However, due to the local-field-like excitonic nonlinearity, the amplitude of the echoes now *increases* with time delay. Thus, for resonant excitation of the lowest exciton and short-range disorder on an energy scale much smaller than the exciton binding energy, the time-dependent Hartree–Fock theory does not result in disorder-induced dephasing, but predicts a signal that increases with increasing time delay.

As in the ordered case (cf. Fig. 12.3), Coulomb correlations are needed to recover nonincreasing echoes, see Figs. 12.4d and e. The modulations of the time-resolved traces that are present in the ordered case, see Fig. 12.3, partly survive in the presence of disorder and induce some polarization-dependent structure on top of the echoes, see Figs. 12.4d and e. As in the ordered case, also with disorder, the modulations are stronger and the time-resolved intensity is weaker for the xy compared to xx configuration. Furthermore, a disorder-induced decrease of the time-resolved intensity with increasing delay is present for xy, but absent for xx. Figure 12.4 gives the first hint towards the existence of polarization-dependent disorder-induced dephasing and this phenomenon is discussed in detail below.

In order to analyze the disorder-induced dephasing, we discuss results on time-integrated four-wave-mixing traces. Having shown that both "CI,1st" and "CI,corr" alone suffer from an unphysical temporal rise, we do not discuss those two terms separately from now on, but combine them into a single term, which is called Coulomb interaction-induced ("CI") nonlinearity. The complete four-wave-mixing signal is thus made up of contributions due to "pb" and "CI".

Numerical results on time-integrated traces with diagonal disorder of three different strengths $W^e = 1.5\,\text{meV}$, $2.5\,\text{meV}$, and $3.5\,\text{meV}$ in the conduction band are displayed in Figs. 12.5a for the xx and (b) for the xy configuration. Furthermore, the results for a diagonal valence-band disorder of $W^h = 1.5\,\text{meV}$ are also included. In all cases, apart from some noise, there is no obvious decay of the time-integrated traces with time delay for xx. For xy, however, clearly the time-integrated intensity decreases with increasing time delay. Furthermore, it can be seen in Fig. 12.5b that the decay for $W^e = 2.5\,\text{meV}$ is faster than for $W^e = 1.5\,\text{meV}$. However, for $W^e = 3.5\,\text{meV}$, the decay of the time-integrated traces does not seem to speed up further. The weak modulation appearing for $W^e = 3.5\,\text{meV}$ could be due to exciton–biexciton beating which becomes visible if the width of the temporal photon echo is comparable to the inverse of the biexciton binding energy. Possibly, in this case, averaging over more disorder realizations is required or the appearance of the beats complicates the extraction of the decay. To obtain an estimate of the strength of the disorder-induced dephasing, we fit the curves in Fig. 12.5b with a single exponential decay $\alpha_0 \exp(-\tau/(T_2/4))$ (see thin lines),

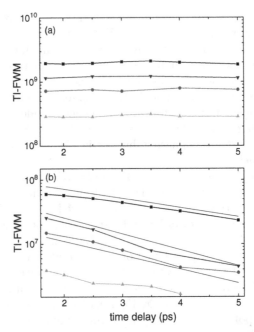

Fig. 12.5. Time-integrated traces versus time delay for three conduction-band disorder strengths of $W^e = 1.5\,\mathrm{meV}$ (*squares*), $2.5\,\mathrm{meV}$ (*circles*), and $3.5\,\mathrm{meV}$ (*triangles up*), respectively, as well as one valence-band disorder strength of $W^h = 1.5\,\mathrm{meV}$ (*triangle down*) (**a**) for xx and (**b**) for xy excitation. The *symbols* indicate the time delays considered in the numerical calculations and the *lines* are guides for the eye. The *thin lines* appearing in (**b**) describe exponential decays $\propto \exp(-\tau/(T_2/4))$, with $T_2 = 12\,\mathrm{ps}$, $7\,\mathrm{ps}$, and $8\,\mathrm{ps}$, from top to bottom, respectively. [From S. Weiser et al., Phys. Rev. **B61**, 13088 (2000)]

where α_0 is a prefactor determining the overall strength. Note that $T_2/4$ would be the decay constant of the time-integrated traces of an inhomogeneously broadened two-level system, if a dephasing time T_2 was introduced phenomenologically. For the case of conduction-band disorder, we obtain $T_2 = 12\,\mathrm{ps}$ for $W^e = 1.5\,\mathrm{meV}$ and $T_2 = 8\,\mathrm{ps}$ for $W^e = 2.5\,\mathrm{meV}$, respectively.

For the valence-band disorder of $W^h = 1.5\,\mathrm{meV}$, we obtain $T_2 = 7\,\mathrm{ps}$. Due to the larger mass of the heavy holes, the same value of disorder considered in the valence band has a stronger effect and thus induces more rapid dephasing in comparison to disorder in the conduction band.

To get insight into the origin of the obtained disorder-induced dephasing, we show the time-integrated traces for conduction-band disorder and xy excitation induced by the "pb" and "CI" terms separately in Fig. 12.6. As expected, "pb" alone shows no dephasing since the present disorder strengths are much weaker than the exciton binding energy. If one suspected that the dephasing of the total time-integrated traces was induced by an inhomogeneous

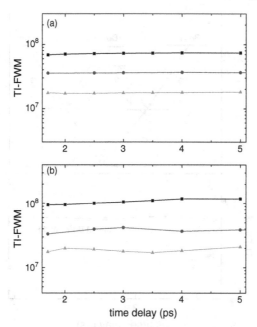

Fig. 12.6. Time-integrated traces versus time delay for three conduction-band disorder strengths of $W^e = 1.5$ meV (*squares*), 2.5 meV (*circles*), and 3.5 meV (*triangles up*) for xy excitation. (a) Only Pauli blocking and (b) only Coulomb interaction-induced nonlinearities are considered. The *symbols* indicate the time delays considered in the numerical calculations and the *lines* are guides for the eye. [From S. Weiser et al., Phys. Rev. **B61**, 13088 (2000)]

distribution of the biexciton binding energy, one would expect that "CI" alone induces the dephasing. Surprisingly, however, we find in Fig. 12.6 that also "CI" alone shows no dephasing. Therefore, we conclude that for the present model with short-range disorder the dephasing is not due to an inhomogeneous distribution of the biexciton binding energy, but instead is the consequence of an interference effect between "pb" and "CI". This strong interference can be inferred from comparing the amplitudes of "pb" and "CI", see Fig. 12.6, to the much smaller amplitude of the total signal, see Fig. 12.5b. For the xx configuration such a strong interference between "pb" and "CI" does not happen, because in this case "CI" is much stronger than "pb", as has been discussed above, see Sect. 11.2.2, and thus the four-wave-mixing signal is dominated by "CI".

To gain further understanding of the disorder-induced dephasing found here, we have performed additional calculations in which we have artificially neglected different biexciton states. The biexciton states that are needed to describe the $\chi^{(3)}$ response are sketched in Fig. 12.7a. Whereas the interaction of two single excitons with opposite spin results in the formation of both a bound biexciton as well as unbound biexciton states, the interaction between

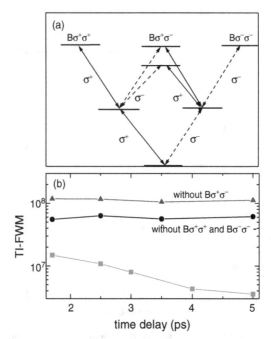

Fig. 12.7. (a) Level scheme representing the single and two-exciton states relevant in $\chi^{(3)}$. (b) Time-integrated four-wave-mixing traces for conduction-band disorder of $W^e = 2.5\,\text{meV}$ (*squares*) for xy excitation. Also shown are results of artificial calculations where either all unbound two excitons states made of two excitons with the same spin ($B\sigma^+\sigma^+$ and $B\sigma^-\sigma^-$) are neglected (*circles*) or all bound and unbound biexciton states made of two excitons with opposite spin ($B\sigma^+\sigma^-$) are neglected (*triangles up*), respectively. [From S. Weiser et al., Phys. Rev. **B61**, 13088 (2000)]

two excitons with the same spin results only in unbound biexciton states. It should be noted that the biexciton has a significant binding energy (about 2.7 meV for the present parameters), whereas the energies of the lowest unbound biexciton states are only slightly larger than twice the energy of the lowest exciton. To demonstrate that unbound biexciton states are indeed important in the context of disorder-induced dephasing, we display in Fig. 12.7b the time-integrated traces for $W^e = 2.5\,\text{meV}$, neglecting all unbound biexcitons made of single excitons with identical spin ($B\sigma^+\sigma^+$ and $B\sigma^-\sigma^-$). As shown by the circles in Fig. 12.7b, the decay of the time-integrated traces is absent if those biexciton states are neglected. Also, if all biexciton states made of two single excitons with opposite spin are neglected, see triangles in Fig. 12.7b, there is no dephasing of the time-integrated traces. The results of these two artificial calculations allow us to conclude that the effect of disorder-induced dephasing (Fig. 12.7b, squares), as analyzed within the present model, is due to destructive interference of contributions induced by different unbound biexciton states.

Besides considering the two extremes of one ordered and one disordered band, we have also performed calculations considering simultaneously disordered valence and conduction bands. For these calculations, we chose $W^e = 1.8\,\text{meV}$ and $W^h = 0.18\,\text{meV}$, i.e., the disorder strengths in the two bands are proportional to their real-space nearest-neighbor couplings, which are inversely proportional to the effective mass at the band extrema. Furthermore, in this case of two disordered bands, we can assume the site disorder to be either correlated, uncorrelated, or even anticorrelated. The results (not shown) demonstrate that there is no dephasing present for xx excitation in any of the cases. The dephasing for the xy configuration is about the same, i.e., it is basically independent of the disorder correlation. We thus do not find any distinct signature of correlations between conduction and valence-band disorder in the present model and the studied parameters. For $W^e = 1.8\,\text{meV}$ and $W^h = 0.18\,\text{meV}$ this is simply due to the fact that the conduction-band disorder is much stronger than the valence-band disorder, which can basically be neglected. Thus, to investigate possible correlations between the conduction- and valence-band disorder, both should be of comparable strength. Additional calculations performed for $W^e = W^h = 1\,\text{meV}$ (not shown) result in very similar dephasing for xy excitation considering correlated and uncorrelated disorder in the conduction and valence bands. For anticorrelated disorder, however, there is basically no dephasing in the numerical results. This can be understood considering that, in this case for $W^e = W^h = 1\,\text{meV}$ on each isolated site, the transition energy is unchanged. Furthermore, it is important to note that within the present model the excitons are strongly bound and thus the relative motion extends only over a few sites. Thus, the effects induced by anticorrelated disorder are rather weak within the present model.

12.2.3 Four-Wave Mixing in the $\chi^{(5)}$ Limit

In Sect. 11.2.2, we studied exciton–biexciton beats in the time-integrated and time-resolved four-wave-mixing traces of the ordered model. It has been shown that within the $\chi^{(3)}$ limit there are no beats in the time-integrated traces for the ordered situation. Such beats become, however, visible in the $\chi^{(5)}$ limit. Besides excitation with higher intensities, also disorder may lead to exciton–biexciton beats in the time-integrated four-wave-mixing traces already in the $\chi^{(3)}$ limit. To explain this effect, it is sufficient to consider a phenomenological disorder-induced inhomogeneous broadening. If this inhomogeneous broadening is assumed to be correlated for the exciton and the exciton to biexciton transitions, it leads to a photon-echo-like time-resolved four-wave-mixing signal. Assuming a Gaussian inhomogeneous broadening thus introduces a Gaussian temporal envelope $\propto \exp(-((t-\tau)/\delta))^2)$ of the time-resolved four-wave-mixing response, where lower limits for the temporal widths of the photon echo δ are given by the inverse of the width of the inhomogeneous distribution and by the temporal width of the exciting pulses.

To phenomenologically include disorder effects into our calculations, we simply assume that such a correlated Gaussian inhomogeneous broadening of the exciton and biexciton energies is present. The spectral width of the inhomogeneous broadening is assumed to be larger than the spectral pulse widths, such that the inhomogeneous broadening simply introduces a photon-echo emission of the time-resolved four-wave-mixing response with a width determined by the incident pulses (δ_1 and δ_2). Therefore, in order to compute polarization-dependent time-integrated four-wave-mixing signals in the presence of inhomogeneous broadening, we simply multiply the time-resolved four-wave-mixing amplitude obtained for the homogeneous system by $\exp(-((t-\tau)/\delta))^2)$ using $\delta = \delta_1 = \delta_2$ and then integrate over time to obtain the time-integrated four-wave-mixing signal.

The time-integrated signals including inhomogeneous broadening are displayed in Fig. 12.8. As shown in Fig. 12.8a, already in $\chi^{(3)}$, inhomogeneous broadening induces exciton–biexciton beats for both xx and xy excitations. Since, in our calculations, we have used rather long pulses of $\delta_1 = \delta_2 = 500$ fs which are not much shorter than the period of the beats (≈ 1.2 ps), the disorder-induced beating and the polarization dependence of the beat phase are not very pronounced. Another result of the inhomogeneous broadening is the strong suppression of four-wave-mixing signals for negative delays, which can be noticed by comparing Figs. 11.10a and 12.8a (note the different time delay axes in these figures). This suppression is a consequence of the fact that for negative time delays τ the maximum of the echo amplitude $\propto \exp(-((t-\tau)/\delta))^2)$ occurs at $t = \tau < 0$, i.e., at a temporal position before the pulses have excited the system.

With increased intensity, we find in Fig. 12.8b that the beats for xx become slightly stronger. For xy, however, one can see in Fig. 12.8b and also in Fig. 12.8c, where the pure $\chi^{(5)}$ contribution is shown, that half-period beats appear. Such half-period beats in the time-integrated four-wave-mixing traces have actually been observed experimentally on disordered samples for xy excitation and increasing excitation intensity. For an explanation of these half-period beats, which only appear for the xy and not for the xx configuration, one should consider the $\chi^{(5)}$ contributions to the time-resolved four-wave-mixing traces, as displayed in Fig. 11.8b and 11.9b. The main difference between xx and xy is that, for xy, the $\chi^{(5)}$ contributions introduce an additional time-delay-dependent phase shift of the time-resolved four-wave-mixing signals, which is not present for xx. Due to the emission of the time-resolved four-wave-mixing signal as a photon echo, in disordered systems, the time-resolved trace may be strongly influenced by the dependence of the phase on the time delay which introduces the half-period beats.

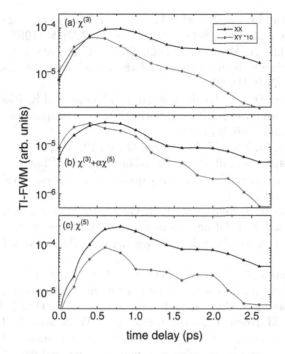

Fig. 12.8. Time-integrated four-wave mixing versus time delay for linear parallel (*triangles up*) and linear perpendicular (*circles*) excitation. Disorder is included as correlated inhomogeneous broadening as discussed in the text: (**a**) pure $\chi^{(3)}$; (**b**) $\chi^{(3)} + \alpha\chi^{(5)}$; and (**c**) pure $\chi^{(5)}$. The *lines* are guides for the eye. [From T. Meier and S.W. Koch, in *Ultrafast Physical Processes in Semiconductors*, special issue of Series *Semiconductors and Semimetals*, Vol. 67, pp. 231–313 (Academic Press, New York, 2001)]

12.3 Suggested Reading

1. T.F. Albrecht, K. Bott, T. Meier, A. Schulze, M. Koch, S.T. Cundiff, J. Feldmann, W. Stolz, P. Thomas, S.W. Koch, E.O. Göbel "Disorder mediated biexcitonic beats in semiconductor quantum wells", Phys. Rev. B **54**, 4436 (1996)
2. D. Bennhardt, P. Thomas, A. Weller, M. Lindberg, and S.W. Koch, "Influence of Coulomb interaction on the photon echo in disordered semiconductors", Phys. Rev. **B43**, 8934 (1991)
3. D. Bennhardt, P. Thomas, R. Eccleston, E.J. Mayer, and J. Kuhl, "Polarization dependence of four-wave-mixing signals in quantum wells", Phys. Rev. **B47**, 13485 (1993)
4. D. Brinkmann, K. Bott, S.W. Koch, and P. Thomas, "Disorder-induced dephasing of excitons in semiconductor heterostructures", phys. stat. sol. b **206**, 493 (1998)

5. O. Carmel and I. Bar-Joseph, "Four-wave-mixing studies of quantum-well excitons in a magnetic field", Phys. Rev. **B47**, 7606 (1993)
6. S.T. Cundiff, H. Wang, and D.G. Steel, "Polarization-dependent picosecond excitonic nonlinearities and the complexities of disorder", Phys. Rev. **B46**, 7248 (1992)
7. R. Eccleston, D. Bennhardt, J. Kuhl, P. Thomas, and K. Ploog, "Intensity dependent four wave mixing polarization rules in quantum wells", Solid State Commun. **86**, 93 (1993)
8. E.O. Göbel, M. Koch, J. Feldmann, G. von Plessen, T. Meier, A. Schulze, P. Thomas, S. Schmitt-Rink, K. Köhler, and K. Ploog, "Time resolved four wave mixing in GaAs/AlAs quantum well structures", phys. stat. sol. b **173**, 21 (1992)
9. P. Grossmann, J. Feldmann, E.O. Göbel, P. Thomas, D.J. Arent, K.A. Bertness, and J.M. Olson, "Homogeneous and inhomogeneous linewidths of excitons in partially ordered $Ga_{0.52}In_{0.48}P$", Appl. Phys. Lett. **65**, 2347 (1994)
10. P. Grossmann, J. Feldmann, E.O. Göbel, P. Thomas, D.J. Arent, K.A. Bertness, and J.M. Olson, "Homogeneous linewidths of excitons in high and low band gap $Ga_{0.52}In_{0.48}P$", phys. stat. sol. b **188**, 557 (1995)
11. F. Jahnke, M. Koch, T. Meier, J. Feldmann, W. Schäfer, P. Thomas, S.W. Koch, E.O. Göbel, and H. Nickel, "Simultaneous influence of disorder and Coulomb interaction on photon echoes in semiconductors", Phys. Rev. **B50**, 8114 (1994)
12. M. Koch, D. Weber, J. Feldmann, E.O. Göbel, T. Meier, A. Schulze, P. Thomas, S. Schmitt-Rink, and K. Ploog, "Subpicosecond photon echo spectroscopy on GaAs/AlAs short period superlattices", Phys. Rev. **B47**, 1532 (1993)
13. Ch. Lonsky, P. Thomas, and A. Weller, "Optical dephasing in disordered semiconductors", Phys. Rev. Lett. **63**, 652 (1989)
14. U. Siegner, D. Weber, E.O. Göbel, D. Bennhardt, V. Heuckeroth, R. Saleh, S. D. Baranovskii, P. Thomas, H. Schwab, C. Klingshirn, J.M. Hvam, and V.G. Lyssenko, "Optical dephasing in semiconductor mixed crystals", Phys. Rev. **B46**, 4564 (1992)
15. C. Sieh, T. Meier, A. Knorr, F. Jahnke, P. Thomas, and S.W. Koch, "Influence of carrier correlations on the excitonic optical response including disorder and microcavity effects", Eur. Phys. J. B **11**, 407 (1999)
16. S. Weiser, T. Meier, J. Möbius, A. Euteneuer, E.J. Mayer, W. Stolz, M. Hofmann, W.W. Rühle, P. Thomas, and S.W. Koch, "Disorder-induced dephasing in semiconductors", Phys. Rev. **B61**, 13088 (2000)

Coherent Excitation Spectroscopy

The discussions of Chap. 12 showed that disorder-induced dephasing is related to optical excitations coupled internally to each other either due to a relaxed selection rule or due to many-particle interactions. As we have explained in detail in Chap. 10, internal coupling leads to quantum beats which can be identified experimentally by analyzing the t-τ plot of the modulus of the third-order optical polarization. The resulting pattern tells us whether we have polarization interference or quantum beats.

In many cases, however, such t-τ plots cannot so easily be interpreted. In particular, the exciton–biexciton beats which are certainly due to internal coupling show a different pattern if compared to the quantum beats of three-level systems. It is also not trivial to record the four-wave-mixing signal in real time. Therefore, one has looked for other experimental approaches which give the required information in the frequency domain.

One such approach is the so-called coherent excitation spectroscopy. It is a variant of the four-wave-mixing experiment. However, the two excitation pulses are no longer identical, as far as their central frequency and spectral width are considered. Instead, the first pulse is long, i.e., spectrally narrow, and centered at an energy called the excitation energy $\hbar\omega_{\mathrm{exc}}$, while the second pulse is short and covers the entire spectral range of the transition energies under study. In coherent excitation spectroscopy, one records the four-wave-mixing signal as a function of energy, which is called the detection energy $\hbar\omega_{\mathrm{det}}$, in the frequency domain. The data are then presented as contour plots in a two-dimensional figure with axes giving the excitation energy $\hbar\omega_{\mathrm{exc}}$ and the detection energy $\hbar\omega_{\mathrm{det}}$. In a sense, a coherent excitation spectroscopy experiment is a combination of four-wave-mixing and pump–probe experiments.

13.1 Level Systems

In order to introduce the coherent excitation spectroscopy technique, it is useful to consider an ensemble of few-level absorbers, see Fig. 13.1, as an

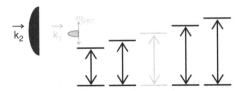

Fig. 13.1. An ensemble of two-level absorbers excited by a long pulse with wave vector k_1, centered at the frequency ω_{exc}, and probed by a short pulse with wave vector k_2. Only the subensemble which is resonant with ω_{exc} gives rise to a four-wave-mixing signal

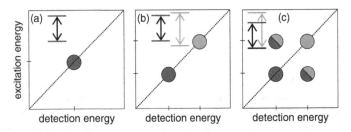

Fig. 13.2. (a) The response in the ω_{exc}–ω_{det} plane for an ensemble of independent two-level absorbers. There is a signal only on the main diagonal at $\omega_{\mathrm{exc}} = \omega_{\mathrm{det}}$. (b) An ensemble consisting or two subensembles having two different resonant frequencies ω_1 and ω_2 leads to two peaks on the main diagonal. (c) On the other hand, an ensemble of three-level systems produces four peaks, two on the main diagonal and two additional signals indicating the internal coupling due to the common ground state

example. Let us start with an ensemble of identical two-level absorbers which all have the same energy difference $\hbar\omega_{\mathrm{exc}}$. The first spectrally narrow (excitation) pulse with frequency ω_{exc} excites this ensemble. The second pulse, applied without any delay ($\tau = 0$), induces a four-wave-mixing signal which in this case contains only the frequency $\omega_{\mathrm{det}} = \omega_{\mathrm{exc}}$. Therefore, we expect a signal only on the main diagonal in the ω_{exc}–ω_{det} plane as a function of ω_{exc} and ω_{det}, see Fig. 13.2a.

We next consider an ensemble of two-level absorbers consisting of two subensembles having different resonance frequencies ω_1 and ω_2. This situation corresponds to polarization interference in four-wave-mixing experiments. Now, we obtain a signal whenever $\omega_{\mathrm{det}} = \omega_{\mathrm{exc}} = \omega_1$ or $\omega_{\mathrm{det}} = \omega_{\mathrm{exc}} = \omega_2$ coincides with one or the other resonance frequency. Thus, we get two peaks on the main diagonal, see Fig. 13.2b. Obviously, the inhomogeneous ensemble shown in Fig. 13.1 leads to signals along the main diagonal over the entire range of transition frequencies.

The signal looks completely different if the ensemble consists of systems that are internally coupled. To illustrate this fact, we choose the ensemble of three-level systems with resonances at ω_1 and ω_2 that is known to produce quantum beats. Now, in addition to the two signals on the main diagonal, we also see a peak at the position $\omega_{\text{exc}} = \omega_1$ and $\omega_{\text{det}} = \omega_2$ and a peak at $\omega_{\text{exc}} = \omega_2$ and $\omega_{\text{det}} = \omega_1$, see Fig. 13.2c. These additional peaks indicate the internal coupling of the two transitions by the common ground state. Once the higher lying resonance is excited by the first pulse, a four-wave-mixing signal is produced with both the higher and the lower resonance frequency. The same is true for this model if the lower resonance is excited by the first pulse. Then, in addition to a signal at this frequency, a signal at the higher frequency is also obtained.

13.2 Ordered Semiconductors

An application to an ordered semiconductor model is shown in Fig. 13.3a. Now the zero of the energy scales has been placed at the lowest excitonic resonance. For parallel linear (xx) polarized pulses, a signal is obtained on the main diagonal for $\omega_{\text{det}} = \omega_{\text{exc}} = \omega_x$, i.e., at the excitonic resonance. The width of this peak is given by the homogeneous broadening introduced by a finite T_2 time into the model, i.e., even if the excitation is centered at a frequency below the maximum of the excitonic line, but still within the homogeneous width, the four-wave-mixing signal has a maximum at exactly the excitonic resonance.

For perpendicularly polarized (xy) pulses, Fig. 13.3b shows a completely different pattern. In addition to the excitonic feature on the main diagonal, there are two peaks off the main diagonal. These are due to transitions to the biexciton. There are also biexcitonic signals present in the situation of Fig. 13.3a, but for xx excitation the excitonic peak completely dominates the pattern, and the biexcitonic features are not visible on the linear grey scale applied in this figure.

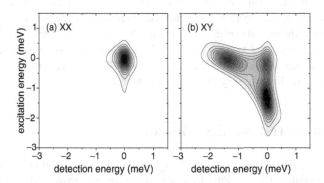

Fig. 13.3. (a) Coherent excitation spectroscopy plot for an ordered semiconductor model for linear parallel polarized pulses. (b) The same for linear perpendicularly polarized pulses showing additional signatures due to transitions to the biexciton. The *grey scale* is linear. The zero of the energy scale corresponds to the lowest excitonic resonance. [After E. Finger et al., phys. stat. sol. b **234**, 424 (2002)]

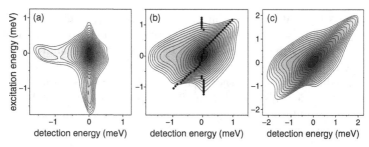

Fig. 13.4. Both pulses are polarized linearly parallel, i.e., in xx configuration. The *grey scale* is logarithmic. (**a**) The ordered semiconductor: A very weak biexciton signature is seen at $\omega_{exc} = 0$ and $\omega_{det} = -1$. The *vertical structure* with maximum at the excitonic resonance $\omega_{exc} = 0$ and $\omega_{det} = 0$ is due to both the homogeneous width of the exciton and a very weak biexcitonic contribution. (**b**) Modeling an inhomogeneous contribution to the line: In addition to the *pattern* seen in (**a**), with the *vertical feature* due to the homogeneous width of the exciton, there is a structure extending along the main diagonal indicating the inhomogeneous contribution to the excitonic line. (**c**) The same as (**b**), but with larger inhomogeneous broadening: Now, the inhomogeneous feature is dominant. Note that the zero of the energy scale corresponds to the lowest excitonic resonance. [After T. Meier et al., phys. stat. sol. b **238**, 537 (2003)]

Even if only the exciton is excited by the first pulse, the second pulse gives rise to a four-wave-mixing signal at the frequency of the exciton to biexciton transition. On the other hand, for excitation at the frequency of the exciton to biexciton transition an excitonic four-wave-mixing signal is also obtained. There is, however, no contribution at the exciton to biexciton transition in this latter situation. The exciton–biexciton system therefore differs from the three-level system, which yields four peaks, see Fig. 13.2c. Our discussion of the temporal dynamics of the third-order polarization has already shown that these two systems are not equivalent. The reason can be traced back to their different structure. If we roughly represent the exciton–biexciton system by a three-level system, see Fig. 11.1, the level common to both transitions is the (excitonic) level in the center, while the common level is the ground state in the ordinary three-level system.

13.3 Disordered Semiconductors

As a second application, we study the influence of disorder on coherent excitation spectroscopy. This is done by calculating the four-wave-mixing signal for an ensemble of ordered semiconductors where the site energies are fixed and equal to each other for any single system, but vary from system to system. The final result is simply the superposition of the uncoupled systems with different transition energies. Therefore, this procedure only leads to inhomogeneous broadening and cannot tell us anything about disorder-induced coupling and dephasing. Parallel linearly polarized pulses are taken, and the gray scale in the coherent excitation spectroscopy plots is now logarithmic in order to present the signal in a larger range of amplitudes.

Figure 13.4a shows the same data as Fig. 13.3a and refers to the ordered situation. Very weak biexcitonic features off the main diagonal are visible for excitation at the lowest exciton. For weak disorder, Fig. 13.4b, one sees features characteristic of a homogeneous line. This is the vertical line of maxima at the excitonic resonance. In addition, there is also a signature of the inhomogeneous nature of the line, given by the maxima along the main diagonal. For larger disorder, Fig. 13.4c, the inhomogeneous feature dominates, and the homogeneous width only leads to a certain broadening of the peak structure along the main diagonal. As is shown in Finger et al., (2002) and Meier et al., (2003), many of the correlation- and disorder-induced signatures to coherent excitation spectroscopy that are discussed here have been observed experimentally.

13.4 Suggested Reading

1. S.T. Cundiff, M. Koch, W.H. Knox, J. Shah, and W. Stolz, "Optical coherence in semiconductors: Strong emission mediated by nondegenerate interactions", Phys. Rev. Lett. **77**, 1107 (1996)
2. A. Euteneuer, E. Finger, M. Hofmann, W. Stolz, T. Meier, P. Thomas, S.W. Koch, W.W. Rühle, R. Hey, and K. Ploog, "Coherent excitation spectroscopy on inhomogeneous exciton ensembles", Phys. Rev. Lett. **83**, 2073 (1999)
3. E. Finger, S.P. Kraft, M. Hofmann, T. Meier, S.W. Koch, W. Stolz, and W.W. Rühle, "Coulomb correlations and biexciton signatures in coherent excitation spectroscopy of semiconductor quantum-wells", phys. stat. sol. b **234**, 424 (2002)
4. X. Li, T. Zhang, C.N. Borca, and S. T. Cundiff, "Many-Body Interactions in Semiconductors Probed by Optical Two-Dimensional Fourier-Transform Spectroscopy", Phys. Rev. Lett. **96**, 057406 (2006)
5. T. Meier, C. Sieh, E. Finger, W. Stolz, W.W. Rühle, P. Thomas, and S.W. Koch, "Signatures of biexcitons and triexcitons in coherent nondegenerate semiconductor optics", phys. stat. sol. b **238**, 537 (2003)

Character of Continuum Transitions

We have already seen that the optical transitions for level systems which have a common ground state lead to quantum beats in four-wave-mixing experiments. In many cases, the discrete excitonic transitions, i.e., the lowest excitonic resonance, and the higher, but weaker, discrete resonances can be described by such simple level systems. Therefore, four-wave-mixing experiments performed on ordered semiconductors with pulses that cover the spectral range of these discrete excitonic resonances show prominent quantum-beat-like features.

On the other hand, if the Coulomb interaction is neglected, the optical transitions in the continuum lead to echo signatures in time-resolved four-wave-mixing traces, since, in this case, the optical line is purely inhomogeneous due to the k-selection rule. Therefore, the question arises, whether the Coulomb interaction, which leads to excitons and thus to the quantum-beat-like features for low-energy excitation, also alters the continuum transitions in such a way that continuum-quantum-beat-like signatures emerge. As we have seen, such continuum-quantum beats lead to a decay, a "dephasing", of the four-wave-mixing traces. Thus, echoes, i.e., signatures of an inhomogeneous line, could be replaced by a simple decay as reproduced by a homogeneous line, if the Coulomb interaction is considered.

In this chapter, we try to answer this question by calculating four-wave-mixing traces and also coherent excitation spectroscopy spectra for continuum transitions for the interacting ordered semiconductor model. Although of fundamental interest, an experimental verification of the results obtained on the basis of our model is probably not easily possible. Optical transitions in the continuum of band–band transitions in ordered semiconductors are known to be extremely susceptible to various kinds of very rapid scattering mechanisms which lead to an extremely fast "true" dephasing, e.g., via carrier LO-phonon and Coulomb scattering. The theoretical description of these processes is beyond the scope of this book. Nevertheless, the calculations yield interesting insights into the Coulomb-interaction-induced coupling of optical transitions in the electron–hole continuum of a semiconductor.

14.1 Four-Wave-Mixing Traces

Figure 14.1 shows the linear spectrum of our model system. The absorption edge of the interaction-free model is at $\hbar\omega = 0$. We see the strong excitonic resonance close to $-20\,\text{meV}$ and a few energetically higher lying discrete resonances, which we have loosely denoted by "1s", "2s", etc. Due to the finite number of sites ($N = 150$), the continuum appears as discrete peaks. We consider a phenomenological dephasing time of $T_2 = 10\,\text{ps}$ here.

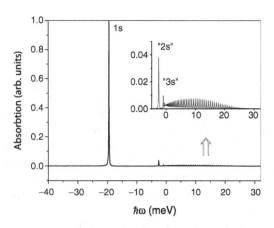

Fig. 14.1. Linear optical spectrum of the model system. The *inset* shows the higher-lying discrete excitonic resonances and the continuum on an enlarged scale

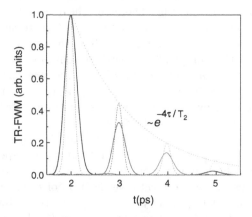

Fig. 14.2. The time-resolved four-wave-mixing trace showing photon echoes which decay with increasing time delay. Without Coulomb interaction (*dashed lines*), the echoes decay according to the phenomenologically introduced dephasing time of $T_2 = 10\,\text{ps}$. With Coulomb interaction included, we still see echoes (*solid lines*). However, their decay is considerably more rapid than in the interaction-free case. Here, echoes generated by a first pulse No. 1 at -2, -3, -4, $-5\,\text{ps}$ and a second pulse No. 2 at $t = 0$ are shown

As we are now interested in optical continuum excitations, we tune the light pulses into the continuum, i.e., $\hbar\omega_L = 16$ meV. If we neglect the Coulomb interaction (dashed lines in Fig. 14.2), we see the expected echoes in the time-resolved traces. Their intensity decays due to the phenomenological dephasing time, as indicated by the dotted line. In the interacting case, the echoes are still present, however, they decay more rapidly (solid lines in Fig. 14.2) than in the case without interaction. We might speculate that this additional decay mechanism is due to the partly homogeneous character of the transitions introduced by internal Coulomb coupling of the continuum transitions.

To substantiate this speculation, we calculate the coherent excitation spectroscopy signals. As we have seen in the previous chapter, these spectra give information about internal couplings between optical excitations in the regime of the discrete excitonic and biexcitonic resonances. The same holds for transitions in the continuum, as is shown below.

14.2 Coherent Excitation Spectroscopy

Figure 14.3 shows how pulse Nos. 1 and 2 are designed. In the following, we concentrate on case (c) with excitation in the continuum. Figure 14.4 shows the spectral four-wave-mixing response as a function of detection energy. Here, both pulses are linearly parallel polarized (xx). We observe two interesting signatures. (i) The spectra are symmetric with respect to the excitation energy (pulse No. 1) if the Coulomb interaction is neglected. In this case, the spectral width is given by the phenomenologically introduced dephasing. For increasing interaction, the spectra become wider. (ii) For sufficiently large interaction strength, the area under the line diminishes considerably.

The first observation (i) is what we expect. A wide spectrum indicates internal coupling. The width of the spectrum shows us how far this coupling

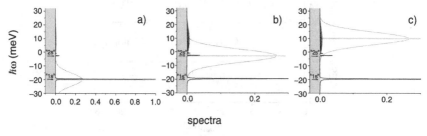

spectra

Fig. 14.3. Coherent excitation spectroscopy for three cases. Here, the linear optical absorption spectrum (*thick line*) and the spectra of pulse No. 1 (*narrow thin line*) and of pulse No. 2 (*broad thin line*) are shown. The central frequencies of both pulses coincide. (**a**) Excitation is at the lowest exciton, (**b**) at the band edge, and (**c**) in the continuum

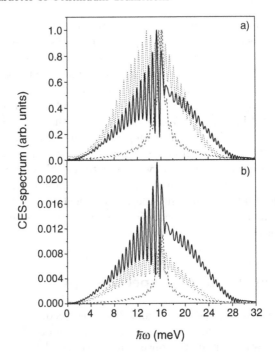

Fig. 14.4. Spectrally resolved four-wave-mixing response for excitation in the continuum with $\hbar\omega_{\mathrm{exc}} = 16\,\mathrm{meV}$. The pulses are linearly parallel polarized (xx). The considered strengths of the Coulomb interaction are $U_0 = 0.01\,\mathrm{eV}$ (*dotted*), $0.005\,\mathrm{eV}$ (*solid*), and 0 (*dashed*). (**a**) All spectra are normalized and (**b**) the spectra are not normalized

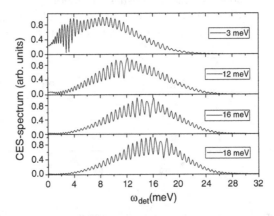

Fig. 14.5. Spectrally resolved four-wave-mixing response for pulses with central energies $\hbar\omega_{\mathrm{exc}} = 3, 12, 16, 18\,\mathrm{meV}$ and Coulomb strength $10\,\mathrm{meV}$. Linearly parallel (xx) polarization has been considered

extends spectrally. Obviously, this coupling is induced by the Coulomb interaction. The second observation (ii) is due to the fact that the Coulomb interaction moves the oscillator strength from the continuum into the discrete excitonic resonances. Compare also Figs. 9.12 and 9.13. Figure 14.5 shows that this spectral feature moves more or less rigidly together with the central energy of the two pulses.

The spectrally resolved response looks quite different if we apply linear cross-polarized pulses (xy). Figure 14.6 shows that the spectra are no longer symmetric. In addition, there is a dip exactly at the excitation energy. This is once more shown in Fig. 14.7, where spectra for both xx and xy polarization are compared directly. It is seen that the Coulomb-induced coupling is more effective for xx polarization, i.e., in this case, the spectrum is much wider compared to the xy situation. Note that both spectra have been normalized in order to be able to compare their widths more easily. Without normalization, the maximum of the xy spectrum is by a factor of about 2 smaller than that of the xx spectrum. Figures 14.7b and c show that the Hartree–Fock and the Pauli-blocking contributions do not depend on polarization and show a narrow, essentially symmetric line.

Therefore, we can conclude that the Coulomb-induced coupling in the continuum is due to Coulomb correlations which are beyond the Hartree–Fock limit. This coupling induces signatures of a homogeneous line. However, the echo-like structure still survives also for the interacting case. The continuum transitions can, therefore, be classified as mixed transitions having both inhomogeneous and homogeneous character. While the former leads to echoes, the latter induces additional dephasing.

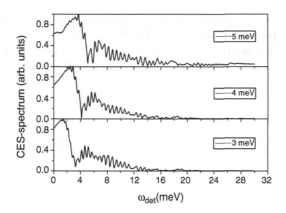

Fig. 14.6. Spectrally resolved four-wave-mixing response for pulses with central energies $\hbar\omega_{\mathrm{exc}} = 3, 4, 5\,\mathrm{meV}$ and Coulomb strength $10\,\mathrm{meV}$. Linear cross-polarization (xy) has been considered

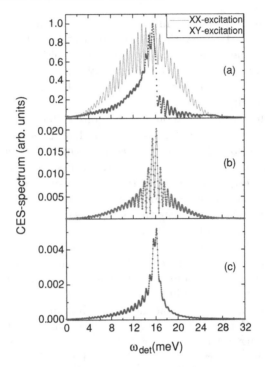

Fig. 14.7. Spectrally resolved four-wave-mixing response for pulses with central energy $\hbar\omega_{exc} = 16\,\mathrm{meV}$ and Coulomb strength $10\,\mathrm{meV}$. (a) Normalized total signal xx polarization (*solid line*) and xy polarization (*dotted line*). (b) Only the Hartree–Fock and (c) only the Pauli-blocking terms are included

The asymmetry and the dip in the xy situation can be understood by solving the simplified equations (8.55) and (8.57) or alternatively (8.60) and (8.61). Such solutions can be used to show that destructive interference among the various nonlinear terms is responsible for these features.

Part III

Applications II

The Semiconductor with Applied Electric Field

In this chapter, we consider a linear chain of N sites which is subject to external electric fields. One might be tempted to include these external electric fields in the direction of the chain by adding a potential term $v_j = eF(t)aj$ to each energy at the site j. The diagonal elements of the Hamiltonian matrices, $T_{jj}^{e,h}$, in this case, read in the electron–hole picture,

$$T_{jj}^{e,h} = \epsilon_j^{e,h} \pm eF(t)aj, \qquad (15.1)$$

while the nondiagonal elements remain unchanged by the electric field. The site energies of electrons and holes are displaced from their field-free values in parallel, i.e., the local energy difference $\hbar\omega_j = \epsilon_j^e + \epsilon_j^h$ is not affected by the external electric field, see Fig. 15.1.

Here, we have to face the problem of the boundary conditions at site 1 and site N. In real life, there are contacts, connecting our finite system to the external world. In other words, we have an open system. Assuming on the other hand that our system can be viewed as a closed system, we have a Hamiltonian that is ill-defined in the limiting case $N \to \infty$, since it does not have a lower bound. This causes problems if one wants to determine the stationary eigenstates of the system (see the literature listed at the end of this chapter).

However, it has to be kept in mind that we study transient phenomena on ultrafast time scales typical for quantum–coherent phenomena. The spectra of the ultrashort optical pulses used to generate the electronic excitations have a certain width inversely proportional to the duration of the pulse. This excitation does not generate stationary populations in single eigenstates but wave packets formed from many eigenstates having different energies. It is the coherent dynamics of these wave packets we are studying. In addition, the unavoidable coupling of these excited species to the outside world leads to dephasing on a time scale T_2. Thus, we are only interested in the spatio-temporal dynamics of the wave packets on this short time scale, which is typically on the order of picoseconds.

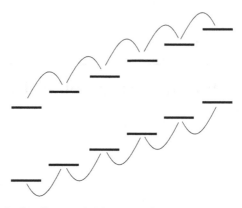

Fig. 15.1. The tight-binding model for an ordered semiconductor with externally applied electric field

Generally, all systems that are investigated experimentally are open systems. Therefore, artifacts which might result from the procedure to render closed systems mathematically tractable, e.g., periodic boundary conditions, are not present in the realistic situations. Whether or not the system is an open system is less important for our situation, which is dictated by experiment.

There are competing dynamical processes in biased systems, like, e.g., field-induced Zener tunneling into other bands. Here, we implicitly assume that these processes occur on a longer time scale than the intraband dynamics of interest to us. Therefore, we use the above Hamiltonian (5.6) with diagonal elements (15.1) in the following discussion of dynamical processes in biased semiconductors.

15.1 Single-Particle Properties

For simplicity of the presentation, we consider in this section a single-band model and an ordered situation. The Hamiltonian reads

$$\hat{H} = \sum_j (\epsilon_0 + eF(t)aj)c_j^\dagger c_j + J \sum_{\langle ij \rangle} c_i^\dagger c_j. \tag{15.2}$$

Let us first concentrate on dc bias fields, i.e., $F(t) = F$. As we show below, for this case, the eigenvalues display a ladder structure (the Wannier–Stark ladder) which can most easily be seen in the limit of extremely strong applied fields. For $eFa \gg J$, the difference in energy of adjacent sites is so large that the coupling J becomes ineffective and can no longer produce states which are extended over more than a single site. Therefore, in this so-called

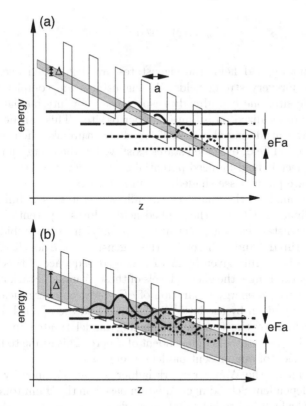

Fig. 15.2. Schematic drawing of the eigenstates with static applied electric field. Every well represents a site and carries a single state. (**a**) For a strong bias field, $eFa > \Delta$, in the regime of Stark localization and (**b**) for an intermediate bias field, $eFa \approx \Delta/3$, in the Wannier–Stark regime. The field-free band widths Δ are indicated by the *shaded areas*. [From T. Meier et al., in *Ultrafast Phenomena in Semiconductors* ed. K.T. Tsen, pp. 1–92 (Springer, New York 2001)]

Stark-localization regime, it is clear that the energy spectrum is discrete and has a ladder structure, i.e., $\epsilon_n = \epsilon_0 + neFa$.

In Fig. 15.2 and some of the following figures, a chain of potential wells is depicted. Every well corresponds to a site in our model. Figure 15.2a shows the Stark-localized wave functions. Due to the small coupling J between the sites, the resulting band width Δ which is indicated by the shaded region in Fig. 15.2a is smaller than eFa, i.e., the field-induced potential change between adjacent sites.

The eigenfunction corresponding to ϵ_n is localized on site n. Consequently, applying a translation in energy and space, one can show that all eigenfunctions Ψ are identical in the sense:

$$\Psi_{n-m}(z) = \Psi_n(z - ma). \tag{15.3}$$

Since without applied field, the eigenstates are delocalized over the entire system and, for very strong fields, the eigenstates are completely localized onto a single site, one can already suspect that, for intermediate fields, the eigenstates are localized on a finite number of sites. This can be understood physically by considering the "tilted" band structure sketched in Fig. 15.2b. Without external field, extended states exist within the energy interval Δ. By adding the electric field-induced potential energy eFz, we arrive at the tilted band-structure picture, see shaded area in Fig. 15.2b.

Next, we analyze the shape of the allowed states as a function of the space coordinate z. Clearly, the applied field induces a potential barrier and an energy eigenstate cannot penetrate appreciably into the forbidden regions outside the tilted band. Therefore, the eigenstates are localized and their spatial width is roughly given by Δ/eF. As a consequence of this localization in a finite spatial range, the energy levels of these Wannier–Stark ladder states are discrete with an energy separation of eFa between energetically adjacent states, i.e., only given by the lattice constant a and the applied field F. Note that, in contrast to the two-level tunneling model treated in Chap. 4, the splitting of the eigenstates is independent of J here. This is due to the spatially infinite extension of our present model system.

More formally, the Wannier–Stark ladder can be obtained by considering the time-independent Schrödinger equation based on the Hamiltonian of (15.2) for $F(t) = F$ which in bra-ket notation reads

$$\hat{H} = \sum_j (\epsilon_0 + eFaj)|j\rangle\langle j| + J \sum_j (|j - 1\rangle\langle j| + |j + 1\rangle\langle j|). \tag{15.4}$$

The eigenstates $|\psi_n\rangle$ and eigenenergies ϵ_n follow from the stationary Schrödinger equation

$$\hat{H}|\psi_n\rangle = \epsilon_n|\psi_n\rangle, \tag{15.5}$$

which in the site representation reads

$$\langle l|\hat{H}|\psi_n\rangle = \epsilon_n\langle l|\psi_n\rangle, \tag{15.6}$$

i.e.,

$$(\epsilon_0 - \epsilon_n + eFal)\langle l|\psi_n\rangle + J(\langle l + 1|\psi_n\rangle + \langle l - 1|\psi_n\rangle) = 0. \tag{15.7}$$

Instead of periodic boundary conditions, the eigenfunctions obey the symmetry relation, (15.3),

$$\langle l|\psi_n\rangle = \langle l + m|\psi_{n+m}\rangle, \tag{15.8}$$

which allows us to rewrite the Schrödinger equation as

$$(\epsilon_0 - \epsilon_n + eFal)\langle l + m|\psi_{n+m}\rangle +$$
$$J(\langle l + m + 1|\psi_{n+m}\rangle + \langle l + m - 1|\psi_{n+m}\rangle) = 0, \qquad (15.9)$$

or

$$(\epsilon_0 - \epsilon_{j-m} + eFa(i - m))\langle i|\psi_j\rangle + J(\langle i + 1|\psi_j\rangle + \langle i - 1|\psi_j\rangle) = 0, \qquad (15.10)$$

where in the last line we have taken $l + m = i$ and $n + m = j$. Comparing with (15.7), we find that the eigenvalues in fact form a ladder

$$\epsilon_{j-m} + eFam = \epsilon_j, \qquad (15.11)$$

or, setting $j = m$,

$$\epsilon_m = \epsilon_0 + eFam. \qquad (15.12)$$

We are now going to determine the eigenfunctions. Using (15.8), the Schrödinger equation (15.7) can be written as

$$-\frac{eFam}{J}\langle m|\psi_0\rangle = \langle m + 1|\psi_0\rangle + \langle m - 1|\psi_0\rangle. \qquad (15.13)$$

This can be solved by applying the recursion relation valid for Bessel functions J_n

$$J_{n-1}(x) + J_{n+1}(x) = \frac{2n}{x}J_n(x). \qquad (15.14)$$

Identifying x with $\frac{-2J}{eFa}$, we find that

$$\langle m|\psi_0\rangle = J_m\left(\frac{-2J}{eFa}\right) \qquad (15.15)$$

solves the Schrödinger equation in the site representation.

The harmonic oscillator is another prototype model in quantum mechanics which has a spectrum of equally spaced energy levels. In contrast to our model in the limit $N \to \infty$, however, the spectrum of the harmonic oscillator has a lower bound. For the harmonic oscillator, it is well known that the spectrum is closely related to the perfect periodic motion of a wave packet in real space. The same is also true for our system. To make this point clear, let us look at the dynamics of a wave packet not in real, but in k space. Without an electric field, the eigenenergies form the cosine band structure $E(k) = \epsilon_0 + 2J\cos(ka)$, $J < 0$, in the Brillouin zone which extends from $-\pi/a$ to π/a. Let us now place a wave packet with central momentum $k = 0$ into the system. Treating the problem on a semiclassical level, Newton's law gives us the rate of change of k as

$$\frac{dk}{dt} = \frac{e}{\hbar} F. \tag{15.16}$$

This "acceleration" in k space arises also in a quantum-mechanical description by transforming the field-induced linear potential given by $eFaj$ into k space.

To show this, we use the time-dependent Schrödinger equation for a time-dependent state $|\psi\rangle$ in k representation, i.e.,

$$\sum_{k'} \langle k|\hat{H}|k'\rangle \langle k'|\psi\rangle = i\hbar \frac{\partial}{\partial t} \langle k|\psi\rangle, \tag{15.17}$$

which, using (15.7), reads

$$\sum_{k'} \sum_{j} \langle k|j\rangle (\epsilon_0 + eFaj)\langle j|k'\rangle \langle k'|\psi\rangle$$
$$+ J \sum_{k'} \sum_{j} (\langle k|j-1\rangle\langle j|k'\rangle + \langle k|j+1\rangle\langle j|k'\rangle)\langle k'|\psi\rangle$$
$$= i\hbar \frac{\partial}{\partial t} \langle k|\psi\rangle. \tag{15.18}$$

With, see (6.1),

$$\langle j|k\rangle = \frac{1}{\sqrt{N}} e^{ikja}, \tag{15.19}$$

we obtain

$$\frac{1}{N} \sum_{k'} \sum_{j} [e^{-ikja} eFaj e^{ik'ja} + (\epsilon_0 + 2J\cos(ka))e^{i(k'-k)ja}]\langle k'|\psi\rangle$$

$$= i\hbar \frac{\partial}{\partial t} \langle k|\psi\rangle. \tag{15.20}$$

The first term can be written as

$$\frac{1}{N} \sum_{k'} \sum_{j} e^{-ikja} j e^{ik'ja} = \frac{i}{Na} \sum_{k'} \sum_{j} e^{ik'ja} \frac{\partial}{\partial k} e^{-ikja}$$

$$= \frac{i}{a} \frac{\partial}{\partial k} \sum_{k'} \delta(k - k'). \tag{15.21}$$

This leads to

$$ieF \frac{\partial}{\partial k} \langle k|\psi\rangle + E(k)\langle k|\psi\rangle = i\hbar \frac{\partial}{\partial t} \langle k|\psi\rangle. \tag{15.22}$$

We now allow k to be explicitly time-dependent

$$k(t) = \frac{1}{\hbar} \int_0^t dt' eF(t') + k(t = 0), \tag{15.23}$$

such that the right-hand side of (15.22) becomes

$$i\hbar\frac{\partial}{\partial t}\langle k(t)|\psi\rangle = i\hbar\frac{\partial}{\partial k}\langle k|\psi\rangle\frac{\partial k(t)}{\partial t} + i\hbar\langle k(t)|\frac{\partial}{\partial t}|\psi\rangle. \tag{15.24}$$

The $\frac{\partial}{\partial k}$ term cancels out and we obtain

$$E(k(t))\langle k(t)|\psi\rangle = \langle k(t)|i\hbar\frac{\partial}{\partial t}|\psi\rangle, \tag{15.25}$$

which means that the temporal dynamics can be described by the field-free solution in k space, however, with a time-dependent $k(t)$. The time dependence is given by (15.23) which is equivalent to the acceleration theorem, i.e., (15.16). Thus, as long as the field-induced coupling between different bands is neglected, the semiclassical acceleration theorem remains valid in a quantum-mechanical treatment.

The acceleration described by (15.16) implies that the wave packet moves undistorted through the first Brillouin zone until it reaches the (e.g., right) zone boundary. Here, due to the periodicity of the crystal, the wave packet is Bragg reflected and reappears at the opposite (e.g., left) zone edge. From there, it continues to move to the zone center and so on, see Fig. 15.3.

The time period of the oscillatory motion is called the Bloch period which is given by

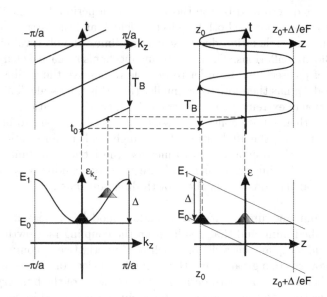

Fig. 15.3. Sketch of the intraband Bloch-oscillation dynamics of a wave packet prepared at the lower band edge at time t_0. *Bottom*: Motion of wave packets in k and real space. *Top*: Time evolution of the center of mass of the k- and real-space wave packets. [From T. Meier et al., in *Ultrafast Phenomena in Semiconductors* ed. K.T. Tsen, pp. 1–92 (Springer, New York 2001)]

$$T_B = \frac{h}{eFa},$$

(15.26)

i.e., it is proportional to the inverse of the splitting between the Wannier–Stark states. The k-space wave packet corresponds to a superposition of Wannier–Stark states in real space. While the k-space packet stays undeformed during its periodic translation through the first Brillouin zone, the real-space packet shows a time-dependent periodic shape deformation.

Under certain circumstances, for example when the wave packet is generated energetically close to the band edges, the center of mass of the wave packet oscillates sinusoidally in real space with the period T_B. This scenario is illustrated in Fig. 15.3. If one does not excite spectrally close to the band edge, but higher in the band, two wave packets around some k_0 and $-k_0$ are initially prepared. Both packets then evolve in time, leading to a more complicated temporal evolution of the center of mass. For example, if one starts in the center of the band, the center of mass shows no dynamics at all since both wave packets move exactly into opposite directions in real space. This exact cancellation of the two wave packets does, however, only happen within the one-dimensional model.

A wave-packet motion representing a charge oscillating in real space gives rise to electromagnetic radiation with the frequency of this oscillation. The frequency is thus determined by the Bloch-oscillation period T_B, (15.26), which is inversely proportional to both the electric field F and the lattice constant a. In real situations, relaxation due to scattering limits the observability of the coherent Bloch oscillations. These scattering mechanisms have a characteristic time scale of picoseconds. In order to be able to observe the oscillations, the Bloch period T_B has therefore to be smaller than this time scale. This requires the application of extremely high electric fields if Bloch oscillations in ordinary crystals with their lattice constant of some tenths of a nm are to be studied. Since in most cases, the application of such high fields is unrealistic, the only way to observe the oscillations is to increase the lattice constant. This can be done by fabricating an artificial solid, e.g., a semiconductor superlattice, which has a lattice constant in the growth direction that can be adjusted at will.

In fact, our one-dimensional tight-binding model is a reasonable model for the so-called miniband that results from the coupling in growth direction of the individual quantum wells of which the superlattice is composed. This miniband governs to a great extent the dynamics of electrons in a superlattice. The two other dimensions with periodicity on the scale of the bulk crystal add additional features to the dynamics, which have been studied in detail on the basis of realistic three-dimensional models of semiconductor superlattices. In contrast to the discussion in the previous chapters, where the the couplings J^e and J^h were chosen to yield realistic effective electron and hole masses close to the band extrema together with a suitable choice of the lattice constant

a, these parameters now have an even more important significance. They directly determine the small electron and hole band widths of the minibands of the semiconductor superlattice. The lattice constant a is now the interwell distance of the superlattice. The Bloch period in such biased superlattices is typically on the order of picoseconds, i.e., the electromagnetic radiation induced by this periodic motion of the wave packet is in the terahertz range.

We now turn to the case of applied ac electric fields and the phenomenon of dynamical localization. A simple way of deriving the band collapse associated with dynamical localization is to consider Newton's law, (15.16), for the rate of change of the wave vector k. Solving this equation for

$$F(t) = F \cos(\omega_L t), \tag{15.27}$$

leads to a time-dependent wave vector

$$k(t) = k_0 + (eF/\hbar\omega_L) \sin(\omega_L t). \tag{15.28}$$

Note that in contrast to previous chapters, ω_L here denotes the frequency of the applied electric field $F(t)$, which is in the THz range, and not the frequency of the optical laser field.

Let us now consider the time-dependent *instantaneous* eigenenergy associated with a k-space state

$$
\begin{aligned}
E(k(t)) &= \epsilon_0 - \frac{\Delta}{2} \cos(k(t)a) \\
&= \epsilon_0 - \frac{\Delta}{2} \left[\cos\left(k_0 a + \frac{eFa}{\hbar\omega_L} \sin(\omega_L t) \right) \right] \\
&= \epsilon_0 - \frac{\Delta}{2} \left[\cos(k_0 a) \cos\left(\frac{eFa}{\hbar\omega_L} \sin(\omega_L t) \right) \right. \\
&\qquad\qquad\qquad \left. - \sin(k_0 a) \sin\left(\frac{eFa}{\hbar\omega_L} \sin(\omega_L t) \right) \right] \\
&= \epsilon_0 - \frac{\Delta}{2} \left[\cos(k_0 a) \left[J_0\left(\frac{eFa}{\hbar\omega_L} \right) + 2 \sum_{n=1}^{\infty} J_{2n}\left(\frac{eFa}{\hbar\omega_L} \right) \cos(2n\omega_L t) \right] \right. \\
&\qquad\qquad\qquad \left. - \sin(k_0 a) \left[2 \sum_{n=1}^{\infty} J_{2n-1}\left(\frac{eFa}{\hbar\omega_L} \right) \sin((2n-1)\omega_L t) \right] \right] \\
&\approx \epsilon_0 - \frac{\Delta J_0\left(\frac{eFa}{\hbar\omega_L}\right)}{2} \cos(k_0 a). \tag{15.29}
\end{aligned}
$$

To reach the last line of (15.29), the time-dependent terms of the energy induced by the ac electric field have been neglected. This is justified in the limit of rapidly oscillating applied fields ($\hbar\omega_L \gg \Delta$), since in this limit $\hbar\omega_L$ dominates over all other energy scales and it is allowed to average over one period of the field. Equation (15.29) implies that in the presence of an applied ac field the band width Δ has to be replaced by $\Delta|J_0(\frac{eFa}{\hbar\omega_L})|$.

Dynamical localization occurs when the argument of the Bessel function in (15.29), $eFa/\hbar\omega_L$, is equal to a zero x_n of $J_0(x)$. For such a ratio between the amplitude and the frequency of the alternating field, the effective band width

$$\Delta(F) = \Delta(F = 0) \left| J_0 \left(\frac{eFa}{\hbar\omega_L} \right) \right| \tag{15.30}$$

vanishes. Vanishing band width implies that there is no coupling between the sites and, thus, that the eigenfunctions are strictly localized onto a single site. Physically, this means that the presence of the time-dependent electrical field introduces coherence into the system, which generally leads to a suppression or even, in special situations, to a complete vanishing of the (coherent) coupling between neighboring sites.

There is no simple classical analog of dynamical localization. We note that this effect is a consequence of a tight-binding dispersion in a single-band model. In more realistic situations, deviations from the clear picture presented here are to be expected.

15.1.1 Linear Optical Response

Here, we extend our discussion to a two-band model. This is necessary if we want to discuss optical properties of biased systems. The Hamiltonian without Coulomb interaction is given by

$$\begin{aligned}
\hat{H} = &\sum_n (\frac{\epsilon_0}{2} + neF(t)a)c_n^\dagger c_n \\
&+ \sum_n (\frac{\epsilon_0}{2} - neF(t)a)d_n^\dagger d_n \\
&+ \sum_{\langle nm \rangle} (J^e c_n^\dagger c_m + J^h d_n^\dagger d_m) \\
&- E(t) \sum_n \mu(c_n d_n + d_n^\dagger c_n^\dagger).
\end{aligned} \tag{15.31}$$

There are several characteristic energies present in our simple model. The largest one is the band-gap energy ϵ_0, which, for typical GaAs-based semiconductor heterostructures, is on the order of $1.5\,\text{eV}$. Another important energy is the miniband width $\Delta^{e,h}$ given by $4|J^{e,h}|$. In the following, the combined miniband width $\Delta = \Delta^e + \Delta^h$ is taken to be $20\,\text{meV}$ (unless specified otherwise), which is about two orders of magnitude smaller than ϵ_0. The applied field $F(t)$ is either static or is assumed to be dominated by frequencies ($\hbar\omega_L \approx 20\,\text{meV}$) which are on the order of the combined miniband width Δ. In the following,

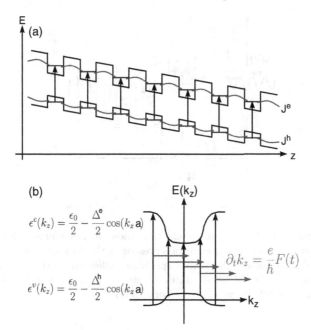

Fig. 15.4. One-dimensional two-band tight-binding model with electric field (a) in real and (b) in k space. [From T. Meier et al., in *Ultrafast Phenomena in Semiconductors* ed. K.T. Tsen, pp. 1–92 (Springer, New York 2001)]

we discuss situations where the field-induced potential drop over one super-lattice period is not too large ($eFa \approx 10\,meV \approx \Delta/2$). Since both $\hbar\omega_L$ and eFa are much smaller than ϵ_0, the interband coupling induced by $F(t)$ can well be neglected.

Within the tight-binding model, $F(t)$ couples to the intrasite optical dipole matrix element μ. A real-space picture of this situation is sketched in Fig. 15.4a. In k space, as in the single-band model, the linear potential induced by the field $F(t)$ results in a time dependence of the wave vector according to $dk/dt = eF/\hbar$. In the two-band model, both types of carriers are accelerated by the field, as shown in Fig. 15.4b. Note that for the holes, the signs of both the charge as well as the wave vector change are opposite to that of the electrons. Thus, consistent with the definition of holes being missing electrons, electrons and holes move uniformly in the same direction in k space.

In addition to the interband coherence p^{he}, we also discuss the intraband dipole moment, which is given by

$$P_{\text{intra}} = \sum_m ema(n^e_{mm}(t) - n^h_{mm}(t)), \qquad (15.32)$$

where $n^e_{mm}(t)$ and $n^h_{mm}(t)$ are the electron and hole populations at site m, respectively. This dipole moment is nothing but the sum over all charges times

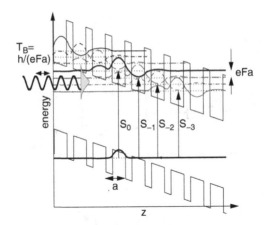

Fig. 15.5. Schematic drawing of optical transitions in the Wannier–Stark regime within a two-band model. The oscillations present in the terahertz signal after excitation with a short optical pulse with a spectrum indicated by the *shaded area* are also envisioned. [From T. Meier et al., in *Ultrafast Phenomena in Semiconductors* ed. K.T. Tsen, pp. 1–92 (Springer, New York 2001)]

their position. In the far field, the emission of terahertz radiation originating from the intraband motion of electrons and holes is proportional to the second derivative of P_{intra} with respect to time and is thus related to the dynamics of $n^e_{mm}(t)$ and $n^h_{mm}(t)$.

Let us now concentrate on the optical response in the presence of applied dc electric fields. In terms of eigenstates with applied dc field, i.e., the Wannier–Stark states, some possible optical transitions are indicated in Fig. 15.5. Since in III–V semiconductor materials, like GaAs, the heavy hole usually has a much larger mass than the electron, Δ^h is much smaller than Δ^e. Therefore, the Wannier–Stark wave functions for the heavy holes are already strongly localized for rather weak applied fields. For the same applied field, the electron Wannier–Stark states can still be delocalized over several sites. Considering a fixed eigenstate of the holes, optical transitions are possible to electronic states which have an appreciable amplitude at the site where the corresponding hole state is localized. In the situation sketched in Fig. 15.5, the strongest optical transition is that to the electron state which is centered at the same site as the hole. Transitions connecting the hole state with electron states that are lower or higher in energy are also possible. However, these transitions are weaker due to the smaller wave function overlap. The "vertical" transition is denoted by S_0, while the other transitions are called $S_{\pm n}$.

Figure 15.6a shows the linear optical absorption for the one-dimensional two-band tight-binding model as a function of applied dc field in a three-dimensional plot. For zero field, we see the typical one-dimensional absorption spectrum with its two van Hove singularities at the band edges which are slightly broadened here due to the phenomenologically introduced interband

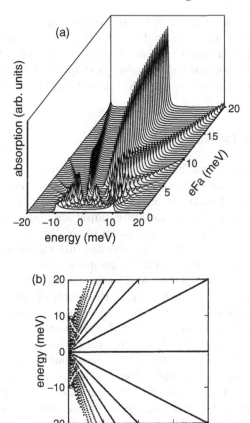

Fig. 15.6. Field-dependent linear optical absorption neglecting Coulomb interaction for a one-dimensional two-band tight-binding model. (a) Three-dimensional plot and (b) the maxima of absorption are displayed in a so-called fan chart. The parameters are: combined miniband width $\Delta = \Delta_c + \Delta_v = 20\,\text{meV}$, interband dephasing time $T_2 = 2\,\text{ps}$. The zero of the energy scale corresponds to the center of the combined miniband. [From T. Meier et al., in *Ultrafast Phenomena in Semiconductors* ed. K.T. Tsen, pp. 1–92 (Springer, New York 2001)]

dephasing time $T_2 = 2\,\text{ps}$. With increasing field, additional peaks develop whose energetic positions shift linearly with the field according to $neFa$. To be able to easily follow the field dependence of the absorption peaks, one often displays the energetic positions of the peaks as a function of applied field in a so-called fan chart, which is plotted in Fig. 15.6b. This fan chart shows the equidistant spacing of the peaks and the linear dependence of their energetic position on the field. When neglecting the Coulomb interaction, as is done here, the S_0 transition is completely independent of field strength. There is some transition region between the continuous one-dimensional miniband absorption and the effectively zero-dimensional peak region in the Stark localized regime that is attributed to Franz–Keldysh oscillations. Within the noninteracting one-dimensional tight-binding description, the optical absorption is completely symmetric with respect to the center of the combined miniband, see Figs. 15.6a and b.

15.1.2 Coherent Density Dynamics

We study the temporal dynamics on the basis of the present model. Figure 15.5 schematically shows the excitation of the system by a short laser pulse that spectrally covers the transitions S_0 to S_{-3}. Initially, the four electron states combine to a wave packet that has finite amplitude mainly at the site where the hole is localized. The temporal evolution of this packet is determined by the phase factors $\exp(i\epsilon_n t)$ of the contributing eigenfunctions where the ϵ_n differ by eFa. This implies that the temporal evolution of the packet is periodic in time with period T_B. As can be concluded from Figs. 15.3 and 15.5, for excitation close to the band gap, the center of mass of the electronic wave packet moves in real space, giving rise to an oscillatory dipole moment. The motion of the hole is more restricted and has opposite direction in real space compared to that of the electronic wave packet.

The situation discussed so far applies to a single transition at a given point in space. However, it is clear that this scenario takes place at all sites in a coherent way. As a result, the electron system moves relative to the hole system giving rise to a macroscopic oscillatory intraband dipole moment, which results in the emission of electromagnetic terahertz radiation.

The origin of the terahertz intraband emission can be understood quite easily. Let us consider that we are able to somehow prepare an electronic wave packet $n(k)$ with an energetic width much smaller than the miniband width around a certain k_0 at time $t = 0$. Due to the applied field, this wave packet is accelerated, i.e., shifted in k space with a velocity proportional to the applied field amplitude F. This motion is described by the acceleration theorem $dk/dt = eF/\hbar$. Assuming a tight-binding dispersion $\epsilon(k) = \epsilon_0 - \Delta/2\cos(ka)$ and using the acceleration theorem results in a time-dependent group velocity, $v(k) = \hbar^{-1}\partial\epsilon(k)/\partial k$, for such a wave packet, i.e.,

$$v(k_0, t) = \frac{\Delta a}{2\hbar} \sin[(k_0 + eF/\hbar t)a]. \tag{15.33}$$

Associated with this group velocity is a current which is given by

$$j(t) = \frac{e}{\hbar} \int n(k, t)v(k)dk \approx \frac{ne\Delta a}{2\hbar} \sin[(k_0 + eF/\hbar t)a], \tag{15.34}$$

where we have assumed that the width of the wave packet is small in k space and where n denotes the total density of the wave packet. Within this simple model, we thus expect sinusoidal oscillations in the current and, since the emitted far field is proportional to the time derivative of the current, also a sinusoidal oscillation in the terahertz signal. The period of the oscillations is just the Bloch period $T_B = h/eFa$.

Alternatively, one can also describe the wave-packet dynamics using the Wannier–Stark states as a basis set. If one prepares a superposition of two neighboring Wannier–Stark states, the oscillation period of the wave packet is given by the inverse of their splitting, which is again the Bloch period T_B.

In a realistic system, the time scale of the coherent oscillatory dynamics is limited by phase-breaking interactions. Their time scale is typically on the order of some picoseconds. If the coherent radiation discussed above is to be observed, the field F and the lattice constant a have to be adjusted, such that several oscillation periods take place within the phase coherence time. To obtain a simple estimate, we take the relevant coherence time for the terahertz emission as the intraband coherence time T_{intra}, which, in terms of the optical Bloch equations, is often parametrized by T_1, where T_1 is the lifetime of the excited states. Terahertz radiation with a well-defined frequency is then observable if $T_B = h/(eFa) < T_1$.

As discussed above, this condition requires a large product of field strength F times lattice constant a. Since the field strength is naturally limited by material properties like breakthrough effects, the lattice constant can be considered as a design parameter in semiconductor superlattices.

Figure 15.7a displays numerically calculated terahertz transients after optical excitation close to the fundamental band gap for various field strengths. These traces were obtained by numerical solutions of the equations of motion neglecting the Coulomb interaction. In our current noninteracting model for all considered fields, the oscillation period is simply given by T_B.

15.1.3 Coherent $\chi^{(3)}$ Processes

While the terahertz emission is induced by the intraband motion of electrons and holes, in nonlinear optical experiments, one is able to temporally resolve the dynamical evolution of interband electron–hole wave packets. These wave packets move in real as well as in reciprocal space. The excitation is vertical in k space, and both packets move with equal velocity rigidly "on top of each other" through the Brillouin zone.

In the following, we consider, in particular, four-wave mixing. Pulse No. 1 excites an electron–hole wave packet that, under the influence of the applied field F, performs Bloch oscillations. When the temporal delay of pulse No. 2 is close to an integer number of Bloch-oscillation periods (nT_B), the electron–hole wave packet has returned to its original position in real and in k space. Loosely speaking, in this situation, there is a strong electron–hole overlap in real space at the arrival of pulse No. 2, which results in a large four-wave-mixing amplitude. If pulse No. 2 hits the sample at a time delay not equal to an integer number Bloch-oscillation periods, the electron–hole overlap is

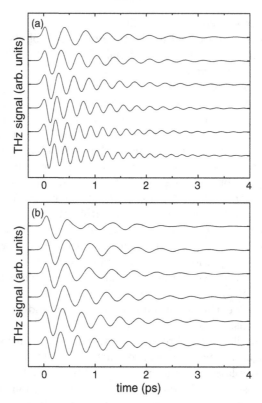

Fig. 15.7. Time-resolved terahertz signals neglecting (**a**) and including (**b**) Coulomb interaction for six dc bias fields using the one-dimensional two-band tight-binding model after excitation with a short laser pulse at $t = 0$. From top to bottom: $eFa = 10\,\text{meV}$, $12\,\text{meV}$, $14\,\text{meV}$, $16\,\text{meV}$, $18\,\text{meV}$, and $20\,\text{meV}$. The parameters are: combined miniband width $\Delta = 20\,\text{meV}$, interband dephasing time $T_2 = 2\,\text{ps}$, intraband dephasing time $T_1 = T_2/2 = 1\,\text{ps}$, laser-pulse width $100\,\text{fs}$, the central laser frequency is tuned close the zero-field $1s$-exciton resonance, and in (**b**) an on-site Coulomb potential of $V = 10\,\text{meV}$ is used. [From T. Meier et al., in *Ultrafast Phenomena in Semiconductors* ed. K.T. Tsen, pp. 1–92 (Springer, New York 2001)]

smaller and, concomitantly, the four-wave-mixing amplitude is weak. Thus, the Bloch-oscillation dynamics should lead to modulations of the amplitude of the time-integrated four-wave-mixing signal with a period given by T_B. This qualitative argument can be confirmed by explicitly solving the Bloch equations.

Figure 15.8 displays both the time-resolved (a) as well as the time-integrated four-wave-mixing signal (c) calculated for the noninteracting model considered here. Temporal modulations with the same period T_B are clearly visible for both types of detection.

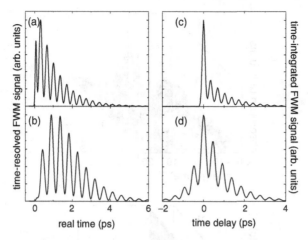

Fig. 15.8. (a) and (b) time-resolved and (c) and (d) time-integrated four-wave-mixing signals using the one-dimensional two-band tight-binding model. Upper plots (a) and (c) are calculated neglecting Coulomb interaction, whereas the lower plots (b) and (d) include Coulomb interaction. The parameters are: combined miniband width $\Delta = 20$ meV, bias field $eFa = 12$ meV, interband dephasing time $T_2 = 2$ ps, intraband dephasing time $T_1 = T_2/2 = 1$ ps, laser pulse width 100 fs, the central laser frequency is tuned close to the flat-band $1s$-exciton resonance, and in (b) and (d) an on-site Coulomb potential of $V = 10$ meV is used. [From T. Meier et al., in *Ultrafast Phenomena in Semiconductors* ed. K.T. Tsen, pp. 1–92 (Springer, New York 2001)]

15.1.4 Dynamical Localization

In this section, we have so far mostly considered external dc fields. The action of a sinusoidally varying field $F(t) = F\cos(\omega_L t)$ has been shown in the previous section to result in a field-dependent band width according to

$$\Delta(F) = \Delta(F = 0)\left| J_0\left(\frac{eFa}{\hbar\omega_L}\right)\right|. \tag{15.35}$$

If the parameter contribution $eFa/\hbar\omega_L$ has a value corresponding to a zero crossing of the Bessel function J_0, the band collapses and dynamical localization occurs. Since this condition does not depend on the width of the band, for our two-band model, the band collapse takes place for both bands simultaneously. Therefore, in this case, the optical spectrum changes from a one-dimensional form (continuum of finite width) to a zero-dimensional one when the band collapses. This is shown in Fig. 15.9a, where the dependence of the linear absorption spectrum on the amplitude of the applied ac field is displayed. The external field can thus be used to control the effective dimensionality of the system. The band collapse is accompanied by a vanishing coupling between adjacent sites. This leads to a suppression of the motion of the excited wave packets in real space and, thus, to a suppression of the terahertz radiation, as shown in Fig. 15.10a (note that 2.4 is a zero of J_0).

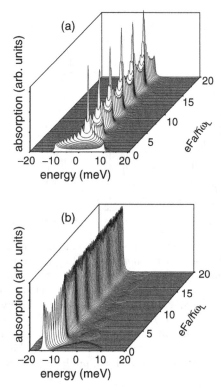

Fig. 15.9. Three-dimensional plots of the linear optical absorption spectra as a function of the amplitude of an ac bias field for the one-dimensional two-band tight-binding model (**a**) without and (**b**) with Coulomb interaction. Parameters: combined miniband width $\Delta = 20\,\text{meV}$, field frequency $\hbar\omega_L = 20\,\text{meV}$, interband dephasing time $T_2 = 2\,\text{ps}$, and in (**b**) an on-site Coulomb potential of $V = 10\,\text{meV}$ is used. The zero of the energy scale corresponds to the center of the combined miniband. [From T. Meier et al., in *Ultrafast Phenomena in Semiconductors* ed. K.T. Tsen, pp. 1–92 (Springer, New York 2001)]

15.2 Interacting System

In order to simplify the calculations, we assume here an on-site Coulomb interaction of the form

$$\hat{H}_C = \sum_m V c_m^\dagger c_m c_m^\dagger c_m + \sum_m V d_m^\dagger d_m d_m^\dagger d_m + \sum_m V c_m^\dagger c_m d_m^\dagger d_m. \quad (15.36)$$

Using an on-site interaction is an approximation which is done for simplicity only, and not meant to quantitatively model an actual system. The δ-like

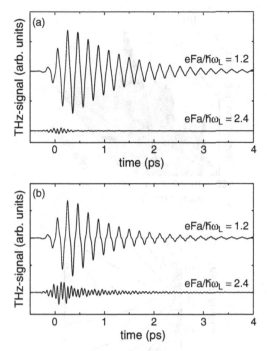

Fig. 15.10. Time-resolved terahertz signals for two amplitudes of the ac bias field using the one-dimensional two-band tight-binding model (**a**) neglecting and (**b**) including Coulomb interaction. Parameters: combined miniband width $\Delta = 20\,\text{meV}$, field frequency $\hbar\omega_L = 20\,\text{meV}$, interband dephasing time $T_2 = 2\,\text{ps}$, intraband dephasing time $T_1 = T_2/2 = 1\,\text{ps}$, on-site Coulomb potential used in (**b**) $V = 10\,\text{meV}$, laser pulse width 300 fs, and the central laser frequencies are tuned in (**a**) close to the band gap and in (**b**) close to the flat-band 1s-exciton resonance, respectively. For $eFa/\hbar\omega_L = 2.4$, dynamical localization occurs, since 2.4 is the first zero of J_0. [From T. Meier et al., in *Ultrafast Phenomena in Semiconductors* ed. K.T. Tsen, pp. 1–92 (Springer, New York 2001)]

potential in real space transforms into a k-space Coulomb potential that is independent of wave vector k. Numerically, it is much simpler to treat such a k-independent potential compared to a more realistic k-dependent potential. In the following, we treat the Coulomb interaction on the Hartree–Fock level.

15.2.1 Linear Optical Response

The linear optical spectrum in a three-dimensional plot and the corresponding fan chart are shown in Fig. 15.11. Comparing this spectrum with that of the noninteracting model, see Fig. 15.6, a number of new features are evident. In the field-free case, the spectrum of the interacting system is dominated by the exciton, which possesses nearly all of the oscillator strength, while the

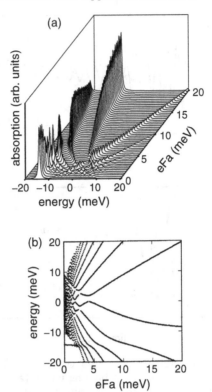

Fig. 15.11. Field-dependent linear optical absorption including Coulomb interaction for the one-dimensional two-band tight-binding model. (**a**) Three-dimensional plot and (**b**) the maxima of the absorption are displayed in a so-called fan chart. The parameters are: combined miniband width $\Delta = \Delta_c + \Delta_v = 20$ meV, on-site Coulomb potential $V = 10$ meV, interband dephasing time $T_2 = 2$ ps and in (**b**) an on-site Coulomb potential of $V = 10$ meV is used. The zero of the energy scale corresponds to the center of the combined miniband. [From T. Meier et al., in *Ultrafast Phenomena in Semiconductors* ed. K.T. Tsen, pp. 1–92 (Springer, New York 2001)]

interband continuum is hardly visible. At high fields, a Wannier–Stark ladder evolves which, however, no longer follows the straight lines given by $neFa$, as can be seen clearly in the fan chart given in Fig. 15.11b. Even the S_0 transition depends on electric field strength, since for strong electric fields the effective dimensionality of the system is reduced to zero which is associated with an increase of the exciton binding energy.

The spectra displayed in Fig. 15.11a show that the oscillator strength is concentrated in the low-energy part of the spectrum, i.e., the S_n transitions with negative n are stronger compared to the ones with positive n. This asymmetry, which is induced by the attractive electron–hole interaction, has to be contrasted to the complete symmetry of the optical absorption with respect to the center of the combined miniband in the noninteracting case, see Fig. 15.6a.

At the energy of the exciton, some avoided crossings are visible in Fig. 15.11. These occur when the energy of a Wannier–Stark state comes close to the energetic position of the field-free exciton. These anticrossings become even more prominent in weakly coupled superlattices, where they lead to drastic deviations from the linear field dependence of excitonic Wannier–Stark ladders.

15.2.2 Coherent Density Dynamics

Terahertz radiation transients calculated for the present model are presented in Fig. 15.7b. The oscillations are characterized by frequencies which correspond to energy differences between excitonic Wannier–Stark ladder states in the fan chart, see Fig. 15.11b. Thus, the period of the oscillations is not just eFa, but is determined by the interplay between the field-induced and the many-body dynamics. For the terahertz traces shown in Fig. 15.7b, the applied field is varied by a factor of two, but the oscillation period does not change that much. This is due to the fact that the energy differences between the excited excitonic Wannier–Stark states is smaller than eFa, i.e., the spacing between the single-particle Wannier–Stark states. If phenomenological damping times are introduced, care should be taken to model the so-called coherent limit, where T_1, i.e., the lifetime of the populations, has to be half the dephasing time T_2 of the interband coherence. Otherwise excitonic frequencies in the terahertz signal frequency can be missed.

15.2.3 Coherent $\chi^{(3)}$ Processes

Four-wave-mixing transients calculated for the interacting case are shown in Figs. 15.8 and 15.12. In calculating these signals, a phenomenological dephasing time T_2 for the interband polarization has been introduced. Consequently, the calculated time-resolved and time-integrated four-wave-mixing signals reflect this damping. We recover here some many-body-induced features which are well-known from four-wave-mixing traces within the Hartree–Fock approximation. The time-resolved traces start in time after both pulses have excited the system. If the many-body interaction is neglected and the system is not inhomogeneously broadened, the time-resolved four-wave-mixing signal reaches, apart from temporal modulations, its maximum immediately after the pulses and subsequently decays as determined by T_2. The Coulomb interaction, treated within the time-dependent Hartree–Fock approximation, induces a renormalization of the driving light field and renormalizes the single-particle energies. This leads to an initial increase of the real-time traces for short times, see Fig. 15.8b (compare to the observations of Fig. 12.3, where $T_2 \to \infty$ has been used). Physically, this can be understood by analyzing the equations of motion for p and n. Without the many-body interaction, only

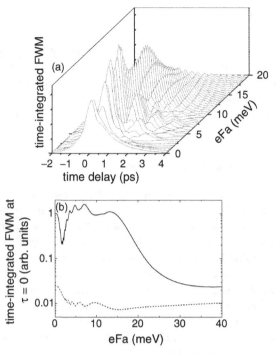

Fig. 15.12. (a) Time-integrated four-wave-mixing signal as a function of the time delay τ for various dc bias fields within the one-dimensional two-band tight-binding model including Coulomb interaction. (b) Field-dependence of four-wave-mixing amplitude at $\tau = 0$. *Solid*: total signal; *dashed*: the Coulomb-induced nonlinearities (as discussed in the text) are neglected. The parameters are: combined miniband width $\Delta = 20$ meV, on-site Coulomb potential $V = 10$ meV, interband dephasing time $T_2 = 2$ ps, intraband dephasing time $T_1 = T_2/2 = 1$ ps, laser pulse width 100 fs, and the central laser frequency is chosen slightly below the flat-band lowest exciton resonance. [After T. Meier et al., Phys. Rev. **B51**, 14490 (1995)]

the short second external laser pulse can interact with the system excited by the first pulse and induce a four-wave-mixing signal. With Coulomb interaction, however, also the induced longer living optical polarization can interact with the previously excited system. This type of many-body-induced optical nonlinearity also leads to four-wave-mixing signals for negative time delays, which are missing in the noninteracting case, see Fig. 15.8d.

The Coulomb-induced many-body nonlinearities are very dominant and lead to a much larger four-wave-mixing signal, as compared to the signal computed with pure Pauli blocking, which is the only nonlinearity present in the noninteracting case. This is demonstrated in Fig. 15.12b, where the field dependence of the four-wave-mixing amplitude for $\tau = 0$ is shown with and without Coulomb interaction. There is a subtle interplay between the many-body-induced nonlinearities corresponding to field and energy renormalization since these two terms enter the equation of motion with different signs. By applying electric fields, these terms can be manipulated. As a consequence, at a certain value of the field, both terms tend to cancel leading to a reduced time-integrated signal, see the dip for $eFa \approx 2\,\mathrm{meV}$ in the solid line of Fig. 15.12b.

Figure 15.12a shows the computed time-integrated four-wave-mixing traces as a function of time delay τ and field strength. For $F = 0$, the well-known behavior is seen with the decay determined by $T_2/2$ for positive delay. The factor $1/2$ arises since the time-integrated signal is given by the integral over the squared modulus of the interband polarization. The signal decay is twice as rapid for negative time delays. With increasing field strength, the signal first becomes weaker for all delay times due to the partial cancellation of the Hartree–Fock terms. At higher field strength, the signal recovers and the Bloch oscillations become visible as modulations of the signal as a function of τ. As expected, the time period of the modulations decreases with increasing field amplitude, but, due to the excitonic interaction, they are not simply given by eFa.

At this point, a caveat has to be mentioned. Even in the coherent limit, one has to be very careful with the interpretation of the numerical results on four-wave mixing obtained within the time-dependent Hartree–Fock approximation. As we have demonstrated in the previous sections, the Hartree–Fock level does not provide a realistic description of some aspects if many-body correlations contribute significantly to the nonlinear optical response. The effects discussed above, however, seem to be described adequately already on the Hartree–Fock level since they are in quite good agreement with experimentally observed results.

15.2.4 Dynamical Localization

Finally, we investigate the effect of dynamical localization on linear optical spectra in the presence of the Coulomb interaction, see Fig. 15.9b. Again, the field-free case is dominated by the excitonic resonance. This remains true also for ac fields with increasing $eFa/\hbar\omega_L$. If this parameter coincides with a zero of the zeroth-order Bessel function, J_0, the miniband collapses and the excitonic resonance is shifted to higher energies. In the present one-dimensional model, already in the field-free case, almost the whole oscillator strength is concentrated in the exciton. Therefore, there is almost no increase of the excitonic absorption occurring when the miniband collapses and the total absorption is concentrated in a single transition. The slight decrease of the oscillator strength between zero field and the first band collapse is due to ac-field-induced exciton ionization.

Figure 15.10b shows the effect of dynamical localization on the amplitude of the terahertz radiation. As in the noninteracting case, compare Fig. 15.10a, the amplitude is greatly reduced if $eFa/\hbar\omega_L$ coincides with the first zero of J_0.

On the basis of the results discussed in this section, we may conclude that the Coulomb interaction introduces important features into the coherent phenomena induced by the externally applied electric dc and ac fields. On the other hand, several qualitative signatures, like the possibility of oscillatory wave-packet motion induced by a dc field or dynamical localization induced by an ac field survive in the presence of the Coulomb interaction.

At this point, however, it still remains to be clarified whether the predicted phenomena persist also in realistic three-dimensional superlattices. In these structures, the motion of the carriers in the planes perpendicular to the growth direction has to be included, making the electron dynamics highly anisotropic. The Coulomb interaction couples the dynamics associated with the different directions and could lead to a strong perturbation of the coherent dynamics predicted by the simple one-dimensional model. Furthermore, realistic superlattices may not be well described by the tight-binding model that is used here in the analysis of both the Bloch oscillations and the dynamical localization. Deviations from this model may induce some perturbations to the picture discussed above. Once a realistic electronic structure is taken into account, one should also treat damping mechanisms due to carrier–carrier and carrier–phonon scattering explicitly. Some of the effects investigated here have been treated for realistic anisotropic three-dimensional systems in the presence of the Coulomb interaction and scattering processes in Meier et al. (2001). As demonstrated there, the complete three-dimensional approach is required in order to quantitatively describe realistic superlattice structures. However, it is further shown that the coherent effects discussed above can be described qualitatively well already within the one-dimensional model.

15.3 Suggested Reading

1. F. Bloch, "Über die Quantenmechanik der Elektronen in Kristallgittern", Z. Phys. **52**, 555 (1928)
2. E.I. Blount, "Formalisms of band theory", in *Solid State Physics*, eds. F. Seitz and D. Turnbull, Vol. 13, p. 305 (Academic Press, New York, 1962)
3. D.H. Dunlap and V.M. Kenkre, "Dynamic localization of a charged particle moving under the influence of an electric field", Phys. Rev. **B34**, 3625 (1986)
4. J. Feldmann, K. Leo, J. Shah, D.A.B. Miller, J.E. Cunningham, T. Meier, G. von Plessen, A. Schulze, P. Thomas, and S. Schmitt-Rink, "Optical investigation of Bloch oscillations in a semiconductor superlattice", Phys. Rev. **B46**, 7252 (1992)
5. S. Glutsch and F. Bechstedt, "Interaction of Wannier–Stark ladders and electrical breakdown in superlattices", Phys. Rev. **B60**, 16584 (1999)

6. S. Glutsch, *Excitons in Low-Dimensional Semiconductors*, Springer Series in Solid-State Sciences, Vol. 141 (Springer, Berlin 2004)

7. J. Hader, T. Meier, S.W. Koch, F. Rossi, and N. Linder, "Microscopic theory of the intracollisional field effect in semiconductor superlattices", Phys. Rev. B **55**, 13799 (1997)

8. M. Holthaus, "Collapse of minibands in far-infrared irradiated superlattices", Phys. Rev. Lett. **69**, 351 (1992)

9. K.-C. Je, T. Meier, F. Rossi, and S.W. Koch, "Theory of quasi-equilibrium nonlinear optical absorption in semiconductor superlattices", Appl. Phys. Lett. **67**, 2978 (1995)

10. J.B. Krieger and G.J. Iafrate, "Time evolution of Bloch electrons in a homogeneous electric field", Phys. Rev. **B33**, 5494 (1986)

11. K. Leo, J. Feldmann, J. Shah, G. von Plessen, P. Thomas, S. Schmitt–Rink, and J. Cunningham, "Optical investigation of Bloch oscillations in semiconductor superlattices", Superlattices & Microstructures **13**, 55 (1993)

12. T. Meier, G. von Plessen, P. Thomas, and S.W. Koch, "Coherent electric-field effects in semiconductors", Phys. Rev. Lett. **73**, 902 (1994)

13. T. Meier, F. Rossi, P. Thomas, and S.W. Koch, "Dynamic localization in anisotropic Coulomb systems: Field induced crossover of the exciton dimension", Phys. Rev. Lett. **75**, 2558 (1995)

14. T. Meier, G. von Plessen, P. Thomas, and S.W. Koch, "Coherent effects induced by static and time-dependent electric fields in semiconductors", Phys. Rev. **B51**, 14490 (1995)

15. T. Meier, H.J. Kolbe, A. Thränhardt, G. Weiser, P. Thomas, S.W. Koch, "Coherent dynamics of photoexcited semiconductor superlattices in homogeneous electric fields", Physica E **7**, 267 (2000)

16. T. Meier, P. Thomas, and S.W. Koch, "Coherent dynamics of photoexcited semiconductor superlattices with applied homogeneous electric fields", in *Ultrafast Phenomena in Semiconductors*, eds. K.T. Tsen (Springer, New York 2001), p. 1

17. E.E. Mendez, F. Agullo-Rueda, J.M. Hong, "Stark Localization in GaAs-GaAlAs superlattices under an electric field", Phys. Rev. Lett. **60**, 2426 (1988)

18. A. Nenciu and G. Nenciu, "On the dynamics of Bloch electrons in constant electric fields", Phys. Lett. **78**, 101 (1980)

19. B. Rosam, D. Meinhold, F. Löser, V.G. Lyssenko, S. Glutsch, F. Bechstedt, F. Rossi, K. Köhler, and K. Leo, "Field-induced delocalization and Zener breakdown in semiconductor superlattices", Phys. Rev. Lett. **86**, 1307 (2001)

20. F. Rossi, T. Meier, P. Thomas, S.W. Koch, P.E. Selbmann, and E. Molinari, "Ultrafast carrier relaxation and vertical-transport phenomena in semiconductor superlattices: A Monte Carlo analysis", Phys. Rev. **B51**, 16943 (1995)

21. A. Thränhardt, H.J. Kolbe, J. Hader, T. Meier, G. Weiser, and S.W. Koch, "Field-dependent absorption in superlattices: Comparison of theory and experiment", Appl. Phys. Lett. **73**, 2612 (1998)

22. P. Voisin, J. Bleuse, C. Bouche, S. Gaillard, C. Alibert, and A. Regreny, "Observation of the Wannier–Stark quantization in a semiconductor superlattice", Phys. Rev. Lett. **61**, 1639 (1988)

23. G. von Plessen and P. Thomas, "Method for observing Bloch oscillations in the time domain", Phys. Rev. **B45**, 9185 (1992)

24. G. von Plessen, T. Meier, J. Feldmann, E.O. Göbel, P. Thomas, K.W. Goossen, J.M. Kuo, and R.F. Kopf, "Influence of scattering on the formation of Wannier–Stark ladders and Bloch oscillations in semiconductor superlattices", Phys. Rev. **B49**, 14058 (1994)

25. G. von Plessen, T. Meier, M. Koch, J. Feldmann, P. Thomas, S.W. Koch, E.O. Goebel, K.W. Goossen, J.M. Kuo, and R.F. Kopf, "Exciton ionization induced by an electric field in a strongly coupled $GaAs/Al_xGa_{1-x}As$ superlattice", Phys. Rev. **B53**, 13688 (1996)

26. J. Zak, "Stark ladder in solids?", Phys. Rev. Lett. **20**, 1477 (1968)

27. J. Zak, "The kq-representation in the dynamics of electrons in solids", in *Solid State Physics*, eds. H. Ehrenreich, F. Seitz, and D. Turnbull, Vol. 27, p. 1 (Academic, New York 1972)

28. C. Zener, "A theory of the electrical breakdown of solid dielectrics", Proc. Roy. Soc. A **145**, 523 (1934)

16

Mesoscopic Semiconductor Rings

After having studied the coherent optical signatures of the ordered and dis-ordered semiconductor model with and without external dc and ac electric fields, we now turn to the discussion of the influence of an external magnetic field on coherent optical response. For a linear chain with an infinite number of sites, i.e., $N \to \infty$, the only influence of a magnetic field is due to the inter-nal degrees of freedom of the individual sites. We expect a Zeeman splitting of the two degenerate upper and lower levels. There are no magnetic field effects related to the motion of the particles in the direction of the chain, i.e., from site to site. Hence, only for finite N, where, due to the periodic boundary con-dition, we have a ring with finite area πR^2, where $R = Na/2\pi$, do magnetic field effects other than intrasite splittings exist.

Therefore, in this chapter, we consider a ring-like arrangement of the sites. It will become clear that the diameter of the ring has to be chosen sufficiently small in order to observe interesting dynamical and magnetic-field-induced signatures in the optical signals. Therefore, these systems are called semicon-ductor nanorings and their dominant optical excitations are magnetoexcitons.

16.1 The Model

16.1.1 No Magnetic Field

The model system is defined by

$$\hat{H} = \hat{H}_0 + \hat{H}_C + \hat{H}_L, \tag{16.1}$$

with, see (5.8),

$$\hat{H}_0 = \sum_{i,j=1}^{N} \sum_e T_{ij}^e c_i^{e\dagger} c_i^e + \sum_{i,j=1}^{N} \sum_h T_{ij}^h d_i^{h\dagger} d_i^h, \tag{16.2}$$

and $\hat{H}_C$ and $\hat{H}_L$ are given by (5.13) with (5.17), and (5.18), respectively.

The physically relevant parameters are the effective masses m_ν and the radius R of the ring. The parameters J_ν and N are chosen in such a way, see (6.19), that (i) the dispersions close to the band extrema are representative for realistic semiconductor nanostructures, and (ii) that the numerics remains manageable.

Perturbation potentials destroying the rotational symmetry of the ring can easily be introduced into our description by taking site-dependent electron and hole energies ϵ_n^e and ϵ_n^h. Disorder in the conduction band is modeled by choosing ϵ_n^e randomly from a Gaussian distribution function $\exp(-(\epsilon_n^e - (\epsilon_0/2))^2/2\sigma_e^2)$, where σ_e characterizes the energetic scale of the disorder. Eventually, we also consider a tilting of the conduction band described by

$$\epsilon_n^e = \sigma_t \sin[2(n-1)\pi/N], \tag{16.3}$$

which corresponds to an additional external potential varying linearly with the y coordinate in the plane of the ring, i.e., the x-y plane, if site $n = 1$ lies on the positive x axis. We further assume that the disorder potentials for electrons and holes or the respective tilting potentials are correlated, as is often the case in semiconductor nanostructures. We take $\delta\epsilon_n^h = (|J^h/J^e|)\delta\epsilon_n^e = (m_e/m_h)\delta\epsilon_n^e$, i.e., we weight the potential perturbations $\delta\epsilon$ inversely with the masses.

The electric light-field pulse is chosen as

$$E(t) = E_0 \frac{1}{\sqrt{\pi}\delta} \exp(-i\omega_L t) \exp(-t^2/\delta^2). \tag{16.4}$$

We study the linear and nonlinear optical response and also the dynamics of the electron–hole dipole moment after optical excitation in the excitonic energy range of the semiconductor nanoring.

The electron and hole dipole moments $\boldsymbol{d}^e$ and $\boldsymbol{d}^h$, defined with respect to the center of the ring, are given by ($\nu = e, h$)

$$d_x^\nu = \pm e \sum_l n_{ll}^\nu R \cos\phi_l, \tag{16.5}$$

$$d_y^\nu = \pm e \sum_l n_{ll}^\nu R \sin\phi_l,$$

$$d_z^\nu = 0, \tag{16.6}$$

where $\phi_l = 2\pi(l-1)/N$ defines the angular position of site l. The signs depend on the charge of the particle in band ν, negative signs correspond to electrons, and positive signs to holes. The total electron–hole dipole moment $\boldsymbol{d}$ is given by

$$\boldsymbol{d} = \boldsymbol{d}^e + \boldsymbol{d}^h. \tag{16.7}$$

The interband polarization is obtained by solving (8.64) in the linear response limit or (8.54) and (8.56) in nonlinear situations. The spectra are calculated assuming a certain finite dephasing time T_2, while we use $T_2 \to \infty$ for the calculation of the coherent dynamics. The densities needed to compute the dynamics of the electron and hole dipole moments, n_{ll}^{ν}, are obtained from (8.14) and (8.15).

The motion of the dipole within the plane of the ring may be considered as a source of electromagnetic radiation. The associated electric field is obtained from the second time derivative of the dipole moment, which constitutes the source term in Maxwell's equations. The corresponding polarization angle is determined by

$$\alpha = \arctan \frac{d^2 d_y/dt^2}{d^2 d_x/dt^2}. \tag{16.8}$$

16.1.2 Magnetic Field Included

We now add the magnetic field to the model. We first consider only a single electron and single hole band. Experimentally, this could be realized by applying circularly polarized light, at least as far as linear optics is concerned. An intrasite Zeeman splitting is neglected here.

We therefore only have the action of the magnetic field on the electron and hole dynamics along the circumference of the ring. This dynamics is determined by the magnetic flux through the area of the ring. We assume that the magnetic field B is perpendicular to the area of the ring and constant. We also assume that $B = 0$ on the circumference of the ring. The magnetic flux is then given by

$$\Phi = \pi R^2 B. \tag{16.9}$$

Since on the circumference of the ring $B = 0$ and

$$B = \nabla \times A(r), \tag{16.10}$$

where $A(r)$ is the vector potential, we can write for r on the circumference

$$A(r) = \nabla \chi(r). \tag{16.11}$$

The scalar potential $\chi(r)$ is, however, not single valued. If we go around the ring once, it changes by an amount $\Delta \chi$, which is given by

$$\begin{aligned} \Delta \chi &= \oint \nabla \chi(r) \cdot dr \\ &= \oint A(r) \cdot dr \\ &= \int_\sigma (\nabla \times A(r)) \cdot df \\ &= \int_\sigma B \cdot df = \Phi, \end{aligned} \tag{16.12}$$

where the contour integral extends over the circumference of the ring and σ means the area of the ring.

We now study how the magnetic field changes the parameters of our tight-binding semiconductor model. Remember that the Hamiltonian matrix $T_{ij}^{e,h}$ is given by the matrix elements of the material Hamiltonian $\hat{H}_0$. Here, we concentrate on the electron part

$$T_{ij}^e = \langle ie|\hat{H}_e|je\rangle \tag{16.13}$$

and ignore for the moment the spin degrees of freedom. The spin can be taken into account easily at the end since, at the beginning of this section, we have made the model assumption that it does not play any role in the interaction with the magnetic field. Furthermore, the magnetic-field-induced terms for the holes can be calculated analogously to those for the electrons.

In order to treat the magnetic field, we apply the continuous real-space representation. In the field-free case, the wave function $\phi(\boldsymbol{r})$ on the ring is the solution of the Schrödinger equation

$$\hat{H}_e\phi(\boldsymbol{r}) = \left[\frac{1}{2m_e}\left(\frac{\hbar}{i}\nabla\right)^2 + V(\boldsymbol{r})\right]\phi(\boldsymbol{r}) = E\phi(\boldsymbol{r}), \tag{16.14}$$

where $V(\boldsymbol{r})$ is the total single-particle potential of the ring.

In the presence of the magnetic field the kinetic energy term contains the vector potential in addition to the canonical momentum. Thus, in this case, the wave functions are now determined by

$$\hat{H}_e(\boldsymbol{A})\,\phi(\boldsymbol{r}) = \left[\frac{1}{2m_e}\left(\frac{\hbar}{i}\nabla + \frac{e}{c}\boldsymbol{A}(\boldsymbol{r})\right)^2 + V(\boldsymbol{r})\right]\phi(\boldsymbol{r}) = E\phi(\boldsymbol{r}), \tag{16.15}$$

where $e > 0$ is the electronic charge and c the velocity of light.

The $\frac{e}{c}\boldsymbol{A}$ term can be formally removed from the Hamiltonian by applying the transformation

$$\phi(\boldsymbol{r}) \rightarrow \tilde{\phi}(\boldsymbol{r}) = \phi(\boldsymbol{r})e^{i\frac{e}{c\hbar}\chi(\boldsymbol{r})} \tag{16.16}$$

to the wave function.

To see this we only have to consider the kinetic-energy term, which can be expressed as

$$\left(\frac{\hbar}{i}\nabla + \frac{e}{c}\boldsymbol{A}(\boldsymbol{r})\right)^2\phi(\boldsymbol{r}) = \left(\frac{\hbar}{i}\nabla + \frac{e}{c}\boldsymbol{A}(\boldsymbol{r})\right)\cdot\left(\frac{\hbar}{i}\nabla + \frac{e}{c}\boldsymbol{A}(\boldsymbol{r})\right)\phi(\boldsymbol{r})$$
$$= -\hbar^2\nabla\cdot\nabla\phi(\boldsymbol{r}) + 2\frac{\hbar e}{ic}\boldsymbol{A}(\boldsymbol{r})\cdot\nabla\phi(\boldsymbol{r})$$
$$+ \frac{\hbar e}{ic}\phi(\boldsymbol{r})\nabla\cdot\boldsymbol{A}(\boldsymbol{r}) + \frac{e^2}{c^2}\boldsymbol{A}^2(\boldsymbol{r})\phi(\boldsymbol{r}). \tag{16.17}$$

On the other hand, starting from the field-free kinetic energy acting on the transformed wave function, we obtain

$$\left(\frac{\hbar}{i}\nabla\right)^2 \tilde{\phi}(\boldsymbol{r}) = -\hbar^2 \nabla \cdot \nabla \left(\phi(\boldsymbol{r})e^{i\frac{e}{\hbar c}\chi(\boldsymbol{r})}\right)$$

$$= -\hbar^2 \nabla \cdot \left(e^{i\frac{e}{\hbar c}\chi(\boldsymbol{r})}\nabla\phi(\boldsymbol{r}) + i\frac{e}{\hbar c}\phi(\boldsymbol{r})e^{i\frac{e}{\hbar c}\chi(\boldsymbol{r})}\nabla\chi(\boldsymbol{r})\right)$$

$$= e^{i\frac{e}{\hbar c}\chi(\boldsymbol{r})}\left[-\hbar^2 \nabla \cdot \nabla\phi(\boldsymbol{r}) + 2\frac{\hbar e}{ic}\boldsymbol{A}(\boldsymbol{r})\cdot\nabla\phi(\boldsymbol{r})\right.$$

$$\left.+\frac{\hbar e}{ic}\phi(\boldsymbol{r})\nabla\cdot\boldsymbol{A}(\boldsymbol{r}) + \frac{e^2}{c^2}A^2(\boldsymbol{r})\phi(\boldsymbol{r})\right], \tag{16.18}$$

where we have used (16.11). Except for the exponential prefactor, we therefore obtain the same result as in (16.17). Thus, if $\phi(\boldsymbol{r})$ solves the Schrödinger equation with field included, then $\tilde{\phi}(\boldsymbol{r})$ is the solution for the Hamiltonian that does not explicitly contain the field, since the exponential factors cancel out after having performed the spatial derivative.

The new function $\tilde{\phi}(\boldsymbol{r})$ has, however, a different boundary condition compared to that of $\phi(\boldsymbol{r})$. The periodic boundary condition of the wave function $\phi(\boldsymbol{r})$ is

$$\phi(\boldsymbol{r}) \rightarrow \phi(\boldsymbol{r}) \tag{16.19}$$

if we go around the ring once. On the other hand, due to the phase factor in (16.16) and because of (16.12), we have instead

$$\tilde{\phi}(\boldsymbol{r}) \rightarrow \tilde{\phi}(\boldsymbol{r})e^{i\frac{e}{\hbar c}\Phi}. \tag{16.20}$$

Surrounding the ring once thus gives a phase factor

$$e^{i\frac{e}{\hbar c}\Phi} = e^{2\pi i\frac{\Phi}{\Phi_0}}, \tag{16.21}$$

where the so-called flux quantum is given by

$$\Phi_0 = \frac{hc}{e}. \tag{16.22}$$

If the ring consists of N sites, then moving one step clockwise (counterclockwise) produces a phase factor

$$e^{\pm 2\pi i\frac{\Phi}{N\Phi_0}}. \tag{16.23}$$

Applying this result to our tight-binding model, we find that the Hamiltonian is given by

$$\hat{H}_0 = \sum_{i,j=1}^{N} \tilde{T}_{ij}^e c_i^\dagger c_i + \sum_{i,j=1}^{N} \tilde{T}_{ij}^h d_i^\dagger d_i, \tag{16.24}$$

where the diagonal elements of the Hamiltonian matrices $\tilde{T}$, which contain the single-particle potential, remain unchanged. However, the nondiagonal elements $\tilde{T}_{ij}^{e,h}$ now contain the additional phase factors given by (16.23), which are called the Peierls phase factors, and read

$$\tilde{T}^e_{i,i+1} = J^e \exp\left[2\pi i \frac{\Phi}{N\Phi_0}\right],$$

$$\tilde{T}^e_{i+1,i} = J^e \exp\left[-2\pi i \frac{\Phi}{N\Phi_0}\right],$$

$$\tilde{T}^h_{i,i+1} = J^h \exp\left[-2\pi i \frac{\Phi}{N\Phi_0}\right],$$

$$\tilde{T}^h_{i+1,i} = J^h \exp\left[2\pi i \frac{\Phi}{N\Phi_0}\right]. \tag{16.25}$$

16.2 Single-Particle Properties

16.2.1 The Single-Particle Spectrum

The eigenvalues of the Hamiltonian $\hat{H}_0$, see (16.24), form two bands

$$E(q_i)^\nu = 2J^\nu \cos(q_i a), \tag{16.26}$$

where $\nu = e, h$ and the wave vectors $q_i, i = 1, \dots, N$ are given by

$$q_i = k_i + \frac{2\pi\Phi}{Na\Phi_0}, \tag{16.27}$$

and k_i are the wave vectors without magnetic field, i.e.,

$$k_i = -\frac{\pi}{a} + \frac{2\pi}{a}\frac{i}{N}. \tag{16.28}$$

As an illustration, we consider a ring with finite (small) radius R and equal electron and hole effective masses $m_e = m_h$. We take $N = 6$ in order to clarify the physical picture. Figure 16.1 shows the valence and conduction bands as well as the relevant optical transitions. The wave vectors k_i for the magnetic-field-free case $\Phi = 0$ are denoted by the vertical dashed lines and those for $\Phi = \Phi_0/4$ by the solid lines. For $\Phi = 0$, there are two two-fold degenerate optical dipole transitions (at $k = \pm\pi/3a$ and $k = \pm2\pi/3a$) and two nondegenerate transitions (at $k = 0$ and $k = \pi/a$). For $\Phi = \Phi_0/4$, the degeneracies are lifted and the spectrum consists of six lines. It can easily be seen that one would obtain three two-fold degenerate transitions for $\Phi = \Phi_0/2$. For $\Phi = m\Phi_0$ with an integer m, the q vectors have moved m steps to the right. For $\Phi = N\Phi_0$, we recover the original picture.

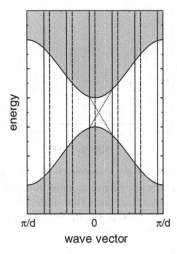

Fig. 16.1. The band structure for zero (*dashed lines*) and a quarter magnetic flux quantum (*solid lines*), i.e., $\Phi = 0$ and $\Phi = \Phi_0/4$, respectively, for a ring having $N = 6$ sites. The eigenvalues are only defined at the *vertical lines*, indicating the allowed k values. For homogeneous optical excitation only transitions along the *vertical lines* are allowed, yielding a discrete spectrum. For inhomogeneous excitation (*broken translational symmetry*), also transitions as indicated by *dotted lines* become allowed. This leads to additional (for equal electron and holes masses degenerate) absorption lines. [From K. Maschke et al., Eur. Phys. J. B **19**, 599 (2001)]

16.2.2 Persistent Currents

Consider an electron generated, under the presence of a flux $\Phi = \Phi_0/4$, in a nondegenerate eigenstate with wave vector q and energy $E(q)^e$ that originated from $k = 0$ in the field-free situation. This excitation carries a *persistent current*, since there is no counterbalance represented by a finite (negative) velocity given by $(1/\hbar)(dE^e(q')/dq')$.

An optical excitation pulse with sufficiently narrow spectral width, such that only the transition at this particular q takes place, generates an electron–hole pair that moves in different directions, i.e., it creates a persistent current, as explained above. This current exists over time scales smaller than those characteristic for coupling to the phonon bath or for radiative recombination. For $\Phi = \Phi_0/2$, on the other hand, no current is generated, since there is an exact counterbalance due to a second transition that has opposite k and is degenerate with the one in question.

The optical absorption spectrum is shown in Fig. 16.2 for $N = 20$. The lines pointing downward show the 11 transitions for $\Phi = 0$ and the upward pointing lines show the 20 nondegenerate transitions for $\Phi = \Phi_0/4$. Note that in this figure the amplitudes indicate the degeneracies of the corresponding transitions. As can be seen in Fig. 16.2, the energies of the two lowest transitions in the spectrum for $\Phi = \Phi_0/4$ depend on the magnetic flux. These

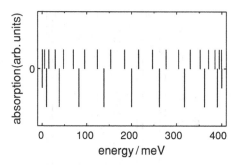

Fig. 16.2. The optical spectrum for a ring with $N = 20$ sites. The *downward lines* indicate transitions in the absence of a magnetic flux. The *upward lines* show the spectrum for a magnetic quarter flux. Note that with flux the degeneracies are lifted. [From K. Maschke et al., Eur. Phys. J. B **19**, 599 (2001)]

transitions merge for $\Phi \rightarrow \Phi_0/2$, which is the situation when no persistent current is present. From (16.26)–(16.28) it is clear that the optical spectrum is periodic in Φ with period $\Phi = \Phi_0$.

16.3 Dynamics of the Electron–Hole Dipole

The following calculations have been performed for a ring with radius $R = 11$ nm containing $N = 20$ sites, which corresponds to a site separation $a = 34$ Å. The chosen coupling parameters $J^e = 50$ meV and $J^h = 10$ meV correspond to effective masses close to GaAs values $m_e = 0.064 m_0$ and $m_h = 0.32 m_0$, where m_0 is the free electron mass. In some calculations, we use values $J^e = J^h = 50$ meV for tutorial reasons. A magnetic flux of $\Phi_0/4$ through the ring would require a magnetic field of about 12 T. As shown in the next section, the relevant time scale for the description of the electron–hole dynamics is on the order of 1–100 ps, which is within the coherent excitonic regime of a semiconductor nanostructure. Larger rings would require smaller magnetic fields, but the relevant time scales also become longer, such that phase-breaking interactions eventually destroy the coherent dynamics one intends to observe. We therefore believe that the chosen parameters are a good compromise.

16.3.1 Noninteracting Particles

In general, a magnetic flux leads to a circular motion of both electrons and holes, see Sect. 16.2.2. For optical excitation close to the gap, electrons and holes move in opposite direction, at least in a situation where the Coulomb interaction can be neglected. The corresponding partial electron or hole currents are largest for fluxes $\Phi = (n + 1/4)\Phi_0$ or $\Phi = (n + 3/4)\Phi_0$. For spatially homogeneous optical excitation, the photogenerated carriers follow the well-known persistent current scenario. In this situation, the optical excitation would just lead to a transient change of the magnetic moment of the ring.

In the following, we assume spatially inhomogeneous excitation, where the photoexcitation takes place at a single site only ($n = 1$, which is located on the positive x axis). In this situation, additional transitions become allowed which are not shown in Fig. 16.2. In order to resolve the spectrum for this situation, we present an enlarged view of the low-energy transitions in Figs. 16.3. Here and in the following, the energy zero corresponds to the lowest transition energy of the perfect noninteracting system for $\Phi = 0$. In the spectrum for $\Phi = \Phi_0/4$, see Fig. 16.3a, we find two absorption lines spaced by about 2.5 meV. The lower peak corresponds to the shifted $k = 0$ transition, the higher one is associated with the energetically lowest indirect transitions, which are two-fold degenerate in the case of equal electron and hole masses, see the dotted lines in Fig. 16.1.

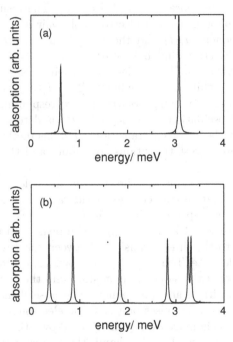

Fig. 16.3. Optical spectrum for a homogeneous ring with $N = 20$ sites close to the fundamental gap without Coulomb interaction, i.e., $U_0 = 0$, and local excitation at site $n = 1$. The magnetic flux is $\Phi = \Phi_0/4$. (a) Equal electron and hole masses $J^e = J^h = 50$ meV. The higher energy peak is due to the two-fold degenerate indirect transitions indicated by *dotted lines* in Fig. 16.1. (b) Different electron and hole masses $J^e = 50$ meV, $J^h = 10$ meV. In this case, the degeneracy is lifted. [From K. Maschke et al., Eur. Phys. J. B **19**, 599 (2001)]

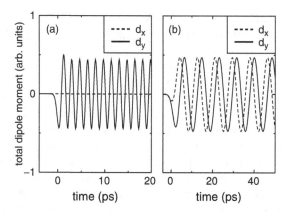

Fig. 16.4. Total dipole moment $d(t) = (d_x(t), d_y(t))$ as a function of time. (**a**) Equal electron and hole masses, see Fig. 16.3a, and (**b**) different masses, see Fig. 16.3b. [From K. Maschke et al., Eur. Phys. J. B **19**, 599 (2001)]

The optical pulse has a central excitation energy $\hbar\omega_L$ close to the fundamental gap and a temporal width of $\delta = 1\,\mathrm{ps}$. The resulting total dipole moment d of the photogenerated electron–hole pair is shown in Fig. 16.4a, where the x component $d_x(t)$ and the y component $d_y(t)$ are shown as a function of time. Electron and hole start at site $n = 1$ and move in opposite directions with equal velocities since their masses are equal. Thus, $d_x(t)$ remains zero, while $d_y(t)$ oscillates. The $d_y(d_x)$ trace in the d_x-d_y plane shows the linearly varying polarization corresponding to this motion, see Fig. 16.5a. The oscillation frequency of $d_y(t)$ is determined by the energetic separation of the involved transitions. For energetically narrow pulse widths, only the two lowest transitions contribute and the motion is purely sinusoidal.

Realistically, in semiconductors, the electron and hole masses are different. In such cases, the situation may change considerably. Figure 16.3b shows the spectrum close to the gap for a ring with $J^e = 50\,\mathrm{meV}$ and $J^h = 10\,\mathrm{meV}$ for excitation at site $n = 1$ and $\Phi = \Phi_0/4$. The peak with the lowest energy corresponds again to the direct transition between the lowest electron state and the highest hole state. The second peak, however, is now due to the indirect transition between the same electron state and the energetically nearest hole state. Both transitions are nondegenerate because of the different masses. As for energetically narrow pulses only a single electron k state contributes, the electron density is homogeneously distributed over the ring after the light pulse, yielding $d^e = 0$. On the other hand, the two contributing hole states form a wave packet that moves around the ring with a velocity given by the energetic separations of the two states involved. This leads to a circularly polarized motion of the total dipole moment $d(t) = d^h(t)$, as seen in Figs. 16.4b and 16.5b, where both d_x and d_y as a function of time as well as the trace $d_y(d_x)$ are shown.

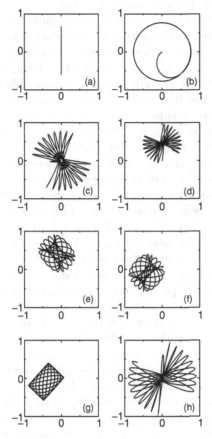

Fig. 16.5. Traces of the total dipole moment $\boldsymbol{d}(t) = (d_x(t), d_y(t))$, (**a–g**), and of the second time derivative of the total dipole moment (**h**) for $\Phi = \Phi_0/4$. *Vertical axes:* y direction; *horizontal axes:* x direction. (**a**) Equal electron and hole masses and inhomogeneous excitation at site $n = 1$ of the homogeneous ring for $U_0 = 0$, see Fig. 16.3a. The polarization is linear, i.e., has only a y component. (**b**) Different masses and inhomogeneous excitation at site $n = 1$ of the homogeneous ring for $U_0 = 0$, see Fig. 16.3b. The polarization is circular. (**c**) Inhomogeneous excitation at site $n = 1$ of the homogeneous ring for $U_0 = 16.5\,\text{meV}$, see Fig. 16.6a. (**d**) Same as (**c**), but homogeneous excitation of a ring with tilted band structure, $\sigma_t = 0.1\,\text{meV}$, see Fig. 16.6d. Note the qualitative similarity with Fig. 16.5c. (**e**) Homogeneous excitation of the ring with tilted band structure and disorder, $\sigma_e = \sigma_t = 0.1\,\text{meV}$, for $U_0 = 16.5\,\text{meV}$. (**f**) Homogeneous excitation of the ring with disorder and without band tilting, $\sigma_e = 0.1\,\text{meV}$, $\sigma_t = 0$ for $U_0 = 16.5\,\text{meV}$. (**g**) Same as (**f**), but without magnetic flux, i.e., $\Phi = 0$. (**h**) Second time derivative of the total dipole moment for the situation of Fig. 16.5d. [From K. Maschke et al., Eur. Phys. J. B **19**, 599 (2001)]

16.3.2 Interacting Particles

It is expected that the Coulomb attraction between electron and hole is important for the wave-packet dynamics. In the following, we assume $U_0 = 16.5\,\text{meV}$, which corresponds to an excitonic binding energy of about $10\,\text{meV}$.

Magnetic field effects in the low-energy excitations can only be expected if the corresponding electron–hole pairs are not too tightly bound, or – in other words – if the excitonic Bohr radius remains comparable with the circumference of the ring. As already mentioned above, small rings are required in any case, since the coherent dynamics can only take place on a time scale smaller than the dephasing time. For given effective masses, the differences in energy leading to the circular motion and, thus, the time scale of the circular dynamics, depend solely on the ring radius R and increase as $1/R^2$ with decreasing R. In particular, the energetic distances depend neither on the number of sites N nor on the site separation a or the coupling parameters J^ν.

We again consider a homogeneous ring and local excitation at site $n = 1$. The lowest transitions are shown in Fig. 16.6a. We see a single peak near $-10\,\text{meV}$ and a split peak near $-9.4\,\text{meV}$. Comparison with the absorption spectrum for homogeneous excitation, see Fig. 16.6b–d, shows that this doublet originates from a higher excitonic bound state which is two-fold degenerate for $\Phi = 0$ and has odd parity, i.e., is optically forbidden for homogeneous excitation. This transition becomes, however, allowed if the symmetry is broken as, e.g., under local excitation conditions, where the circular symmetry of the system is perturbed by the excitation itself. Alternatively, the symmetry may be broken by a perturbing single-particle potential. This is also evident from Figs. 16.6c and d, which show the absorption spectra for homogeneous excitation for $\Phi = 0$ and for $\Phi = \Phi_0/4$, but where now additional band-tilting potentials $\delta\epsilon_n^e = \sigma_t \sin[2(n-1)\pi/N]$ and $\delta\epsilon_n^h = (J^e/J^h)\sigma_t \sin[2(n-1)\pi/N]$ are added to the conduction and valence band site energies, respectively. In this case, the symmetry is broken by the additional potentials and the optical transition into these states becomes allowed even under homogeneous excitation conditions.

The doublet splitting in Fig. 16.6a is due to the magnetic flux, as can also been seen in Fig. 16.6d. This splitting is periodic with period $\Phi_0/2$ and we could call this effect the "quantum confined Zeeman effect". The period halfing is due to the fact that the two peaks cross at $\Phi = \Phi_0/2$.

Similar to the situation treated in the previous section, the optical excitation of the doublet structure leads to electron and hole wave packets moving in opposite directions, but now with different velocities due to the different tight-binding coupling. The time scale of the rotation is determined by the rather small splitting of the doublet. The rotation frequency is thus considerably smaller than the superimposed linear oscillation which is due to the contribution of the lowest excitonic transition that scales with the energetic distance between the first absorption peak and the doublet. This explains the rosette-like trace of the total dipole moment shown in Fig. 16.5c. With our

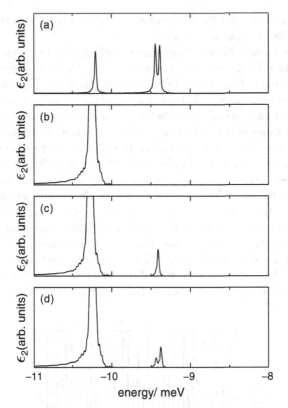

Fig. 16.6. Optical spectrum close to the fundamental gap for the ring with $N = 20$ sites for different electron and hole masses, $J^e = 50\,\text{meV}$, $J^h = 10\,\text{meV}$, with Coulomb interaction of $U_0 = 16.5\,\text{meV}$. (**a**) Homogeneous ring and local excitation at site $n = 1$ and magnetic flux $\Phi = \Phi_0/4$. (**b**) Homogeneous ring, homogeneous excitation, and no magnetic flux, i.e., $\Phi = 0$. The peak is the lowest excitonic resonance. (**c**) Tilted band structure, $\sigma_t = 0.2\,\text{meV}$, homogeneous excitation, and magnetic flux $\Phi = 0$. In addition to the lowest excitonic resonance, the next higher excitonic states become optically allowed. (**d**) Same as (**c**), but with magnetic flux $\Phi = \Phi_0/4$. The second resonance is split. [From K. Maschke et al., Eur. Phys. J. B **19**, 599 (2001)]

parameters, the circular component has a period of 100 ps, while that of the linear component has a period of the order of 5 ps.

The dynamics of the total dipole moment is shown in Fig. 16.5d for $\sigma_t = 0.1\,\text{meV}$. It closely resembles that of Fig. 16.5c for the perfect ring with single-site excitation. The additional stationary dipole moment seen in Fig. 16.5d is caused by the tilting of the valence and conduction bands. The stationary dipole moment increases with the amplitude of the band-tilting potential, whereas the amplitude of the dynamic motion decreases. The latter feature is due to the fact that the excitonic wave function becomes more confined for larger amplitudes σ_t and finally localizes in the region where the differences between the site energies are smallest.

The source term in Maxwell's equations, which is given by the second time derivative of the dipole moment, is shown in Fig. 16.5h. It essentially reflects the dynamical part of the total dipole moment shown in Fig. 16.5d.

Similar effects as discussed above may be expected for a disorder potential, which also breaks the circular symmetry of the ring, thus changing transitions into higher excitonic states, which are forbidden in the homogeneous situation, into allowed ones. A random potential, however, leads to a somewhat more complicated trace of the total dipole, as shown in Fig. 16.5e, where, in addition to the previously used sinusoidal potential, we have added a random potential with $\sigma_e = \sigma_t = 0.1$ meV. Again, a stationary dipole is found which results from a fluctuation of the local band gap, but it now points into a slightly different direction. Figure 16.5f shows the situation without sinusoidal potential. In this case, the stationary dipole has changed its direction and is now solely determined by the disorder potential. The dynamical part, however, looks still very similar. The magnetic flux $\Phi = \Phi_0/4$ leads to a circular component in the dynamical trace, as can be seen in the upper panel of Fig. 16.7, where the

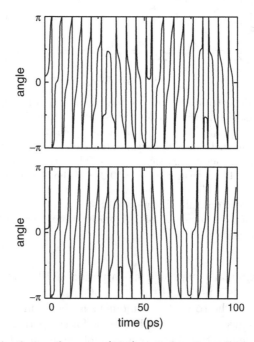

Fig. 16.7. Polarization angle α, see (16.8), as a function of time for homogeneous excitation of the disordered ring. *Upper panel*: with magnetic flux $\Phi = \Phi_0/4$, see Fig. 16.5f. There is no time reversal. The circular character is clearly seen. *Lower panel*: without magnetic flux, see Fig. 16.5g. Two perfect time reversals can be identified. [From K. Maschke et al., Eur. Phys. J. B **19**, 599 (2001)]

polarization angle α defined in (16.8) is shown. The motion is in general not periodic. In fact, a periodic evolution would require that the involved energy differences have rational ratios. The dynamics of the total dipole moment in the absence of the magnetic field is shown in Fig. 16.5g and the corresponding angle α is presented in the lower panel of Fig. 16.7. It is seen that the motion periodically changes its direction and that there is no circular component on the temporal average.

Experimentally, the dipole motion could be detected by measuring the corresponding electromagnetic field, which follows the underlying polarization pattern. Alternatively, one could also use pump–probe or four-wave-mixing experiments to observe the predicted effects, as has been successfully done in the case of Bloch oscillations discussed in the previous chapter.

In order to obtain observable signals, it is usually necessary to perform the experiments on ensembles of rings. The excitation then has to be homogeneous over the entire ensemble and a gradient of composition or strain should be applied to produce the sinusoidal symmetry breaking potential.

Disorder is always present to a certain degree in real samples. Our results show, however, that it does not disturb the dynamics of a single ring severely. For an ensemble of many nanorings, the disorder-induced changes are expected to cancel out at times longer than the pulse duration. As an alternative to such an ensemble, one might fabricate a quantum well with closely spaced holes of diameters in the 10–100 nm range. This arrangement should also facilitate the creation of a band gap gradient over the sample. Although, in this case, the different rings have slightly different optical gaps, the dynamics of the dipole moment is identical for all rings.

16.4 Linear and Nonlinear Magneto-Optics

Here, we consider the linear and, in particular, nonlinear optical properties of interband transitions in semiconductor nanorings with magnetic field and disorder. So far, we have treated the special case of magnetic flux Φ being one quarter of the flux quantum Φ_0 which gives rise to nonvanishing electron and hole currents along the circumference of the ring. Now, we take magnetic fluxes of variable strength.

We find periodic magnetic-field-induced modifications of the spectral positions of bound and unbound excitons and biexcitons. The findings demonstrate that the Aharonov–Bohm effect, known to exist for charged particles moving along the circumference of a ring penetrated by a magnetic flux, also influences quasi-particles like excitons and biexcitons. At first sight, this may seem surprising, as these particles are neutral.

We now allow for two (in the magnetic field-free case) spin-degenerate bands for electrons and holes in order to be able to treat biexcitons. We still neglect, however, the intraband Zeeman splitting. This would lead to additional splittings of the excitonic resonances but is not expected to alter the features discussed here substantially. Numerically computed non-linear pump–probe spectra for counterclockwise polarized pulses demonstrate that the magnetic field strongly influences Coulombic many-particle correlations. In this geometry, the differential absorption spectra of an ordered system are purely induced by many-body correlations as long as only heavy-hole and no light-hole transitions are relevant as has been shown in Chap. 11.

As realistic nanostructures exhibit some amount of structural disorder on mesoscopic scales, we study the influence of energetic disorder on the magnetic-field-induced changes of the linear and nonlinear absorption spectra. It is shown that the excitonic shift survives in the presence of energetic disorder, even if the disorder-induced inhomogeneous broadening of the absorption lines exceeds the magnetic-field-induced shift of the exciton line. Furthermore, we show and discuss results on linear and nonlinear optical spectra of single disordered rings.

The number of sites is taken to be $N = 10$. In one set of calculations, we use $N = 20$. We here use $J^e = 28.05\,\mathrm{meV}$ and $J^h = 4.95\,\mathrm{meV}$. The corresponding effective masses can be evaluated from (6.18). The ratio of the conduction- and valence-band couplings J^e/J^h matches the inverse ratio of the effective conduction- and valence-band masses $(m_c^*/m_v^*)^{-1}$ for GaAs parameters. For the Coulomb interaction, we use $U_0 = 15\,\mathrm{meV}$ (in two sets of calculations we use $U_0 = 0$ and $U_0 = 22.5\,\mathrm{meV}$) and $a_0/a = 0.5$. This results in exciton binding energies of a reasonable magnitude. The values used here were chosen in order to clearly demonstrate how a magnetic field qualitatively influences exciton and biexciton states.

16.4.1 Linear Optical Spectra

We start by investigating the influence of the enclosed magnetic flux on the excitonic line in the linear absorption spectrum. Since all magnetic-field-induced changes are periodic and proportional to $\cos(2\pi\Phi/\Phi_0)$, we limit our investigations to $0 \leq \Phi/\Phi_0 \leq 0.5$. Figure 16.8a displays the excitonic linear absorption for different flux ratios

$$\mathcal{R} = \Phi/\Phi_0. \tag{16.29}$$

Note that the zero of the energy scale corresponds to the lowest optical transition without Coulomb interaction and without magnetic field, i.e., $\mathcal{R} = 0$ in Fig. 16.8b. In Fig. 16.8a we have the exciton line at $E_X = -11.86\,\mathrm{meV}$ for $\mathcal{R} = 0$ which directly corresponds to an exciton binding energy of 11.86 meV. With increasing flux, Fig. 16.8a shows that the exciton line shifts towards higher energies and that the oscillator strength is increasing.

An increased excitonic oscillator strength means that the part of the exciton wave function describing the relative motion of the electron and hole must have an increased value at the origin, i.e., at $r_e = r_h$. Therefore, such an increase would imply an increased exciton binding energy. Since the exciton

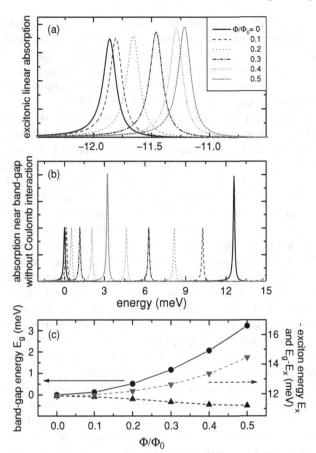

Fig. 16.8. (a) Excitonic linear optical absorption spectra and (b) interband absorption without Coulomb interaction close to the band gap for various magnetic fields corresponding to $\mathcal{R} = \Phi/\Phi_0 = 0, 0.1, 0.2, 0.3, 0.4,$ and 0.5. In both cases Lorentzian homogeneous broadening is introduced by inserting a dephasing time of $T_2 = 10$ ps into (8.54). (c) Band-gap energy E_g (*circles*) and the negative of the exciton energy $-E_X$ (*triangles up*), as well as difference between both $E_g - E_X$ (*triangles down*). The symbols correspond to calculated values, the lines are guides for the eye. [From T. Meier et al., Eur. Phys. J. B **22**, 249 (2001)]

binding energy is defined with respect to the lowest optical transition of the same system without Coulomb interaction, we have to calculate $E_g(\mathcal{R})$, see Fig. 16.8b.

   Without magnetic field, the lowest optical transition corresponds in k space to a transition between the $k = 0$ electron and hole states. The phase factor introduced by the magnetic field in real space results in an effective shift of the allowed values of the wave vectors in k space. Therefore, $E_g(\mathcal{R})$ increases monotonically when the enclosed flux increases from $\Phi = 0$

to $\Phi = 0.5\Phi_0$, see Fig. 16.8b. The second allowed interband transition, which is two-fold degenerate without magnetic field corresponding to $k = 2\pi/Na$ and $k = -2\pi/Na$, splits into two lines with increasing magnetic field. Due to the magnetic-field-induced shift of the k values, one of these lines increases and one decreases in energy. At $\Phi = 0.5\Phi_0$, the lowest transition is two-fold degenerate corresponding to $k = \pi/Na$ and $k = -\pi/Na$.

As shown in Fig. 16.8c, $-E_X(\mathcal{R})$ is cosine-like and decreases slightly, whereas $E_g(\mathcal{R})$ increases quadratically since the cosine tight-binding dispersion in the vicinity of the band gap is quadratic. Due to the fact that $E_g(\mathcal{R})$ increases more strongly than $-E_X(\mathcal{R})$ decreases, the exciton binding energy defined as $\Delta E_X(\mathcal{R}) = E_g(\mathcal{R}) - E_X(\mathcal{R})$ increases as a function of $\mathcal{R}$. Thus, in agreement with the increase in oscillator strength shown in Fig. 16.8a, we find that the exciton binding energy increases when going from $\mathcal{R} = 0$ to $\mathcal{R} = 0.5$.

An interesting qualitative analogy exists with the shift of the excitonic absorption in a superlattice biased by an ac electric field. As the magnetic field considered here, also an ac electric field changes the single-particle dispersion and moves, due to dynamical localization, the lowest optical transition without Coulomb interaction towards higher values, see Sect. 15.2.4. Since this dynamical localization is accompanied by a change of the effective dimension of the exciton from one to close to zero, the binding energy of the exciton increases if the ac electric field is applied. As for dynamical localization, we here also see simultaneously a blue shift and increases of the oscillator strength and the binding energy of the exciton, see Fig. 16.8a.

The change of the exciton energy due to the Aharonov–Bohm effect occurs only in mesoscopic systems. Since the effect relies on the finite probability of the electron and hole forming the exciton to tunnel around the ring, the exciton Bohr radius needs to be comparable to the diameter of the ring. If we increase the Coulomb interaction by a factor of 1.5, i.e., we use $U_0 = 22.5\,\text{meV}$, the exciton is more strongly bound and the Bohr radius is smaller. Concomitantly, the shift of the exciton energy in the range $\mathcal{R} = 0$ to $\mathcal{R} = 0.5$ is reduced in Fig. 16.9a as compared to Fig. 16.8a. Another clear demonstration of the mesoscopic nature of the excitonic Aharonov–Bohm effect is given in Fig. 16.9b, where it is shown that when using the original parameters, i.e., $U_0 = 15\,\text{meV}$, the effect has almost completely disappeared for a doubled number of sites $N = 20$ corresponding to a doubled diameter of the ring.

16.4.2 Nonlinear Optical Spectra

The nonlinear optical response of the ring is analyzed by calculating the differential absorption as obtained in pump–probe spectroscopy performed with counterclockwise polarized pulses, i.e., with $\sigma^+\sigma^-$ excitation. In this geometry, the differential absorption spectra of an ordered system are purely induced by many-body correlations as long as only heavy-hole and no light-hole transitions are relevant. This is due to the fact that the two degenerate optically

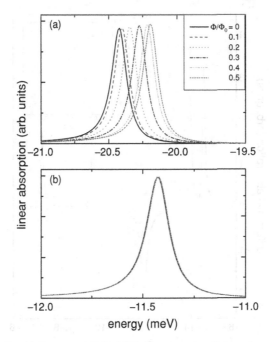

Fig. 16.9. Excitonic linear optical absorption spectra for various magnetic fields corresponding to $\mathcal{R} = \Phi/\Phi_0 = 0$, 0.1, 0.2, 0.3, 0.4, and 0.5. (a) For an increased Coulomb interaction of $U_0 = 22.5$ meV and (b) for an increased system size of $N = 20$. In both cases, Lorentzian homogeneous broadening is introduced by inserting a dephasing time of $T_2 = 10$ ps into (8.54). [From T. Meier et al., Eur. Phys. J. B **22**, 249 (2001)]

allowed heavy-hole excitons that can be excited with σ^+ and σ^- polarized light do not share a common state.

The spectra displayed in Fig. 16.10 show negative contributions at the exciton energy, which correspond to the pump-induced bleaching of the exciton transition. Positive induced absorption appears spectrally below and above the exciton due to transitions to bound and unbound two-excitons, respectively. As the exciton, the spectral position of the exciton to bound biexciton transition, which appears at $E_{BX} - E_X$, also shifts towards higher energies if $\mathcal{R}$ increases from 0 to 0.5. The fact that the exciton to bound biexciton transition shifts more strongly with $\mathcal{R}$ than the exciton line, corresponds to a *decrease* of the biexciton binding energy $\Delta E_{BX} = 2E_X - E_{BX}$, see inset of Fig. 16.10. So, comparing the cases $\mathcal{R} = 0$ with $\mathcal{R} = 0.5$, we find that the increase of the exciton binding energy is accompanied by a decrease of the biexciton binding energy of about 15%. This difference can be understood to be due to the weaker interaction among the more tightly bound excitons. If the excitons were very strongly bound, i.e., if the electron and hole were confined to the same site as is the case for a Frenkel exciton, the biexciton

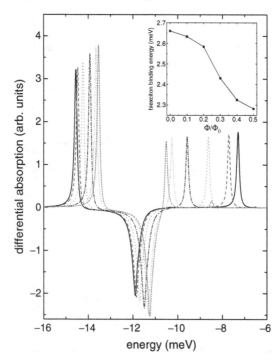

Fig. 16.10. Differential absorption spectra using a σ^+ polarized pump and a σ^- polarized probe pulse for various magnetic fields corresponding to $\mathcal{R} = \Phi/\Phi_0 = 0$, 0.1, 0.2, 0.3, 0.4, and 0.5. The pump pulse arrives 2 ps before the probe, its duration is 1.18 ps (FWHM of pulse intensity), and it is spectrally tuned to the exciton resonance for each value of the field. A Lorentzian homogeneous broadening is introduced by inserting a dephasing time of $T_2 = 10$ ps into (8.54). No broadening is used in (8.56) in order to obtain spectrally narrow two-exciton lines. Zero differential absorption is indicated by the *horizontal line*. The *inset* shows the flux dependence of the biexciton binding energy obtained as $\Delta E_{BX} = 2E_X - E_{BX}$. [From T. Meier et al., Eur. Phys. J. B **22**, 249 (2001)]

binding energy would vanish and only unbound two-excitons would appear in the spectra. Thus, with the magnetic-field-induced stronger exciton binding we go from an extended Wannier towards a more tightly bound Frenkel exciton. A decrease of the biexciton binding energy comparing the cases $\mathcal{R} = 0$ with $\mathcal{R} = 0.5$ is observed for all parameters that have been investigated, i.e., this effect seems to be robust and especially survives for larger as well as weaker Coulomb interaction (not shown in figure).

As clearly shown in Fig. 16.10, the magnetic field shifts not only change the resonances of bound biexcitons to higher energies, but also introduce an even stronger red shift of the transitions to the lowest unbound two-exciton resonance, i.e., the stronger binding of the excitons introduces a weaker repulsive interaction among them, which is consistent with the explanation given above.

16.4.3 Influence of Disorder

Figure 16.11a displays the excitonic linear optical absorption spectra for $\mathcal{R} = 0$ (solid) and $\mathcal{R} = 0.5$ (dashed) for the ordered system. The spectra displayed in Figs. 16.11b and c were calculated by averaging over 10 000 disorder realizations (corresponding to a measurement on an ensemble of 10 000 rings) using $\sigma_e = 2\,\text{meV}$ and $4\,\text{meV}$, respectively. They show some fluctuations due to the finite number of disorder realizations and a much broader and red shifted exciton absorption than for the ordered system. This can be understood to be the result of the disorder-induced inhomogeneous broadening of the excitonic transition energy. Even for these quite broad spectra which are characterized by an inhomogeneous linewidth that exceeds the magnetic-field-induced shift

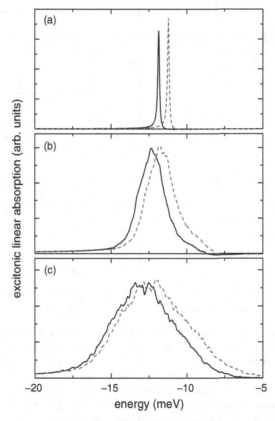

Fig. 16.11. Excitonic linear optical absorption spectra for magnetic fields corresponding to $\mathcal{R} = \Phi/\Phi_0 = 0$ (*solid*) and 0.5 (*dashed*). (a) Without disorder, (b) with disorder described by $\sigma^c = 2\,\text{meV}$, and (c) for $\sigma^c = 4\,\text{meV}$. Lorentzian homogeneous broadening is introduced by inserting a dephasing time of $T_2 = 10\,\text{ps}$ into (8.54). For (b) and (c) the spectra have been averaged over 10,000 disorder realizations. [From T. Meier et al., Eur. Phys. J. B **22**, 249 (2001)]

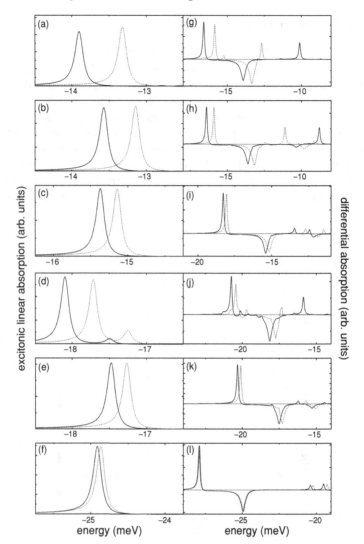

Fig. 16.12. *Left panels*: excitonic linear optical absorption spectra for magnetic fields corresponding to $\mathcal{R} = \Phi/\Phi_0 = 0$ (*solid*) and 0.5 (*dashed*). (**a–c**) Three individual disorder realizations with $\sigma^c = 4\,\mathrm{meV}$ and (**d–f**) the same realizations for $\sigma^c = 10\,\mathrm{meV}$. *Right panels*: differential absorption spectra using a σ^+ polarized pump and a σ^- polarized probe pulse for $\mathcal{R} = \Phi/\Phi_0 = 0$ (*solid*) and 0.5 (*dashed*). (**g–i**) For the same disorder realizations as (**a–c**) and (**j–l**) for the same disorder realizations as (**d–e**). The pump pulse arrives 2 ps before the probe, its duration is 1.18 ps (FWHM of pulse intensity), and it is spectrally tuned to the exciton resonance for each value of the field. A Lorentzian homogeneous broadening is introduced by inserting a dephasing time of $T_2 = 10\,\mathrm{ps}$ into (8.54). No broadening is used in (8.56) in order to obtain spectrally narrow biexciton lines. Zero differential absorption is indicated by the horizontal line. [From T. Meier et al., Eur. Phys. J. B **22**, 249 (2001)]

of the exciton line, one can still see an effect of the magnetic field. Thus, the shift of the exciton line due to the magnetic field seems not to be destroyed by energetic disorder on the scale considered here and even survives the inhomogeneous broadening.

Even more interesting than the disorder-induced inhomogeneous broadening is the question of how the disorder influences the relative motion of the electron–hole pair forming the exciton. Note that in order to find any Aharonov–Bohm effect, the relative motion of the exciton needs to extend over the entire ring. This condition is not fulfilled if the disorder is strong enough to localize the relative motion to a small portion of the ring. In order to address this question, we investigate linear and nonlinear absorption spectra for individual disorder realizations.

For $\sigma^e = 4\,\mathrm{meV}$, the magnetic-field-induced shift of the exciton energy is still present, see Figs. 16.12a–c, however, it is smaller than without disorder, cf. Fig. 16.8a, and its precise value depends on the disorder realization. Whereas for one realization, Fig. 16.12a, there is still some observable increase of the oscillator strength of the exciton, this effect is much smaller for another realization, Fig. 16.12b, and is absent in a third case, Fig. 16.12c. Consequently, only for one particular disorder realization, see Fig. 16.12a, does the magnetic field still significantly influence the relative motion of the exciton. This results in a clear reduction of the interaction among excitons as demonstrated by the corresponding nonlinear pump–probe spectrum shown in Fig. 16.12g. As in the ordered case, the energy difference between the exciton and the bound biexciton as well as between the exciton and the lowest unbound biexciton state is smaller for $\mathcal{R} = 0.5$ than for $\mathcal{R} = 0$. In Figs. 16.12h and i on the other hand, the biexciton binding energy is unaffected by the magnetic field. One can, however, even for the cases of Fig. 16.12h, still see some reduction of the energy difference between the exciton and the lowest unbound biexciton state. Additionally, the disorder is responsible for the weak additional lines appearing in the pump–probe spectra.

Even if the disorder is increased to $\sigma^c = 10\,\mathrm{meV}$ (note that this value is close to the exciton binding energy) all three realizations still show some reduced shift of the exciton energy, see Figs. 16.12d–f. The weak absorption line above the lowest exciton appearing in Fig. 16.12d is induced by the rather strong disorder. For $\sigma^c = 10\,\mathrm{meV}$, however, no realization shows any increase of the exciton oscillator strength. Consequently, also the biexciton binding energy is unaffected by the magnetic field, see Figs. 16.12j–l.

16.5 Suggested Reading

1. F. Bloch, "Josephson effect in a superconducting ring", Phys. Rev. **B2**, 109 (1970)

2. N. Byers and C.N. Yang, "Theoretical considerations concerning quantized magnetic flux in superconducting cylinders", Phys. Rev. Lett. **7**, 46 (1961)

3. K. Maschke, T. Meier, P. Thomas, S.W. Koch, "Coherent dynamics of magnetoexcitons in semiconductor nanorings", Eur. Phys. J. B **19**, 599 (2001)

4. T. Meier, P. Thomas, S.W. Koch, "Linear and nonlinear optical properties of semiconductor nanorings with magnetic field and disorder – Influence on excitons and biexcitons", Eur. Phys. J. B **22**, 249 (2001)

5. T. Meier, P. Thomas, S.W. Koch, and K. Maschke, "Signatures of trions in the optical spectra of doped semiconductor nanorings in a magnetic field", phys. stat. sol. b **234**, 283 (2002)

6. T. Meier, C. Sieh, E. Finger, W. Stolz, W.W. Rühle, P. Thomas, and S.W. Koch, "Signatures of biexcitons and triexcitons in coherent nondegenerate semiconductor optics", phys. stat. sol. b **238**, 537 (2003)

7. R.A. Römer and M.E. Raikh, "Aharonov–Bohm effect for an exciton", Phys. Rev. **B62**, 7045 (2000)

8. R.A. Römer and M.E. Raikh, "Aharonov–Bohm oscillations in the exciton luminescence from a semiconductor nanoring", phys. stat. sol. (b) **221**, 535 (2000)

Coherent Density Dynamics in Disordered Semiconductors

In this chapter, we study some aspects of transport phenomena related to the possibility of particle localization by disorder. The classical theories of disorder-driven localization, the Anderson localization, have been formulated based on noninteracting models. For a certain time, the predictions of these theories were generally accepted. However, experiments on two-dimensional systems gave evidence that the understanding of Anderson localization is far from complete. It was argued that the Coulomb interaction might be responsible for the signatures seen in those experiments which disagreed with theoretical predictions.

It was therefore necessary to develop a theory of electronic transport, in particular dc transport, that treats disorder and many-particle interaction on an equal footing. However, such a theory is extremely hard to evaluate, since it requires us to study a dc (zero-frequency, i.e., long-time) property in the thermodynamic limit (i.e., for large systems). Both analytical and numerical approaches remained rather limited in their success. It is therefore helpful to reduce the model to a very schematic scale such that particle–particle interaction can be described and the numerical evaluation is still feasible.

17.1 The Two-Interacting-Particles Problem

The minimal model that allows us to treat particle–particle interaction is given by exactly two electrons. In fact, the problem of two interacting particles in a random potential is an excellent paradigm for the general question of the interplay of disorder and interactions in many-body systems. Considering the two-interacting-particles localization length l_2, one finds an interaction-induced increase $l_2 > l_1$ over the single-particle localization length l_1. Here, l_1 and l_2 are measured in units of the lattice constant of a one-dimensional chain.

Due to this artificial model, existing works comprise purely theoretical case studies. In fact, the model of just two particles in a single band does not correspond to any real physical situation. On the other hand, optical excitations in semiconductors quite naturally produce pairs of charged particles, either in bound states (excitons) or in the continuum of Coulomb-correlated electron–hole pairs, see Chap. 14. Provided the excitation intensity is not too strong, the dynamics of the optical excitation can be viewed as a superposition of that of isolated pairs of mutually interacting particles, namely electrons and holes. Therefore, the experimental and theoretical study of the spatio-temporal dynamics of optically excited electron–hole pairs in disordered semiconductors should yield insights into the notoriously difficult problem of disorder-driven localization of interacting particles.

17.2 The Semiconductor Two-Interacting-Particles Problem

In the following, we investigate the spreading of an electron–hole wave packet after local excitation by an optical pulse. The interaction is given by the long-range Coulomb potential which, besides producing bound states (excitons) near the edges of the excitation spectrum, also correlates the electrons and holes in the pair continuum. Theoretical studies of the spatio-temporal dynamics of excitonic wave packets show that their motion is rather limited in the presence of scattering. Here, we focus our interest on the dynamics of optically generated wave packets in the pair continuum. We find that the excitation conditions in the presence of particle–particle interaction influence the carrier dynamics dramatically.

17.2.1 The Model

We consider the two-band Hamiltonian, (5.8), with a single conduction and a single valence band, nearest-neighbor couplings J^e and J^h, and site energies ϵ_i^e and ϵ_i^h which are randomly distributed over the interval $[(\epsilon_0/2) - W/2, (\epsilon_0/2) + W/2]$ and can be chosen to be uncorrelated, correlated, or anticorrelated. In the following we start by considering uncorrelated disorder and analyze the other cases in Sect. 17.2.5. We use the Coulomb interaction in its monopole–monopole form, (5.13), with the matrix elements given by (5.17). The regularization constant a_0 has been chosen here to be five times the lattice constant a.

In the following, we assume a *weak and local* initial excitation at the central site $i = 0$, which is modeled by setting $\mu_i = \delta_{i,0}$ for the local dipole matrix element. The interband coherences $p_{ij}^{he}(t)$ are obtained from the linear-response equation of motion, (8.64), and the intraband quantities n_{ij}^e and n_{ij}^h from the conservation laws, i.e., (8.14) and (8.15), which are valid in the low-excitation coherent limit. We study the influence of disorder, for various correlation properties and interaction strengths, on the localization of the electron–hole pair. The equation of motion is solved numerically for a number of realizations (typically 20–40) of the disorder drawn from the distribution of site energies. The observables are then configurationally averaged over these realizations.

Instead of considering the localization length which describes the asymptotic behavior of wave functions, we calculate the experimentally more relevant participation number $\Lambda(t)$, see (6.38), which is given by

$$\Lambda(t) = \frac{[\sum_i n_{ii}(t)]^2}{\sum_i n_{ii}^2}. \tag{17.1}$$

Here, n_{ii} stands for either n_{ii}^e or n_{ii}^h. A wave packet localized at site 0, i.e., $n_{ii} = \delta_{i0}$, leads to $\Lambda = 1$, while an excitation uniformly extended over the sample of N sites, i.e., $n_{ii} = 1/N$, corresponds to $\Lambda = N$. The calculations were performed for chains containing $N = 240$ sites. Boundary effects can easily be identified in the temporal evolution of Λ and do not play a role as long as $\Lambda < N/2$. Thus, the data presented are free of finite-size effects.

The optical pulse is defined by its mean energy $\hbar\omega$ and the temporal width δ of the Gaussian envelope which is proportional to $\exp(-(t/\delta)^2)$. We define an excitation energy E_{exc} referring to the bottom of the (ordered) absorption band, i.e., $E_{exc} = \hbar\omega - E_{gap}$. All results have been calculated using $\delta = 100\,\text{fs}$, which corresponds to a spectral width of 22 meV (FWHM).

17.2.2 Electron–Hole Symmetry

Here, we consider the situation of a symmetric band structure with $J^e = J^h = 20\,\text{meV}$. The absorption spectra with and without Coulomb interaction are shown in Fig. 17.1 for the ordered case. Note that in this chapter we use $U = \pm 1$ as a multiplication factor of the Coulomb matrix elements i.e., $V_{ij} \rightarrow UV_{ij}$ determining the sign of the interaction. The interaction is attractive if $U < 0$ or repulsive if $U > 0$. The peak structure near the absorption edge in Fig. 17.1 is due to the formation of excitons. Upon changing the sign of the Coulomb interaction, the bound states are shifted from the bottom to the top of the absorption spectrum.

As the dynamics of electrons and holes are the same for the assumed symmetric band structure, we restrict our discussion to the electrons and write $\Lambda = \Lambda^e$. We first discuss the situation in the absence of Coulomb interaction. Figure 17.3, which is repeated here for convenience, see Fig. 6.7, shows the corresponding $\Lambda(t)$ for different disorder parameters W after excitation by a pulse at $E_{exc} = 80\,\text{meV}$. The excitation is centered in the absorption spectrum, as indicated in Figs. 17.1 and 17.2. It is seen in Fig. 17.3 that $\Lambda(t)$ saturates within 1 ps after the excitation. Here and below, we take the saturation value Λ_{sat} as a measure of localization. As expected, this value decreases rapidly with increasing disorder. We find $\Lambda_{sat} \approx W^{-1.3}$ as W is varied over the range 20–240 meV for $J = 20\,\text{meV}$.

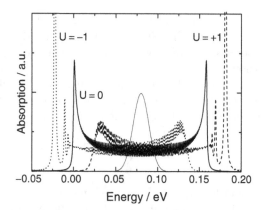

Fig. 17.1. Absorption spectrum of the ordered chain for $U = 0$, -1, $+1$ and for equal electron and hole masses, i.e., $J^e = J^h = 20\,\mathrm{meV}$. The spectrum of the optical pulse with $E_{\mathrm{exc}} = 80\,\mathrm{meV}$ is also given. [From D. Brinkmann et al., Eur. Phys. J. B **10**, 145 (1999)]

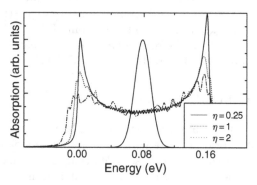

Fig. 17.2. Absorption spectrum of the disordered chain for $U = 0$ and equal electron and hole masses, i.e., $J^e = J^h$. Three different disorder parameters $\eta = W/J$ are used. The spectrum of the optical pulse with E_{exc} in the center of the absorption continuum is also given

Figure 17.4 shows $\Lambda(t)$ in the presence of an attractive electron–hole interaction of $U = -1$. Comparison of Figs. 17.3 and 17.4 reveals three remarkable features.

(i) The interaction clearly leads to a reduction of the localization of the particles. It has been carefully checked that the long-time saturation value of $\Lambda(t)$ is not due to a finite size effect; values $\leq N/2$ are fully converged with respect to the sample size.

(ii) While the extension of the wave packets in the noninteracting situation saturates quickly ($\leq 1\,\mathrm{ps}$), the interacting wave packets reach their saturation value at much longer times.

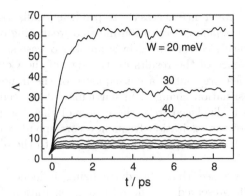

Fig. 17.3. The participation number $\Lambda(t)$ for the noninteracting case $U = 0$ for various disorder parameters W. Note that after a very rapid increase, the participation number, i.e., the extension of the wave packet, saturates. This indicates Anderson localization

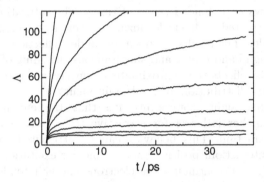

Fig. 17.4. The participation number for electrons $\Lambda(t)$ for the interacting model $U = -1$ and for the same disorder strength W as in Fig. 17.3. Only for the largest disorder, is a saturation due to localization seen. Note the different scale of both the *horizontal* and the *vertical axis* as compared with Fig. 17.3

(iii) The sign of the Coulomb interaction ($U = \pm1$) has no influence on the propagation of the particles if we excite in the center of the band. The same is true if we apply a very short excitation pulse which spectrally covers the entire band. In this case, the spectral position of the central pulse frequency within the band is then completely irrelevant and the excited particle pair wave packet is initially generated exclusively at site $i = 0$.

17.2.3 Dynamic-Correlation-Induced Delocalization

In all cases where J^e and J^h are of comparable magnitude, we find that the participation number is enhanced by the interaction. In a mean field picture, it is the temporal fluctuations of the field originating from the partner particle which destroy the coherence necessary to produce localization. This explanation in terms of a *dynamic*-correlation-induced weakening of the influence of disorder can be nicely corroborated by a number of case studies.

The independence of the results on the sign of the Coulomb interaction is not a general feature, but is a consequence of the imposed electron–hole symmetry. This situation changes, in particular, if the central frequency of excitation pulses is displaced from the center of the absorption band. Note that this choice of the excitation frequency corresponds to the realistic situation where electron–hole pairs are excited close to the absorption edge in semiconductors.

In Fig. 17.5, we see that the delocalizing influence of the Coulomb interaction is still preserved for the off-center excitation. The central excitation energy of the pulse is placed in the lower part of the pair continuum at $E_{\mathrm{exc}} = 40\,\mathrm{meV}$. In Fig. 17.6, we plot the participation number Λ for light electrons and heavy holes using $J^e = 2J^h = 20\,\mathrm{meV}$. Results averaged over 60 realizations are shown for a large disorder of $W = 80\,\mathrm{meV}$ and $U = 0, \pm 1$. This figure shows an extended range of times. The saturation of the traces can well be interpreted as due to localization, even in the interacting case, for this large disorder. The results are invariant under reflection of the excitation frequency through band center and simultaneous switching of the sign of the interaction. This reflects the approximate symmetry, apart from fluctuations in the site energy distribution, of the Hamiltonian.

It is, at first sight, counterintuitive that the enhancement of the participation number is larger for attractive ($U = -1$) than for repulsive ($U = +1$) interaction. This behavior can be attributed to the fact that for attractive interaction, positive masses, and for excitation into the lower half of the excitation continuum the electron–hole pair tends to stay closer together. The fluctuating field due to the accompanying particle is then more

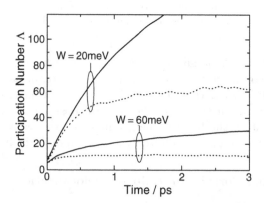

Fig. 17.5. Temporal evolution of the participation number $\Lambda(t)$ for excitation in the lower half of the continuum and for $U = 0$ (*dotted line*) and $U = -1$ (*solid line*). [From D. Brinkmann et al., Eur. Phys. J. B **10**, 145 (1999)]

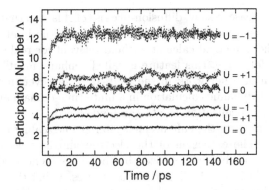

Fig. 17.6. Temporal evolution of the participation number $\Lambda(t)$ for electrons (*upper traces*) and for holes (*lower traces*) for excitation in the lower half of the continuum and $U = 0, \pm 1$, $J^e = 2J^h = 20\,\mathrm{meV}$, $E_{\mathrm{exc}} = 40\,\mathrm{meV}$, and $W = 80\,\mathrm{meV}$. [From D. Brinkmann et al., Eur. Phys. J. B **10**, 145 (1999)]

pronounced as compared to the case of repulsive interaction, where the mutually repulsive particle pair tends to be separated. Hence, the *dynamic-correlation-induced* weakening of the influence of disorder is less effective for repulsive than for attractive interaction.

A completely different behavior is found for a *static* field. We consider now an infinitely heavy hole, i.e., $J^h = 0$, which produces a static field and excitation at the (interaction-free) band center. For both attractive and repulsive interactions, the participation number is *decreased* with respect to the noninteracting case. This result is easily understood without invoking fluctuating fields since at the band center electron states have their maximal extension. In the presence of interaction, off-center states are admixed, which leads to greater confinement. The effect of the static interaction is thus opposite to that of a fluctuating field.

The strong retardation of the saturation in the interacting case can also be understood in our picture. With or without interaction, the electron and hole wave packets spread over a range determined by the extension of the single-particle levels involved in the optical transition just after the short excitation pulse. The fluctuating Coulomb field due to the partner particle then leads to an increase of the spread of the wave packets. As a consequence, the average fluctuating field acting on a given particle is reduced, which in turn tends to slow further spreading, eventually leading to the observed saturation at long times.

17.2.4 Finite-Time Scaling

We have demonstrated that compared to the noninteracting case the interaction induces a significant change of the dynamics. Instead of an exponential rise of $\Lambda(t)$ towards a saturation value on the time scale of the excitation pulse, in the interacting case, a slow, diffusion-like rise of Λ is seen that does not seem to saturate, at least for not too strong disorder, in the limited time regime accessible to the numerical calculation (and relevant for real situations). For brevity, in the following, this feature is called "enhancement" in accordance with the notion of other work in the field. It is, however, not implied that from the present dynamic calculation in a finite time-domain anything like an enhancement of a localization length can be deduced.

In order to quantify the enhancement, finite-time scaling is applied in the following sense. The extension of the electron and hole wave packets as a function of time can be characterized by the above-mentioned participation numbers $\Lambda^{e,h}$ which are single-particle quantities. Using the pair amplitude p_{ij}^{he}, two-particle quantities, i.e., the center-of-mass coordinate

$$R = \left(\sum_{i,j} |p_{ij}^{he}|^2 (i+j)^2/2 \right)^{1/2} \tag{17.2}$$

and the relative coordinate

$$\rho = \left(\sum_{i,j} |p_{ij}^{he}|^2 (i-j)^2/2 \right)^{1/2} \tag{17.3}$$

can be defined. Their ratio R/ρ is a measure of the interaction-induced enhancement.

Although, by a dynamic calculation for finite times, it is impossible to decide whether the two-particle packet is localized or not, a long-time saturation value of the center-of-mass coordinate R_∞ is assumed for practical purposes. While the temporal rise of the interaction-free traces is exponential with a time scale determined by the pulse duration, the short-time behavior in the interacting case looks like a diffusive process. So the following interpolation relation is used

$$R(t) = ((Dt)^{-1/2} + R_\infty^{-1})^{-1}. \tag{17.4}$$

Figure 17.7 shows the very good quality of typical fits for times larger than the pulse width. It is found that, while the diffusivity D is weakly dependent on the disorder and the interaction, R_∞ shows a much stronger dependence. Also R_∞ can be taken to characterize the enhancement.

17.2.5 Influence of Correlated Versus Anticorrelated Disorder

Next, we study the effect of correlated versus anticorrelated disorder and of the interaction strength on the enhancement in more detail. We show data for $J^e = J^h = 20\,\text{meV}$ exclusively, i.e., equal electron and hole masses are

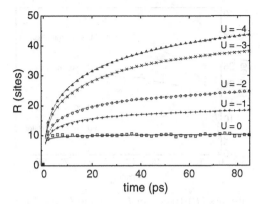

Fig. 17.7. Fit according to (17.4) of the center-of-mass trace for different interaction strengths U. [From P. Thomas et al., phys. stat. sol. b **218**, 125 (2000)]

assumed. The lattice constant, equal to the disorder length scale, is taken to be $a = 20$ Å. The central frequency ω of the excitation pulse is chosen to be situated below the center of the absorption band calculated without interaction. Specifically, the gap of the noninteracting case is the origin of the energy axis, the band center is at $\Delta = 80$ meV and we use $\hbar\omega = 40$ meV. As we have seen, for this excitation condition, the sign of the interaction is not irrelevant. Here, we use an attractive interaction throughout ($U < 0$). A temporal pulse width $\delta = 100$ fs is used corresponding to a spectral width of 22 meV. The number of sites N is taken large enough, typically $N = 240$, such that, for not too small disorder, the locally excited wave packet does not reach the boundary of the sample in times of typically some ten picoseconds. After this time acoustic phonon scattering leads to effective dephasing and the coherent phenomena studied here are destroyed. For principal studies, however, the calculation is sometimes performed for much longer times, although the results are no longer relevant to experiments.

While the previous results have been obtained using uncorrelated disorder, here, the influence of correlation of the disorder is investigated. Spatial correlation enters the dynamics of locally generated excitations through the presence of the Coulomb interaction and also through the spectral response to the excitation pulse. The influence of disorder correlations on the optical spectra is obtained using optical dipole matrix elements that are equal for all sites. If the dynamics of the local excitation is calculated, on the other hand, we only take, as above, the central matrix element to be nonzero.

In the case of correlated disorder, the electron and hole wave functions are pairwise identical, since both particles have the same environment. For $U = 0$, all optical matrix elements connecting these pairs of states are equal, all others are zero. Therefore, the optical spectrum for $U = 0$ resembles the density of single-particle states, however, with a band width being the sum of that of the two bands.

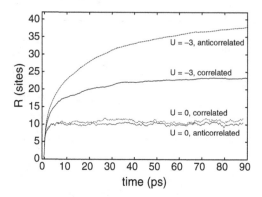

Fig. 17.8. Center-of-mass (R) traces for $U = 0$ and $U = -3$ for anticorrelated and correlated disorder of strength $W^e = W^h = W = 60\,\mathrm{meV}$. [From P. Thomas et al., phys. stat. sol. b **218**, 125 (2000)]

Figure 17.8 shows center-of-mass traces for correlated and anticorrelated disorder for $U = 0$ and $U = -3$. Remarkably, the enhancement for anticorrelated disorder is much more pronounced than in the correlated case. This is also visible in Fig. 17.9, which shows the same findings as contour plots of the two-particle wave function, $|p_{ij}^{he}|^2$, for anticorrelated and correlated disorder at time $t = 165\,\mathrm{ps}$ for the same parameters used in Fig. 17.8. Again, the anticorrelated situation shows much larger enhancement.

Analyzing the optical spectra for the anticorrelated case, Fig. 17.10, one realizes a peak which, for $U = 0$, lies in the center of the band, i.e., at $\Delta = 80\,\mathrm{meV}$ and shifts towards lower energy for increasing attractive (i.e., negative) U. For the excitation condition leading to the large center-of-mass enhancement at $U = -3$, pairs of states in the vicinity of this peak are excited. The origin of the peak can be traced back to optical transitions connecting states in the tails of the single-particle bands. Even for moderate disorder, these pairs of states with nearly identical transition energies are strongly localized at the same spatial position and therefore their optical matrix elements are large. Exact eigenvectors calculated numerically for a short sample ($N = 10$) and for $U = 0$ and $U = -3$ confirm the strong overlap within the contributing pairs. Obviously, these strongly localized tail states cannot be responsible for the large enhancement.

The fact that the enhancement for anticorrelated disorder is larger than that for correlated disorder is not fully understood yet. A prominent difference exists in the equations of motion for p_{ij}^{he} for the two cases. For the anticorrelated case, the disorder is absent in the equations for the diagonal elements p_{ii}^{he}, while it is present in the respective equation for the correlated case.

The behavior of excitons in disordered systems is usually discussed in terms of the relative coordinate $r = i - j$ and the center-of-mass coordinate $x = (i+j)/2$ instead of the indices i and j (for equal masses, not to be confused with R and ρ, cf. (17.2) and (17.3)). After the transformation $(i, j) \rightarrow (r, x)$,

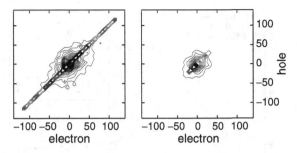

Fig. 17.9. Contour plots of $|p_{ij}^{he}|^2$ at 165 ps for anticorrelated (*left*) and correlated (*right*) disorder and $U = -3$. The other parameters are the same as in Fig. 17.8. [From P. Thomas et al., phys. stat. sol. b **218**, 125 (2000)]

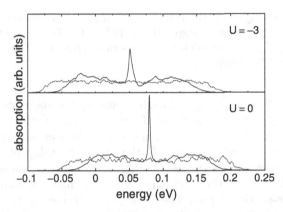

Fig. 17.10. Absorption spectra for $U = 0$ and $U = -3$ for correlated (*dotted line*) and anticorrelated (*solid line*) disorder. Other parameters as above. [From P. Thomas et al., phys. stat. sol. b **218**, 125 (2000)]

the relative coordinate r can be integrated out in the equation of motion if, e.g., the disorder W is smaller than the exciton binding energy and if discrete excitonic resonances are considered. The equation of motion in terms of x would not contain any disorder in the anticorrelated case, in contrast to the correlated case. Although, for the parameters used in Fig. 17.8, this approach is not strictly valid, it suggests a possible solution for the problem at hand.

In the literature, a so-called duality has been discussed that relates the enhancement for large U to that for small U. The present calculations yield similar results, see Fig. 17.11 for anticorrelated disorder. The ratio R_∞/ρ_∞ is close to unity for small U, while it has a maximum around $U = -3$ to -4 which increases with increasing disorder. Of course, both R_∞ and ρ_∞ decrease with increasing disorder, however, ρ_∞ faster than R_∞. Remarkably, the position of the maximum does not depend on the amount of disorder. This behavior of R_∞/ρ_∞ is apparently related to the fact that, for the chosen excitation energy

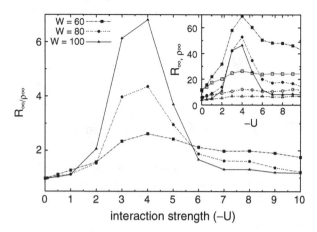

Fig. 17.11. Dependence of the enhancement R_∞/ρ_∞ on interaction strength U for three different disorder strengths $W^e = W^h = W$. The *inset* shows R_∞ (*full symbols*) and ρ_∞ (*open symbols*) versus U separately. [From P. Thomas et al., phys. stat. sol. b **218**, 125 (2000)]

$\hbar\omega$, the center of the optical spectrum, indicated by the above mentioned peak, shifts through the spectral position of the laser pulse with changing U, see Fig. 17.10. Around $U = -3$ to -4, the peak coincides with the excitation energy ω, while at larger negative U the excitation takes place again in a region where states outside the center of the single-particle bands are optically coupled. We have seen that the enhancement is more pronounced for less localized single-particle states. These states are situated in the center of the single-particle bands. The Coulomb enhancement is therefore largest if the electron–hole packet is generated in the center of the optical spectrum since the interaction couples these weakly localized states most effectively.

17.3 Suggested Reading

1. D. Brinkmann, J.E. Golub, S.W. Koch, P. Thomas, K. Maschke, and I. Varga, "Interaction-assisted propagation of Coulomb-correlated electron–hole pairs in disordered semiconductors", Eur. Phys. J. B **10**, 145 (1999)
2. D.L. Shepelyansky, "Coherent propagation of two interacting particles in a random potential", Phys. Rev. Lett. **73**, 2607 (1994)
3. P. Thomas, I. Varga, T. Lemm, J.E. Golub, K. Maschke, T. Meier, S.W. Koch, "Propagation of Coulomb-correlated electron–hole pairs in semiconductors with correlated and anticorrelated disorder", phys. stat. sol. b **218**, 125 (2000)

18

Current Echoes

As another coherent transport phenomenon, we describe here the theoretical prediction of the current echo. As we have seen, the prototype of a system exhibiting echo features is realized by an ensemble of independent two-level resonances where the transition frequencies are distributed according to a certain distribution function. The spectral response of such a system is characterized by an inhomogeneous line. After short-pulse excitation at time $t = 0$ the field-induced macroscopic coherence of the ensemble decays due to the randomization of the individual phases on a time scale determined by the reciprocal spectral width of the distribution of states excited within the inhomogeneous line. Provided that there are no other phase-destroying interactions, excitation of the system with a second delayed pulse at $t = \tau$ leads to phase conjugation of the coherence and, as a consequence, to its rephasing at $t = 2\tau$. At this time, a spontaneous field, i.e., the echo, is emitted. However, echoes are not restricted to uncoupled two-level quantum systems. Already above, we have discussed echoes in interacting many-particle systems like photoexcited semiconductors.

Generally, the ingredients for the existence of an echo are a sufficiently wide spectrum of levels, inhomogeneous character of the transitions, coherent preparation of the excited ensemble, and the absence of rapid dephasing. In the case of the photon echo that has been treated above, an interlevel or interband coherence is excited. Therefore, the rephased coherence is the source of the observed signal.

In the following, we investigate another kind of echo. Let us consider the situation of a degenerate electron system in a disordered potential landscape. A short voltage pulse at $t = 0$ generates a current pulse, which decays within the elastic scattering time. For a system without Coulomb interaction and in the absence of dephasing interactions, it has been shown that a second voltage pulse at $t = \tau$ leads to a reconstruction of a macroscopic current burst at $t = 2\tau$. Later, such current echoes were also predicted in models which include the Coulomb interaction.

Experimentally, the generation of such an echo requires current pulses on a time scale shorter than the elastic scattering time, which in turn has to be shorter than the inelastic scattering time, i.e., on the order of or shorter than picoseconds. This poses a challenge to the experimentalists. In particular, the generation of current pulses on such short time scales is not trivial.

Using optical coherent control techniques, it has, however, been demonstrated that it is possible to optically generate intraband currents in semiconductors on a femto- to picosecond time scale. Moreover, the detection of ultrashort current pulses should be possible using terahertz detection schemes.

In order to demonstrate the feasibility of a current echo experiment using such an optical coherent control excitation scheme, we show that, on the basis of our disordered two-band tight-binding Hamiltonian, currents can be generated in a disordered semiconductor. Furthermore, interesting new intraband echoes appear for a suitably chosen sequence of optical pulses. Despite the superficial similarities, it is also pointed out that this echo phenomenon profoundly differs from the conventional photon echo.

In addition, we would like to mention that one can, by making use of the spin dependence of the optical dipole matrix elements, see, e.g., Sect. 5.4, also generate spin current by the excitation with properly polarized coherent control pulses. For the description of the generation of such spin currents, it is sufficient to use the semiconductor Bloch equations (with or without Coulomb interaction) including the spin-dependent optical selection rules. This topic is not described in the following, but additional information is provided in some of the publications listed under "Suggested Reading".

18.1 Current Generation by Coherent Control

As we are interested in fundamental questions concerning the generation and the general properties of the current echo, we adopt the simplest model that allows us to incorporate disorder in a natural way. This is our one-dimensional two-band tight-binding Anderson Hamiltonian with diagonal disorder, as introduced in (5.8). Here, we treat nondegenerate electron and hole bands. Disorder is characterized by the disorder parameter $\eta^{e,h} = W^{e,h}/J^{e,h}$.

The light field $E(t)$ couples to the total dipole moment

$$\hat{d} = \sum_{\substack{i,j \\ \lambda,\lambda'=e,h}} d_{ij}^{\lambda,\lambda'} a_i^{\lambda\dagger} a_i^{\lambda'}, \tag{18.1}$$

where

$$d_{ij}^{\lambda,\lambda'} = -e\left(R_i \delta_{\lambda\lambda'} + r_{\lambda\lambda'}\right)\delta_{ij} \tag{18.2}$$

and λ and λ' take the values e, h. The second term in (18.2) is the interband dipole moment responsible for interband optical transitions known from (5.22), whereas the first term in (18.2) is the intraband dipole moment responsible for transport processes. The intraband dipole matrix elements are a necessary ingredient of the present theory, since the dynamics of the intraband coherences and of intraband currents are relevant. We assume that the electric light field is parallel to the intrasite optical dipole matrix element and also to the chain of sites. The Coulomb interaction is ignored in this model.

The total polarization (density) is given by

$$P = \frac{\langle \hat{d} \rangle}{V} \tag{18.3}$$

and the total current by

$$
\begin{aligned}
\mathcal{J} &= \langle \dot{\hat{d}} \rangle \\
&= \frac{1}{i\hbar} \langle [\hat{d}, \hat{H}_0] \rangle \\
&= \frac{ie}{\hbar} \sum_{\substack{\lambda \\ ij}} (R_i - R_j) T_{ij}^\lambda \langle a_i^{\lambda\dagger} a_j^{\lambda'} \rangle \\
&\quad + \frac{ie}{\hbar} \sum_{\substack{\lambda\lambda' \\ ij}} r_{\lambda\lambda'} \left(T_{ij}^{\lambda'} - T_{ij}^\lambda \right) \langle a_i^{\lambda\dagger} a_j^{\lambda'} \rangle \\
&= \frac{ie}{\hbar} R \sum_{\lambda i} J^\lambda \left(\langle a_i^{\lambda\dagger} a_{i+1}^\lambda \rangle - \langle a_i^{\lambda\dagger} a_{i-1}^\lambda \rangle \right) \\
&\quad - \frac{ie}{\hbar} \sum_{\lambda\lambda' i} r_{\lambda\lambda'} \left(\epsilon_i^\lambda - \epsilon_i^{\lambda'} \right) \langle a_i^{\lambda\dagger} a_i^{\lambda'} \rangle \\
&\quad + \frac{ie}{\hbar} \sum_{\lambda\lambda' i} r_{\lambda\lambda'} \left(J^\lambda - J^{\lambda'} \right) \left(\langle a_i^{\lambda\dagger} a_{i+1}^{\lambda'} \rangle + \langle a_i^{\lambda\dagger} a_{i-1}^{\lambda'} \rangle \right), \tag{18.4}
\end{aligned}
$$

where $R = R_{i+1} - R_i$. These two observables P and $\mathcal{J}$ have both intra- and interband contributions given by the first and second term in (18.2), respectively.

We adopt the notation of (5.7) for the electron and hole operators and that of (8.5) and (8.10) for the interband polarization (and the interband current) and for the intraband coherences related to the intraband current. The interband current $\mathcal{J}^{\text{inter}}$, the intraband current $\mathcal{J}^{\text{intra}}$, and the total polarization P are given by

$$\mathcal{J}^{\text{intra}} = \frac{2eR}{\hbar} \left[J^e \sum_i \Im n_{i+1,i}^e + J^h \sum_i \Im n_{i+1,i}^h \right], \tag{18.5}$$

$$\mathcal{J}^{\text{inter}} = -\frac{2er_{cv}}{\hbar} \sum_i \left(\epsilon_i^e + \epsilon_i^h \right) \Im p_{ii}^{he}$$

$$+ \frac{2er_{cv}}{\hbar} \left(J^e - J^h \right) \sum_i \Im \left(p_{i-1,i}^{he} + p_{i+1,i}^{he} \right), \tag{18.6}$$

$$P = -\frac{e}{V} \sum_i \left(R_i \left(n_{ii}^e - n_{ii}^h \right) + r_{cv} \Re p_{ii}^{he} \right), \tag{18.7}$$

respectively. In (18.7), the first and second terms refer to the intra- and interband polarization, respectively.

Using the Heisenberg equation of motion and taking the expectation values everywhere, we obtain the following set of (extended) optical Bloch equations

$$\frac{\partial}{\partial t} p_{ij}^{he} = -\frac{i}{\hbar} \left(\epsilon_i^h + \epsilon_j^e \right) p_{ij}^{he}$$

$$-\frac{i}{\hbar} J^h \left(p_{i-1j}^{he} + p_{i+1j}^{he} \right) + \frac{i}{\hbar} J^e \left(p_{ij-1}^{he} + p_{ij+1}^{he} \right)$$

$$+ \frac{i}{\hbar} eE(t) \cdot \left[(R_i - R_j) p_{ij}^{he} + r_{cv} \left(n_{ij}^e + n_{ji}^h \right) \right], \tag{18.8}$$

$$\frac{\partial}{\partial t} n_{ij}^e = \frac{i}{\hbar} \left(\epsilon_i^e - \epsilon_j^e \right) n_{ij}^e$$

$$+ \frac{i}{\hbar} J^e \left(n_{ij-1}^e + n_{ij+1}^e - n_{i-1j}^e - n_{i+1j}^e \right)$$

$$+ \frac{i}{\hbar} eE(t) \cdot \left[(R_i - R_j) n_{ij}^e + r_{cv} \left(p_{ij}^{he} - (p_{ji}^{he})^* \right) \right], \tag{18.9}$$

$$\frac{\partial}{\partial t} n_{ij}^h = \frac{i}{\hbar} \left(\epsilon_i^h - \epsilon_j^h \right) n_{ij}^h$$

$$- \frac{i}{\hbar} J^h \left(n_{ij-1}^h + n_{ij+1}^h - n_{i-1j}^h - n_{i+1j}^h \right)$$

$$- \frac{i}{\hbar} eE(t) \cdot \left[(R_i - R_j) n_{ij}^h + r_{cv} \left((p_{ij}^{he})^* - p_{ji}^{he} \right) \right]. \tag{18.10}$$

This set of equations differs from the conventional optical Bloch equations for the noninteracting tight-binding model, see (8.9), (8.12), and (8.13), by the terms containing the positions R_i of the sites. Below, (18.8)–(18.10) are solved numerically in all orders of the external light fields.

We use the following coherent control scheme in order to optically generate short intraband current pulses in both bands. The first excitation at time $t = 0$ is chosen to be due to a light field

$$E(t) = E_1 e^{-(t/\delta_1)^2} \cos\left((\omega/2)t\right) + E_2 e^{-(t/\delta_2)^2} \cos\left(\omega t + \phi_{1,2}\right), \quad (18.11)$$

where E_i are the amplitudes, δ_i the temporal widths, and $\phi_{1,2}$ is the relative phase of the two contributions which have frequencies ω (called full-gap pulse) and $\omega/2$ (called half-gap pulse) which satisfy $\hbar\omega/2 < \epsilon_0 < \hbar\omega$; i.e., only the full-gap pulse can generate resonant interband optical excitations. An example of such a time-dependent electric field is shown in Fig. 18.1. Note that the displayed field is not symmetric with respect to zero field. This asymmetry is the origin for the excitation of the intraband currents in this nonlinear process.

18.2 Current Decay in Disordered Semiconductors

For an ordered situation ($\eta^e = \eta^h = 0$), intraband currents are generated on the time scale of the optical pulse. The direction of the current depends on the relative phase $\phi_{1,2}$ of the two contributions, $E^{(\omega/2)}(t)$ and $E^{(\omega)}(t)$. Following an ultrafast rise, the current stays constant in the ordered case.

If disorder is introduced, the current is expected to decay due to elastic scattering at the disorder. However, one finds an extremely noisy current trace for a single realization of disorder. This noise can, in principle, be reduced by taking a configurational average. It turns out, however, that one has to average over an extremely large number of realizations in order to achieve a smooth trace.

Unfortunately, the dominant contribution to the current in the disordered case is not given by the same mechanism as in the ordered case, where the current is of at least third order in the external light field. Instead, in the presence of disorder, there is an additional trivial contribution that is at least of second order in the field. Responsible for this contribution are quantum-beat-like

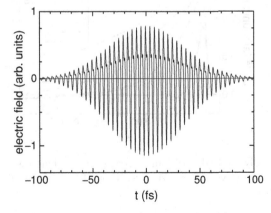

Fig. 18.1. Electric field of the coherent control light pulse consisting of two frequencies ω and $\omega/2$. Note the asymmetric form around zero field strength

excitations due to the spatial inhomogeneity of the disordered sample. In a self-averaging situation these contributions vanish. The convergence is, however, extremely slow. In order to improve the convergence, one may consider two different ensembles. (i) For any given realization of disorder, the spatially inverted realization is also considered in the ensemble. Then, the unwanted contributions cancel each other. (ii) Correlated disorder is used. In this case, the quantum-beat-like contributions are mutually compensated already in a given realization, provided that the effective masses of electrons and holes are identical ($|J^e| = |J^h|$). Otherwise, the widths of the disorder energy distributions have to scale with the couplings. In realistic disordered heterostructures, the disorder is expected to be more or less correlated in this sense. Therefore, the echo traces shown here are obtained using the second method. The remaining nontrivial current traces still show fluctuations which are partly due to the small number of sites of the sample. Moreover, even for a very long chain the inherent localization effectively leads to a fragmentation of the system into pieces which have lengths on the order of the localization length. The fluctuations are reduced by averaging over sufficiently many different realizations of the correlated disordered samples (usually several thousands).

In Fig. 18.2, decaying currents following excitation with coherent-control pulses are shown for different disorder parameters η and for an excitation where the frequency ω is in the center of the optical spectrum. Here, the total optical band width is 272 meV and the pulse duration is 50 fs. Only for the smallest disorder parameter $\eta = 0.5$, is the decay approximately exponential. For larger disorder, the decay is faster than exponential and has oscillatory character, showing the effect of localization. The Fourier transform for $\eta = 4$

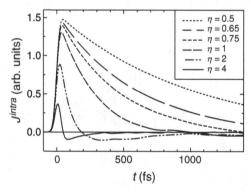

Fig. 18.2. Optically excited intraband currents for various disorder strengths of $\eta = \eta^e = \eta^h$ for a chain with $N = 151$ sites. [From J. Stippler et al., phys. stat. sol. b **221**, 379 (2000)]

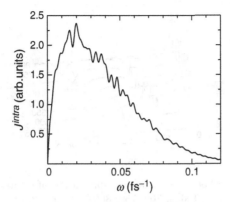

Fig. 18.3. Fourier transform of the intraband current for $\eta = 4$ from Fig. 18.2. This graph shows the localization-induced deviation from the Drude-like Lorentzian at low frequencies. [From J. Stippler et al., phys. stat. sol. b **221**, 379 (2000)]

is given in Fig. 18.3, which exhibits the localization-induced deviation from a Drude-like Lorentzian at low frequencies.

18.3 Equal Electron and Hole Masses

The current echo appearing spontaneously after two delayed current excitations is demonstrated in Fig. 18.4 for a chain of 71 sites and disorder parameter $\eta = 2$. It is obvious that the echo appears at $t = 2\tau$ if the delay of the excitation pulses is τ. As already stated above, the process of generating a current is of at least third order in the field amplitudes. Thus, the current echo is expected to be at least of sixth order. Therefore, it is found to be advantageous to use relatively high light intensities for our model. All light pulses have the area (half the Rabi angle) of $0.32\,\pi$, the central frequency (i.e., ω) coincides with the center of the absorption spectrum, and the spectral width is $66\,\text{meV}$. No truncation in terms of orders of the external pulses has been applied. For this excitation condition, it is sufficient to average over 150 realizations of the disorder. Note that the amplitude of the current echo is inverted relative to that of the two primary currents, see Fig. 18.4.

18.4 Different Electron and Hole Masses

There are three findings which point out that the dynamics initiated by the coherent control excitation characterizes a new coherent phenomenon.

(i) For equal electron and hole masses, a spontaneous delayed signal pulse is completely absent if, instead of correlated disorder, we consider a model

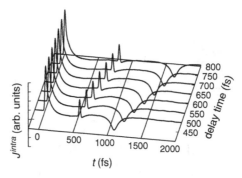

Fig. 18.4. Current echoes after repeated application of excitation pulses at $t = 0$ and $t = \tau = 500, 550, 600, 650, 700, 750, 800$ fs for $\eta = 2$, correlated disorder, and optical excitation in the band center with pulse areas of 0.32π. An average over 150 disorder realizations has been taken and $N = 71$. [From J. Stippler et al., phys. stat. sol. b **221**, 379 (2000)]

with uncorrelated disorder. Remember that, in this case, current fluctuations are excited even by normal band-to-band excitation which can be suppressed only by extensive configurational averaging over a large number of disorder realizations. After this averaging, no detectable signal can be seen in the resulting current traces, while the echoes are rather prominent for correlated disorder and an averaging over only 150 realizations.

(ii) For the case of equal electron and hole masses, the spontaneous delayed response is obtained even for a simplified excitation sequence. The first pulse at $t = 0$ is taken to generate a current, i.e., it is given by (18.11). However, for the second pulse at $t = \tau$, we take a simple full-gap pulse with only the frequency ω. This pulse has to enter the equations of motion at least twice. As a result, we still see a spontaneous response at the time $t = 2\tau$, see upper panel in Fig. 18.5. We therefore conclude that the echo signal is not due to two externally excited current pulses, but instead only the first pulse has to be a current-generating pulse. The reverse order of pulses does not yield a delayed response.

(iii) For different electron and hole masses, $J^h/J^e < 1$, we find that, surprisingly, there are two separate intraband responses, one preceding $t = 2\tau$ and one following $t = 2\tau$. In particular, the preceding contribution is due to the valence band and appears at

$$t_v = \left(1 + \frac{J^h}{J^e}\right) \tau, \qquad (18.12)$$

while the delayed contribution is due to the conduction band and appears at

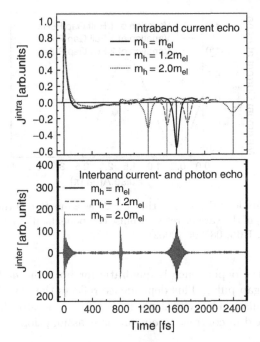

Fig. 18.5. Intra- and interband dynamics for different electron and hole masses. The excitation at $t = 0$ is via a coherent control pulse. The excitation at $t = \tau = 800\,\text{fs}$ is only via a full-gap pulse, which does not generate a current. While for $m_h \neq m_{el}$ the intraband dynamics shows two distinct responses, one preceding and one following $t = 2\tau$ for different masses, the interband echo (photon echo) always appears exactly at $t = 2\tau$. [From C. Schlichenmaier et al., Phys. Rev. **B65**, 085306 (2002)]

$$t_c = \left(1 + \frac{J^e}{J^h}\right)\tau, \tag{18.13}$$

see upper panel in Fig. 18.5. Note that the interband photon echo appears in all these cases at $t = 2\tau$, as shown in the lower panel of Fig. 18.5.

18.5 Excitation Intensity Dependence

The amplitude of the spontaneous signal depends on the excitation intensity of the various pulses in different ways. Figure 18.6 shows the amplitude as a function of the pulse areas (half the Rabi angles Θ_n) of the pulse No. n. $\Theta_n = \pi$ corresponds to complete inversion of a two-level absorber excited resonantly by pulse No. n.

It is shown in Fig. 18.6 that in order to obtain large echoes, in particular, the half-gap contribution has to be sufficiently strong, while the first full-gap pulse does not need to have a very high intensity. In the limit of low excitation intensity, the amplitude of the signal depends linearly on the area of the first

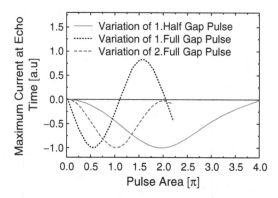

Fig. 18.6. Amplitude of the echo at time $t = 2\tau$ as a function of the pulse area (half of the Rabi angle) of the various relevant pulse components. [From C. Schlichenmaier et al., Phys. Rev. **B65**, 085306 (2002)]

full-gap pulse, the dependence is quadratic for both the half-gap pulse and the second full-gap pulse. This dependence reflects the lowest relevant order of the various pulses. The echo signal is therefore at least of fifth order in the external light field in contrast to our previous assumption.

18.6 Intraband Dynamics

The appearance of the intraband echoes can be understood on the basis of a simplified model. Let us assume that we were able to diagonalize the conduction- and the valence-band Hamiltonian, resulting in eigenstates $|h\nu\rangle$ and $|e\nu'\rangle$ having energies ϵ_ν^h and $\epsilon_{\nu'}^e$ for the valence and the conduction band, respectively. In the eigenbasis, the Hamiltonian is given by

$$\hat{H}_0 = \sum_{\nu=1}^{M} \epsilon_\nu^h a_\nu^{h\dagger} a_\nu^h + \sum_{\nu'=1}^{M'} \epsilon_{\nu'}^e a_{\nu'}^{e\dagger} a_{\nu'}^e, \tag{18.14}$$

where $a_\nu^{e,h}$ ($a_\nu^{e,h\dagger}$) annihilates (creates) an electron or a hole in the eigenstate $|(e,h)\nu\rangle$ of $\hat{H}_0$. The light–matter interaction is described by, see (3.14),

$$\hat{H}_L = -E(t)\hat{P}, \tag{18.15}$$

with the electric field $E(t)$ and the polarization operator, see (3.15),

$$\hat{P} = \sum_{\nu=1}^{M} \sum_{\nu'=1}^{M'} \mu_{\nu\nu'}^{he} a_\nu^{h\dagger} a_{\nu'}^e + H.C., \tag{18.16}$$

where $\mu_{\nu\nu'}^{he}$ is the interband dipole matrix element between pairs of upper and lower eigenstates.

Using this Hamiltonian, the optical Bloch equations in the rotating-wave approximation for the M-M' system are given by (setting $\hbar = 1$)

$$\frac{dp_{\nu\nu'}^{he}}{dt} - i(\omega_{\nu'\nu}^{he} - \omega)p_{\nu\nu'}^{he} = -iE(t)\left(\sum_{\mu=1}^{M}\mu_{\mu\nu'}^{he}n_{\nu\mu}^{hh} - \sum_{\mu'}^{M'}\mu_{\nu\mu'}^{he}n_{\mu'\nu'}^{ee}\right),$$

$$\frac{dn_{\nu\mu}^{hh}}{dt} - i(\epsilon_{\mu}^{h} - \epsilon_{\nu}^{h})n_{\nu\mu}^{hh} = iE(t)\sum_{\kappa'=1'}^{M'}\mu_{\nu\kappa'}^{he}(p_{\mu\kappa'}^{he})^* - iE(t)\sum_{\kappa'=1'}^{M'}(\mu_{\mu\kappa'}^{he})^*p_{\nu\kappa'}^{he},$$

$$\frac{dn_{\nu'\mu'}^{ee}}{dt} - i(\epsilon_{\mu'}^{e} - \epsilon_{\nu'}^{e})n_{\nu'\mu'}^{ee} = iE(t)\sum_{\kappa=1}^{M}(\mu_{\kappa,\nu'}^{he})^*p_{\kappa\mu'}^{he} - iE(t)\sum_{\kappa=1}^{M}\mu_{\kappa\mu'}^{he}(p_{\kappa\nu'}^{he})^*,$$

$$(18.17)$$

where $\omega_{\nu\nu'}^{eh} = \epsilon_{\nu'}^{e} - \epsilon_{\nu}^{h}$.

Here, we are interested in the response of $n_{\nu\mu}^{hh}$ and $n_{\nu'\mu'}^{ee}$ to the sequence of the first coherent-control pulse and the second full-gap pulse. For simplicity, we only consider two states in each band, i.e., $|v1\rangle$, $|v2\rangle$ and $|c1\rangle$, $|c2\rangle$. The equations of motion for $n_{\nu\mu}^{hh}$ and $n_{\nu'\mu'}^{ee}$ for $\nu, \mu = 1, 2$ and $\nu', \mu' = 1, 2$ read

$$\frac{dn_{12}^{hh>}}{dt} + i\delta_v n_{12}^{hh>} = iE(t)\mu\left((p_{21}^{he})^* - p_{12}^{he}\right),$$

$$\frac{dn_{12}^{ee>}}{dt} + i\delta_c n_{12}^{ee>} = iE(t)\mu\left(p_{12}^{he} - (p_{21}^{he})^*\right),$$

$$(18.18)$$

where $\delta_v = \epsilon_2^h - \epsilon_1^h$ and $\delta_c = \epsilon_2^e - \epsilon_1^e$. Note that because of the correlated disorder, only optical dipole matrix elements μ between pairs of corresponding states are nonzero, i.e., there is a strict selection rule. Equations (18.18) describe the dynamics after the second full-gap pulse. Just before the arrival of this second pulse, the $n_{\nu\mu}^{hh<}$ and $n_{\nu'\mu'}^{ee<}$ have acquired phases according to δ_v and δ_c, respectively, due to their free motion in the interval between the first pulse and the second pulse. Just after the first pulse, their phases were such that (for the total ensemble of M and M' levels) a macroscopic intraband current is present. Our aim is to find the time at which the phases of $n_{\nu\mu}^{hh>}$ and $n_{\nu'\mu'}^{ee>}$ have the exact same values which the phases of $n_{\nu\mu}^{hh<}$ and $n_{\nu'\mu'}^{ee<}$ had just after the first pulse.

We assume, for simplicity, that the second pulse has the form $\delta(t - \tau)$, i.e., it arrives after a delay time τ following the first pulse and is extremely short. From the equation of motion for p, we find the values of this variable at time $t = \tau$ which enter the driving term on the right-hand side of (18.18)

$$\frac{dp_{12}^{he}}{dt} + i\delta_v p_{12}^{he} = -iE(t)\mu(n_{12}^{hh<} - n_{12}^{ee<}),$$

$$\frac{dp_{12}^{eh}}{dt} + i\delta_c p_{12}^{eh} = iE(t)\mu(n_{12}^{hh<} - n_{12}^{ee<}).$$

$$(18.19)$$

If the solutions of these equations, taken at time $t = \tau$, are inserted into (18.18), we find that the equation for $n_{12}^{hh>}$ has a driving term proportional to $-n_{21}^{ee<}(\tau)$, while that for $n_{12}^{ee>}$ has a driving term proportional to $-n_{21}^{hh<}(\tau)$.

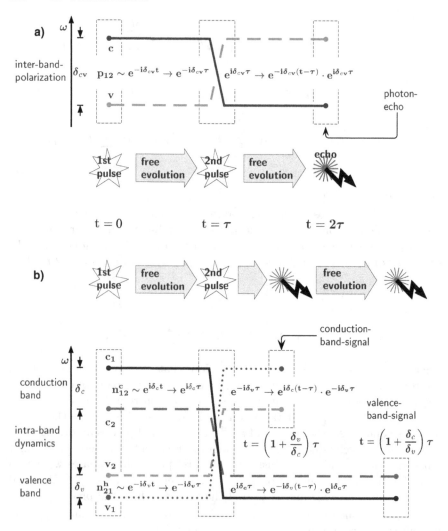

Fig. 18.7. Diagram showing the inversion of intra- and interband phases. (a) The photon echo scenario. (b) The intraband echo scenario. [From C. Schlichenmaier et al., Phys. Rev. **B65**, 085306 (2002)]

Note that due to correlated disorder the dynamics in the upper and lower pair of states is identical up to a global time-scaling factor given by δ_v/δ_c. Consequently, $n_{12}^{hh>}(t)$ has acquired the initial phase of $-n_{21}^{ee<}(\tau)$ at time $t_v = (1 + \delta_c/\delta_v)\tau$ and $n_{12}^{ee>}(t)$ has acquired the initial phase of $-n_{21}^{hh<}(\tau)$ at time $t_c = (1+\delta_v/\delta_c)\tau$. This explanation is schematically displayed in Fig. 18.7.

Turning now back to the ensemble of more than two eigenstates in the bands, we see that all these particular terms add up at times t_v and t_c to an intraband dynamics showing the initially generated intraband current, while all other terms interfere destructively. Thus, these considerations explain the predicted intraband echo phenomenon.

In contrast, the interband photon echo relies on the phase conjugation of the interband polarizations $p_{\nu\nu'}^{he}$, as shown in the upper part of Fig. 18.7. This interband phase conjugation always leads to a restoration of the initial phases at time $t = 2\tau$. If, however, we are interested in the phases of the intraband quantities, we have to consider that not only the interband phase factors are conjugated by the second pulse, but in addition also the intraband phases.

18.7 Suggested Reading

1. R. Atanasov, A. Haché, J.L.P. Hughes, H.M. van Driel, and J.E. Sipe, "Coherent control of photocurrent generation in bulk semiconductors", Phys. Rev. Lett. **76**, 1703 (1996)
2. R.D.R. Bhat and J.E. Sipe, "Optically injected spin currents in semiconductors", Phys. Rev. Lett. **85**, 5432 (2000)
3. A. Haché, Y. Kostoulas, R. Atanasov, J.L.P. Hughes, J.E. Sipe, and H.M. van Driel, "Observation of coherently controlled photocurrent in unbiased bulk GaAs", Phys. Rev. Lett. **78**, 306 (1997)
4. J. Hübner, W.W. Rühle, M. Klude, D. Hommel, R.D.R. Bhat, J.E. Sipe, and H.M. van Driel, "Direct observation of optically injected spin-polarized currents in semiconductors", Phys. Rev. Lett. **90**, 216601 (2003)
5. Huynh Thanh Duc, T. Meier, and S.W. Koch, "Microscopic analysis of the coherent optical generation and the decay of charge and spin currents in semiconductor heterostructures", Phys. Rev. Lett. **95**, 086606 (2005)
6. W. Niggemeier, G. von Plessen, S. Sauter, and P. Thomas, "Current echoes", Phys. Rev. Lett. **71**, 770 (1993)
7. S. Sauter-Fischer, E. Runge, and R. Zimmermann, "Partial dephasing in interacting many-particle systems and current echo", Phys. Rev. **B57**, 4299 (1998)
8. C. Schlichenmaier, I. Varga, T. Meier, P. Thomas, and S.W. Koch, "Optically induced coherent intraband dynamics in disordered semiconductors", Phys. Rev. **B65**, 085306 (2002)
9. M.J. Stevens, A.L. Smirl, R.D.R. Bhat, A. Najmaie, J.E. Sipe, and H.M. van Driel, "Quantum interference control of ballistic pure spin currents in semiconductors", Phys. Rev. Lett. **90**, 136603 (2003)
10. J. Stippler, C. Schlichenmaier, A. Knorr, T. Meier, M. Lindberg, P. Thomas, S.W. Koch, "Current echoes induced by coherent control", phys. stat. sol. b **221**, 379 (2000)

19

Problems

Problem 1. Derive (4.29) from (4.28) using (4.3).

Problem 2.
Add a "dephasing" term

$$-i\hbar \frac{\rho_{rl}}{T_2} \tag{19.1}$$

to the third equation in (4.29). Follow the approach leading to (4.44) and solve for $j(t)$, $p(t)$, and $I(t)$ either analytically or numerically. Discuss, in particular, the case of long times.

Problem 3. Rewrite (5.13) in normal order, i.e., all creation operators appear on the left-hand side of all annihilation operators. Thus identify the contributions of the Coulomb interaction to the single-particle energies.

Problem 4. Generate a spatially correlated disorder potential numerically. Consider a single-band one-dimensional system with lattice constant a. Plot the disorder potential for length $L = na$ with $n = 1, 3, 5, 10, 50$ for a system of $N = 200$ sites. Start from (i) a uniform random (box) and (ii) a Gaussian distribution of width W of site energies ϵ_i of site i.

Hint: Look for the Box-Muller transformation to generate the Gaussian distribution.

Problem 5. Calculate the correlation functions of the above determined disorder potentials. The correlation function is defined by

$$\Phi(x) = \langle \epsilon_{i+x} \epsilon_i \rangle. \tag{19.2}$$

The brackets denote the configurational average. Discuss the resulting correlation function with respect to the length L. When using a Gaussian window for the sliding average make sure that its full width at half maximum should be as large as the width of the box window.

Problem 6. Derive (6.24) and (6.25) for a density of states of the entire band and close to the band edge of a one-dimensional single-band system with a dispersion relation given by (6.13).

Problem 7. Derive the optical spectrum $\chi''(\omega)$ for transitions close to the absorption edge of a one-dimensional two-band semiconductor with dispersion relations for electrons and holes given by (6.13). Use (6.17) to obtain an analytical expression close to the absorption edge. Calculate the spectrum numerically for the full dispersion of the bands and plot the result.

Problem 8. Determine the participation number Λ for coherent tunneling as a function of time. Consider the particle being placed at the left-hand side at time $t = 0$. Use (4.27) and (6.38) and identify $n_{ll} = \rho_{ll}$ and $n_{rr} = \rho_{rr}$.

Problem 9. Derive the Kramers–Kronig relations. The Fourier transform of $\chi(t)$ is defined by

$$\chi(\omega) = \int_{-\infty}^{\infty} \chi(t) e^{i\omega t} dt \tag{19.3}$$

and is a complex function, i.e.,

$$\chi(\omega) = \chi'(\omega) + i\chi''(\omega). \tag{19.4}$$

In our case $\chi'(\omega)$ and $\chi''(\omega)$ are real functions and coincide with the real and imaginary part, respectively, of the Laplace-transform (9.2) for $\Im z \to 0^+$.

The Kramers–Kronig relations follow quite generally from the causality of the linear response function. They read

$$\chi'(\omega) = -\frac{1}{\pi} P \int_{-\infty}^{\infty} \frac{\chi''(\omega')}{\omega - \omega'} d\omega', \tag{19.5}$$

$$\chi''(\omega) = \frac{1}{\pi} P \int_{-\infty}^{\infty} \frac{\chi'(\omega')}{\omega - \omega'} d\omega' \tag{19.6}$$

where $P \int$ denotes the principal value of the integral, i.e.,

$$P \int_{-\infty}^{\infty} \frac{f(x)}{x - x_0} dx = \lim_{\epsilon \to 0+} \left(\int_{-\infty}^{x_0 - \epsilon} \frac{f(x)}{x - x_0} dx + \int_{x_0 + \epsilon}^{\infty} \frac{f(x)}{x - x_0} dx \right).$$

Causality means that

$$\chi(t) = \theta(t) f(t). \tag{19.7}$$

Here, $\theta(t)$ is the unit step function which can be expressed as

$$\theta(t) = \frac{1}{2}(u(t) + 1), \tag{19.8}$$

with

$$u(t) = \begin{cases} -1 & \text{if } t < 0 \\ 1 & \text{if } t > 0. \end{cases}$$

Hint: The function $u(t)$ can be written in the form

$$u(t) = \lim_{\epsilon \to 0^+} \frac{i}{2\pi} \int_{-\infty}^{\infty} \left(\frac{1}{\omega' + i\epsilon} + \frac{1}{\omega' - i\epsilon} \right) e^{-i\omega' t} d\omega'. \tag{19.9}$$

Use the Dirac identity

$$\lim_{\epsilon \to 0} \frac{1}{\omega \pm i\epsilon} = \frac{P}{\omega} \mp i\pi\delta(\omega) \tag{19.10}$$

and prove that

$$\chi(\omega) = \frac{i}{\pi} P \int_{-\infty}^{\infty} \frac{\chi(\omega')}{\omega - \omega'} d\omega'. \tag{19.11}$$

Problem 10. From linear response theory the complex response function $\chi(z)$ is given by

$$\chi(z) = -\frac{i}{\hbar} \int_0^{\infty} \langle [\hat{P}(t), \hat{P}(0)] \rangle_0 e^{izt} dt. \tag{19.12}$$

Evaluate this expression for an ensemble of identical two-level absorbers and show that the term in $\chi''(\omega)$ describing emission coincides with (9.13).
Here $\hat{P}(t)$ is

$$\hat{P}(t) = e^{-i\hat{H}_0 t/\hbar} \hat{P} e^{i\hat{H}_0 t/\hbar} \tag{19.13}$$

and

$$\langle \cdots \rangle_0 = Tr(\rho_0 \cdots) \tag{19.14}$$

with the equilibrium density matrix

$$\langle 1v|\rho_0|1v \rangle = 1,$$
$$\langle 1c|\rho_0|1c \rangle = 0,$$
$$\langle 1v|\rho_0|1c \rangle = 0,$$
$$\langle 1c|\rho_0|1v \rangle = 0.$$

Problem 11. Calculate the optical polarization in the time domain for a Fano system with a single discrete state by Fourier-transforming (9.35). The result is given in (9.37) and (9.38).

Problem 12. Calculate the optical polarization in the time domain for a Fano system with two discrete states by Fourier-transforming (9.34). The result is given in (9.37) and (9.43). Use the following parameters: $q_1 = q_2 = q$ and $\Gamma_1 = \Gamma_2 = \Gamma$.

Problem 13. Plot the optical dipole matrix elements $\mu(k)$ and $\bar{\mu}(k)$ given in (9.74) for various values of the ratios δ_ν/J^ν ($\nu = e, h$) and discuss their k-dependence.

Problem 14. Four-wave-mixing in tunneling structures: Consider a two-level absorber having a lower level at energy ϵ_1^v and an upper level at energy ϵ_1^c. The upper level is coherently coupled through some barrier to a second level having energy ϵ_2^c. The coupling matrix element is given by the real energy J. A two-beam degenerate four-wave-mixing experiment is performed by applying two identical pulses with relative delay τ to the system from directions $\boldsymbol{k}_1$ and $\boldsymbol{k}_2$. Their photon energies $\hbar\omega_L$ are close to $\epsilon_1^c - \epsilon_1^v$. In the ground state the electron sits in the lower level with energy ϵ_1^v.

Calculate the time-resolved and time-integrated four-wave-mixing signals for very short pulses, i.e., the envelopes are given by $\delta(t)$ and $\delta(t-\tau)$-functions, for various upper-level energy differences $\epsilon_1^c - \epsilon_2^c$. Apply (10.64).

Hint: Diagonalize the upper-level part of the Hamiltonian and consult Chap. 4.

Show that the temporal four-wave-mixing traces describe the tunneling dynamics of the electron between the upper-level subsystem.

How does the situation change if pulse-envelopes of finite width $\delta_1 = \delta_2$ are applied such that $\delta_i^{-1} \ll |J|$? Treat this problem qualitatively and also quantitatively by solving the optical Bloch equations for the system at hand numerically for various values of δ_i and system parameters J and $\epsilon_1^c - \epsilon_2^c$.

Hint: Use, e.g., the fourth-order Runge–Kutta method to solve the relevant equations of motion.

Problem 15. Consider a pump–probe experiment performed with ultra-short optical pulses. Approximate the pump and probe pulse envelopes by δ-functions and calculate the differential absorption spectrum $\delta\alpha(\omega)$ as a function of the time delay τ by solving (8.55) and (8.57) supplemented by dephasing times, i.e. (using a slightly modified notation)

$$\frac{d}{dt}p = \frac{i}{\hbar}\left(-\hbar\omega_p p + \mu_p^* E(1 - |p|^2) + V_p p|p|^2 + V_B p^* \bar{B}\right) - \frac{p}{T_{2,p}},$$

$$\frac{d}{dt}\bar{B} = \frac{i}{\hbar}\left(-\hbar\omega_B \bar{B} + pp\right) - \frac{\bar{B}}{T_{2,B}}.$$

up to third order in the field.

Hint: Follow the steps presented in Chap. 10 for level systems and write $\delta\alpha$ according to (11.5), i.e.,

$$\delta\alpha(\omega) = \delta\alpha_{\text{pb}}(\omega) + \delta\alpha_{\text{CI,1st}}(\omega) + \delta\alpha_{\text{CI,corr}}(\omega), \qquad (19.15)$$

as a sum of three additive contributions.

Discuss the line shape of the three contributions for positive and negative delays τ.

(The obtained results may be compared with Appendix of C. Sieh et al., Eur. Phys. J. B **11**, 407 (1999).)

Problem 16. Repeat problem 15 using (8.60) and (8.61), i.e. (using a slightly modified notation)

$$\frac{d}{dt}p = \frac{i}{\hbar}\left(-\hbar\omega_p p + \mu_p^* E(1 - |p|^2) + \tilde{V}_B p^* B\right) - \frac{p}{T_{2,p}},$$

$$\frac{d}{dt}B = \frac{i}{\hbar}\left(-\hbar\omega_B B - \mu_B^* Ep\right) - \frac{B}{T_{2,B}},$$

instead of (8.55) and (8.57). These equations allow one to write $\delta\alpha$ as a sum of two additive terms

$$\delta\alpha(\omega) = \delta\alpha_{\mathrm{pb}}(\omega) + \delta\alpha_{\mathrm{CI}}(\omega), \tag{19.16}$$

where the Coulomb-interaction-induced contribution is given by $\delta\alpha_{\mathrm{CI}}(\omega) = \delta\alpha_{\mathrm{CI,1st}}(\omega) + \delta\alpha_{\mathrm{CI,corr}}(\omega)$.

What can one learn about the compensation between $\delta\alpha_{\mathrm{CI,1st}}(\omega)$ and $\delta\alpha_{\mathrm{CI,corr}}(\omega)$ by comparing the results of problem 15 with the present one?

Problem 17. Repeat problem 15 using a cw-pump field with frequency ω_L and an ultrashort δ-function probe pulse. Discuss the optical Stark effect, in particular, the line shape and the amplitudes of $\delta\alpha_{\mathrm{pb}}(\omega)$, $\delta\alpha_{\mathrm{CI,1st}}(\omega)$, and $\delta\alpha_{\mathrm{CI,corr}}(\omega)$ as a function of the detuning between ω_L and ω_p.

(Some of the obtained results may be compared with P. Brick et al., Phys. Rev. B **64**, 075323 (2001).)

Problem 18. To describe the dependence of the differential absorption on the polarization directions of the pump and probe pulses, the equations considered in problem 15 have to be supplemented by the dipole matrix elements describing the circularly polarized heavy-hole to electron transitions in III-V semiconductors, see (5.23). Since these transitions are spin-degenerate, the total polarization is determined by two interband coherences, p^+ and p^-, i.e.,

$$\mathbf{P} = \phi(0)(\boldsymbol{\sigma}^+ p^+ + \boldsymbol{\sigma}^- p^-), \tag{19.17}$$

where p^+ and p^- can be excited by clock- and counterclockwise (σ^+ and σ^-) polarized light fields, respectively.

Considering the spin-degenerate optical selection rules in the microscopic equations, (8.54) and (8.56), one finds that three different types of biexciton coherences, i.e., B^{++}, B^{--}, B^{+-} are relevant. Whereas B^{++} and B^{--} describe interacting excitons of the same spin, B^{+-} describes the interactions of excitons with different spin which may lead to the formation of a bound biexciton state.

The dynamics of these quantities can be described by the following set of schematic equations

$$\frac{d}{dt}p^+ = \frac{i}{\hbar}\left(-\hbar\omega_p p^+ + \mathbf{E}\cdot(\boldsymbol{\sigma}^+)^*(1-|p^+|^2) + V_p p^+ |p^+|^2 + \tilde{V}_p p^+ |p^-|^2\right.$$
$$\left. + V_B(p^+)^*\bar{B}^{++} + \tilde{V}_B(p^-)^*\bar{B}^{+-}\right) - \frac{p^+}{T_{2,p}},$$

$$\frac{d}{dt}p^- = \frac{i}{\hbar}\left(-\hbar\omega_p p^- + \mathbf{E}\cdot(\boldsymbol{\sigma}^-)^*(1-|p^-|^2) + V_p p^- |p^-|^2 + \tilde{V}_p p^- |p^+|^2\right.$$
$$\left. + V_B(p^-)^*\bar{B}^{--} + \tilde{V}_B(p^+)^*\bar{B}^{+-}\right) - \frac{p^-}{T_{2,p}},$$

$$\frac{d}{dt}\bar{B}^{++} = \frac{i}{\hbar}\left(-\hbar\omega_B\bar{B}^{++} + p^+ p^+\right) - \frac{\bar{B}^{++}}{T_{2,B}},$$

$$\frac{d}{dt}\bar{B}^{--} = \frac{i}{\hbar}\left(-\hbar\omega_B\bar{B}^{--} + p^- p^-\right) - \frac{\bar{B}^{--}}{T_{2,B}},$$

$$\frac{d}{dt}\bar{B}^{+-} = \frac{i}{\hbar}\left(-\hbar\omega_B\bar{B}^{+-} + p^+ p^-\right) - \frac{\bar{B}^{+-}}{T_{2,B}}.$$

Approximate the pump and probe pulse envelopes by δ-functions and use the equations given above to calculate the differential absorption $\delta\alpha(\omega)$, see (11.4), up to third order in the field.

Discuss the influence of the polarization directions of the pump and probe pulses.

Investigate for which cases excited state absorption induced by transitions to a bound biexciton is visible and compare with the findings of Fig. 11.4.

(Some of the obtained results may be compared with Appendix of C. Sieh et al., Eur. Phys. J. B **11**, 407 (1999).)

Problem 19. Consider the schematic equations presented in problem 15 and use them to calculate the time-resolved four-wave-mixing signal in self-diffraction geometry, i.e., emitted in the direction $2\mathbf{k}_2 - \mathbf{k}_1$, up to third order in the fields by approximating the envelopes of the incident pulses by δ-functions.

Write the four-wave-mixing polarization $P_{\mathrm{FWM}}(t,\tau)$ according to (11.12) as a sum of three terms, i.e.,

$$P_{\mathrm{FWM}}(t,\tau) = P_{\mathrm{FWM}}^{\mathrm{pb}}(t,\tau) + P_{\mathrm{FWM}}^{\mathrm{CI,first}}(t,\tau) + P_{\mathrm{FWM}}^{\mathrm{CI,corr}}(t,\tau). \qquad (19.18)$$

Which terms contribute to the four-wave-mixing response for negative delays, i.e., $\tau < 0$?

What are the decay times of the time-resolved four-wave-mixing signal, i.e., $|P_{\mathrm{FWM}}(t,\tau)|^2$, for positive and negative delays?

Show that exciton–biexciton beats are present in the time-resolved but not in the time-integrated signal.

Calculate $P_{\mathrm{FWM}}(\omega,\tau)$, which is the Fourier transform of $P_{\mathrm{FWM}}(t,\tau)$. Discuss biexciton-induced signatures in the spectrally resolved four-wave-mixing signal, i.e., $|P_{\mathrm{FWM}}(\omega,\tau)|^2$

Consider a *correlated* inhomogeneous broadening of the exciton and biexciton frequencies, i.e., $\delta\omega_B = 2\delta\omega_p$, and show that this leads to photon echoes. How does the result change if the inhomogeneous broadening of the the exciton and biexciton frequencies is not fully correlated, e.g., $\delta\omega_B = \alpha\delta\omega_p$ with $\alpha \neq 2$?

Show that exciton–biexciton beats may appear in the time-integrated signal if inhomogeneous broadening is considered.

What happens to the signal for negative time delays in the presence of inhomogeneous broadening?

Problem 20. Consider the schematic equations presented in problem 18 and use them to calculate the time-resolved four-wave-mixing signal up to third order in the fields by approximating the envelopes of the incident pulses by δ-functions.

Discuss the dependence of $\boldsymbol{P}_{\mathrm{FWM}}(t, \tau)$ on the polarization directions of the incident pulses.

Compare, in particular, the strength of the spectrally-resolved signal $|\boldsymbol{P}_{\mathrm{FWM}}(\omega, \tau)|^2$ at $\omega = \omega_p$ and at $\omega = \omega_B - \omega_p$, i.e., at the exciton and at the exciton to biexction transition frequencies, when exciting with linearly parallel and orthogonal polarized pulses, i.e., the so-called xx and xy excitation configurations.

Problem 21. Consider the schematic equations presented in problem 15 and use them to calculate the coherent-excitation-spectroscopy signal up to third order in the fields. Treat pulse No. 1 as a cw-field with frequency ω_{exc} and pulse No. 2 as a δ-pulse and calculate the spectrally-resolved four-wave-mixing signal $|P_{\mathrm{FWM}}(\omega, \omega_{\mathrm{exc}})|^2$.

Show that the biexciton-induced terms explain the two off-diagonal peaks ($\omega \neq \omega_{\mathrm{exc}}$) that show up in Fig. 13.3b.

Problem 22. Extend problem 21 and analyze the dependence of the coherent excitation spectroscopy signal on the polarization directions of the incident beams by using the schematic equations presented in problem 18. Treat pulse No. 1 as a cw pulse with frequency ω_{exc} and pulse No. 2 as a δ-pulse and calculate the spectrally resolved four-wave-mixing signal $|P_{\mathrm{FWM}}(\omega, \omega_{\mathrm{exc}})|^2$.

Consider, in particular, excitation with linearly parallel and orthogonal polarized pulses, i.e., the so-called xx and xy excitation configurations. Compare the analytical results with those presented in Fig. 13.3.

Problem 23. Solve the Wannier equation (9.45) supplemented with a dephasing time, i.e. (using a slightly modified notation)

$$\frac{d}{dt}p_{12} = \frac{i}{\hbar}\left(-\sum_j T^e_{2j}p_{1j} - \sum_i T^h_{i1}p_{i2} + V_{12}p_{12} + E(t)\cdot(\mu_{12})^*\right) - \frac{p_{12}}{T_2},$$

with

$$\mu_{12} = \mu_0\delta_{12}, \tag{19.19}$$

numerically using a very short pulse. Determine the time-dependent polarization, i.e.,

$$P(t) = \sum_{ij} \mu_{ij}^{he} p_{ij}(t). \qquad (19.20)$$

Obtain $P(\omega)$ via a Fourier transformation and plot its imaginary part, i.e., $\Im P(\omega)$, which is proportional to the absorption $\alpha(\omega)$.

Start by neglecting the Coulomb interaction, i.e., set $V = 0$, and reproduce, e.g., Figs. 6.4 and 9.12. Then add the interaction and reproduce, e.g., Figs. 9.13 and 9.14. Vary the effective masses, the strength of the Coulomb attraction, and the dephasing time and discuss the changes in the calculated spectra.

Hint: Use, e.g., the fourth-order Runge–Kutta method to solve the equations of motion.

Problem 24. Solve the Wannier equation, see problem 23, numerically and obtain linear optical absorption spectra of mesoscopic semiconductor rings with magnetic field. As described in Chap. 16, one has to consider complex magnetic-field-dependent phase factors in the tight-binding couplings, see (16.25), i.e.,

$$\tilde{T}_{i,i+1}^e = J^e \exp\left[2\pi i \frac{\Phi}{N\Phi_0}\right],$$

$$\tilde{T}_{i+1,i}^e = J^e \exp\left[-2\pi i \frac{\Phi}{N\Phi_0}\right],$$

$$\tilde{T}_{i,i+1}^h = J^h \exp\left[-2\pi i \frac{\Phi}{N\Phi_0}\right],$$

$$\tilde{T}_{i+1,i}^h = J^h \exp\left[2\pi i \frac{\Phi}{N\Phi_0}\right]. \qquad (19.21)$$

Reproduce Figs. 16.8 and 16.9 and vary the parameters to investigate under which conditions significant magnetic-field-dependent modifications of the linear absorption exist.

Problem 25. Use the results of Sect. 8.2.2 to reproduce the equations of motion describing the optical response in the coherent $\chi^{(3)}$ limit as presented in Sect. 8.2.3.

Problem 26. Follow the derivations presented in Sects. 8.2.2 and 8.2.5 and verify (8.77).

Problem 27. The wave function of the central Wannier–Stark state with energy ϵ_0 in site representation is given by (15.15), i.e.,

$$\psi_0(m) = \langle m|\psi_0\rangle = J_m\left(\frac{-2J}{eFa}\right). \qquad (19.22)$$

Plot the wave function for different ratios $\frac{-2J}{eFa}$.

Analyze the extension of the wave function in the Wannier–Stark regime, where eFa is finite and comparable to but smaller than the band width $4|J|$.

Discuss the Stark localization, i.e., the localization of the wave function in the limit of strong fields F.

Problem 28. Assume that a wave packet consisting of electronic Wannier–Stark states, see problem 27, is generated via optical excitation with a short pulse at $t = 0$, i.e.,

$$\Psi(m, t = 0) = \sum_{n=-\infty}^{\infty} a_n \psi_n(m). \tag{19.23}$$

The state $\psi_n(m)$ has the energy $\epsilon_n = \epsilon_0 + neFa$ and can be obtained from $\psi_0(m)$ via the relation $\psi_n(m) = \psi_0(m + n)$, see (15.3).

If the field F is sufficiently large such that the Wannier–Stark states of the holes are localized (note that due to their larger mass, i.e., smaller J, already moderate electric field F are able to localize the hole Wannier–Stark states) and if the optical excitation is local at site $m = 0$, the amplitudes of the wave packet are given by $a_n = \psi_n(0) = J_n\left(\frac{-2J}{eFa}\right)$.

Analyze the dynamical evolution ("breathing") of the wave packet which is described by

$$\Psi(m, t) = \sum_{n=-\infty}^{\infty} a_n \psi_n(m) e^{-i\epsilon_n t}, \tag{19.24}$$

by plotting the time-dependent electron probability, i.e., $|\Psi(m, t)|^2$.

An oscillatory motion of the wave packet in real-space characteristic of the Bloch oscillation dynamics as shown in Fig. 15.3 is obtained if, e.g., predominantly Wannier–Stark states close to the lower band edge ($\epsilon_0 - 2|J|$) are excited. Visualize Bloch oscillations by considering only a few states in the vicinity of $\epsilon_0 - 2|J|$ when calculating the wave-packet dynamics.

Problem 29. In the presence of an ac electric field, the effective band width is given by (15.30), i.e.,

$$\Delta(F) = \Delta(F = 0) \left| J_0\left(\frac{eFa}{\hbar\omega_L}\right) \right|. \tag{19.25}$$

Plot $J_0(x)$ for $x > 0$ and determine the smallest arguments for which $J_0(x)$ vanishes.

Assuming $a = 10\,\text{nm}$ and $\hbar\omega_L = 20\,\text{meV}$, what is the smallest amplitude of the ac field F which leads to dynamical localization, i.e., a complete collapse of the (mini-)band?

Index

About the Authors

Torsten Meier obtained his Ph.D. (Dr. rer. nat., 1994) in physics at the Philipps-Universität Marburg. From 1995 to 1997 he worked as a Post-Doc in the group of Prof. Shaul Mukamel, Department of Chemistry, University of Rochester, New York, USA. Afterwards he returned to Marburg, received his Habilitation in theoretical physics in December 2000, and became Privatdozent in April 2001. Since April 2002 he has been a Fellow in the Heisenberg program of the Deutsche Forschungsgemeinschaft.

Peter Thomas obtained his Ph.D. (Dr. rer. nat., 1969) in physics at the Philipps-Universität Marburg. In 1973 he became professor of theoretical physics in Marburg. Since 1975 he has established fruitful cooperation with various groups in Budapest, Hungary, and also teaches as guest professor at the Budapest University of Technology and Economics.

Stephan W. Koch obtained his Ph.D. (Dr. phil. nat., 1979) in physics at the Universität Frankfurt. Until 1993 he was a full professor both in the Department of Physics and at the Optical Sciences Center at the University of Arizona, Tucson, USA. In the fall of 1993, he joined the Philipps-Universität Marburg where he is a full professor of theoretical physics. He is a Fellow of the Optical Society of America. He has received the Leibniz prize of the Deutsche Forschungsgemeinschaft (1997) and the Max-Planck Research Prize of the Humboldt Foundation and the Max-Planck Society (1999).